Bruno Apolloni, Witold Pedrycz, Simone Bassis, and Dario Malchiodi

The Puzzle of Granular Computing

Studies in Computational Intelligence, Volume 138

Editor-in-Chief
Prof. Janusz Kacprzyk
Systems Research Institute
Polish Academy of Sciences
ul. Newelska 6
01-447 Warsaw
Poland
E-mail: kacprzyk@ibspan.waw.pl

Further volumes of this series can be found on our homepage:
springer.com

Vol. 117. Da Ruan, Frank Hardeman
and Klaas van der Meer (Eds.)
Intelligent Decision and Policy Making Support Systems, 2008
ISBN 978-3-540-78306-0

Vol. 118. Tsau Young Lin, Ying Xie, Anita Wasilewska
and Churn-Jung Liau (Eds.)
Data Mining: Foundations and Practice, 2008
ISBN 978-3-540-78487-6

Vol. 119. Slawomir Wiak, Andrzej Krawczyk
and Ivo Dolezel (Eds.)
Intelligent Computer Techniques in Applied Electromagnetics, 2008
ISBN 978-3-540-78489-0

Vol. 120. George A. Tsihrintzis and Lakhmi C. Jain (Eds.)
Multimedia Interactive Services in Intelligent Environments, 2008
ISBN 978-3-540-78491-3

Vol. 121. Nadia Nedjah, Leandro dos Santos Coelho
and Luiza de Macedo Mourelle (Eds.)
Quantum Inspired Intelligent Systems, 2008
ISBN 978-3-540-78531-6

Vol. 122. Tomasz G. Smolinski, Mariofanna G. Milanova
and Aboul-Ella Hassanien (Eds.)
Applications of Computational Intelligence in Biology, 2008
ISBN 978-3-540-78533-0

Vol. 123. Shuichi Iwata, Yukio Ohsawa, Shusaku Tsumoto, Ning
Zhong, Yong Shi and Lorenzo Magnani (Eds.)
Communications and Discoveries from Multidisciplinary Data, 2008
ISBN 978-3-540-78732-7

Vol. 124. Ricardo Zavala Yoe
Modelling and Control of Dynamical Systems: Numerical Implementation in a Behavioral Framework, 2008
ISBN 978-3-540-78734-1

Vol. 125. Larry Bull, Bernadó-Mansilla Ester
and John Holmes (Eds.)
Learning Classifier Systems in Data Mining, 2008
ISBN 978-3-540-78978-9

Vol. 126. Oleg Okun and Giorgio Valentini (Eds.)
Supervised and Unsupervised Ensemble Methods and their Applications, 2008
ISBN 978-3-540-78980-2

Vol. 127. Régie Gras, Einoshin Suzuki, Fabrice Guillet
and Filippo Spagnolo (Eds.)
Statistical Implicative Analysis, 2008
ISBN 978-3-540-78982-6

Vol. 128. Fatos Xhafa and Ajith Abraham (Eds.)
Metaheuristics for Scheduling in Industrial and Manufacturing Applications, 2008
ISBN 978-3-540-78984-0

Vol. 129. Natalio Krasnogor, Giuseppe Nicosia, Mario Pavone
and David Pelta (Eds.)
Nature Inspired Cooperative Strategies for Optimization (NICSO 2007), 2008
ISBN 978-3-540-78986-4

Vol. 130. Richi Nayak, Nikhil Ichalkaranje
and Lakhmi C. Jain (Eds.)
Evolution of the Web in Artificial Intelligence Environments, 2008
ISBN 978-3-540-79139-3

Vol. 131. Roger Lee and Haeng-Kon Kim (Eds.)
Computer and Information Science, 2008
ISBN 978-3-540-79186-7

Vol. 132. Danil Prokhorov (Ed.)
Computational Intelligence in Automotive Applications, 2008
ISBN 978-3-540-79256-7

Vol. 133. Manuel Graña and Richard J. Duro (Eds.)
Computational Intelligence for Remote Sensing, 2008
ISBN 978-3-540-79352-6

Vol. 134. Ngoc Thanh Nguyen and Radoslaw Katarzyniak (Eds.)
New Challenges in Applied Intelligence Technologies, 2008
ISBN 978-3-540-79354-0

Vol. 135. Hsinchun Chen and Christopher C. Yang (Eds.)
Intelligence and Security Informatics, 2008
ISBN 978-3-540-69207-2

Vol. 136. Carlos Cotta, Marc Sevaux
and Kenneth Sörensen (Eds.)
Adaptive and Multilevel Metaheuristics, 2008
ISBN 978-3-540-79437-0

Vol. 137. Lakhmi C. Jain, Mika Sato-Ilic, Maria Virvou,
George A. Tsihrintzis, Valentina Emilia Balas
and Canicious Abeynayake (Eds.)
Computational Intelligence Paradigms, 2008
ISBN 978-3-540-79473-8

Vol. 138. Bruno Apolloni, Witold Pedrycz, Simone Bassis
and Dario Malchiodi
The Puzzle of Granular Computing, 2008
ISBN 978-3-540-79863-7

Bruno Apolloni
Witold Pedrycz
Simone Bassis
Dario Malchiodi

The Puzzle of Granular Computing

Bruno Apolloni
Dip. Scienze dell'Informazione
Università degli Studi di Milano
Via Comelico, 39/41 20135 Milano
Italy

Witold Pedrycz
University of Alberta
Dept. Electrical & Computer Engineering
9107 116 Street
Edmonton AB T6G 2V4 Canada
& Systems Research Institute
Polish Academy of Sciences
Warsaw, Poland

Simone Bassis
Dip. Scienze dell'Informazione
Università degli Studi di Milano
Via Comelico, 39/41
20135 Milano
Italy

Dario Malchiodi
Dip. Scienze dell'Informazione
Università degli Studi di Milano
Via Comelico, 39/41
20135 Milano
Italy

ISBN 978-3-540-79863-7

e-ISBN 978-3-540-79864-4

DOI 10.1007/978-3-540-79864-4

Studies in Computational Intelligence   ISSN 1860949X

Library of Congress Control Number: 2008926081

© 2008 Springer-Verlag Berlin Heidelberg

This work is subject to copyright. All rights are reserved, whether the whole or part of the material is concerned, specifically the rights of translation, reprinting, reuse of illustrations, recitation, broadcasting, reproduction on microfilm or in any other way, and storage in data banks. Duplication of this publication or parts thereof is permitted only under the provisions of the German Copyright Law of September 9, 1965, in its current version, and permission for use must always be obtained from Springer. Violations are liable to prosecution under the German Copyright Law.

The use of general descriptive names, registered names, trademarks, etc. in this publication does not imply, even in the absence of a specific statement, that such names are exempt from the relevant protective laws and regulations and therefore free for general use.

Typeset & Cover Design: Scientific Publishing Services Pvt. Ltd., Chennai, India.

Printed in acid-free paper

9 8 7 6 5 4 3 2 1

springer.com

To those who are called upon to make decisions, practically the whole of mankind, politicians included. Faced with necessarily granular information, we dont expect people to arrive at the optimum decision. But we demand that they make reasonable choices.

Image created by Guglielmo Apolloni

Preface

The computer era is characterized by a massive shift in problem solving from the search of a suitable solving procedure to the careful exploitation of data information. If the most common recipe to solve a problem forty years ago was "look for the right formula", the common today's direction is "consider what suggestions data tell you about the problem solution". This data driven goal intensively exploited in the eighties in the connectionist paradigm, but even early adopted in the sixties by the fuzzy set paradigm, opened the door of an ideal megastore where a plenty of methods are promoted, each claiming high percentage of success in a vast variety of problems. Depending on the operational context and on the peculiarities that are enhanced of, they are alternatively declared methods of computational intelligence, to denote loose axiomatic premises compensated by the analogies with procedures that are supposed to be followed by intelligent (human or animal) beings and the appeal of the results they promise, or more recently methods of granular computing to root their rationale on the data information content. These looseness and approaching bivalence are not a defect *per se*, but very often produce an overlap in solving procedures leaving the user dubious on the solution way he has to follow in a specific problem.

Aim of this monograph is to help the reader to compose the puzzle pieces represented by these procedures into a high diversified mosaic of the modern methods for solving computational problems, within which the localization of a single tile is highly informative of its proper use. The strategy we adopted is to privilege the information management aspects in respect to the algorithmic sophistication. This is why we opted in the title for the second categorization of the methods in terms of information granules. Since our goal is rather ambitious, we will start with a very fundamental knowledge phenomenology with the commitment of

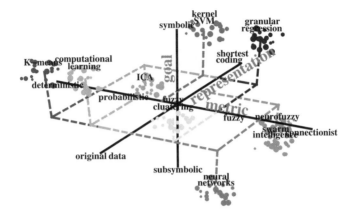

Fig. 0.1. Three dimensional framework representing an inference task. Clusters: some among the tasks afforded in the monography.

moving to more complex phenomena just as it becomes needed. Nevertheless the book is hard to read. It does not mean that it requires the reader to posses an extensive training in mathematics or in computer science. To a significant extent the book is self contained, apart some elementary mathematics and probability notions. Nevertheless, a coherent line of reasoning is developed along all the book demanding an accurate consideration on the part of the reader. With this caveat and guidelines, we expect the book to be beneficial both to the intellectually curious student and to the researcher open to modern conceptual paradigms.

We frame the matter of the book within the general goal of inferring rules from a suitable reading of a sample of data the rules refer to. The whole relies on an intelligent interplay between the properties we may observe in the data and the task we are called to solve on the basis of them, hence in a selection and management of data features that are highly dependent on their operational context and processing goal. The entire study can be cast in a three-dimensional framework (see Fig. 0.1), whose individual axes come with a well-defined semantics:

- The level of capability we have of stating a *metric* in terms of some order, or at least a certain degree of similarity between the data. This gives rise to a hierarchy of the methods in terms of available information content. Here the richest to poorest direction moves from deterministic, to probabilistic, then fuzzy and finally connectionist methods. We will follow this broadly defined hierarchy allowing for any kind of inversions and contaminations.

- The amount of structure we have embedded in these data, i.e., the features forming a *representation* of them. We will mainly quantify this information in terms of their distribution law or, dually, of the degree of structural connections existing therein.
- The typology of operational results we are expecting in face of the data at hand, i.e., the *goal* we aim to achieve. The taxonomy concerns both the kind of rule we want to discover and its description. As for the former, we distinguish between tasks such as classification, regression, etc. As for the latter, we mainly distinguish the cases where we are interested in a formal description of the rule, say through a formula or a procedure, from those where we are satisfied of having a computational device, for instance a set of instructions and parameters, producing correct solutions to a vast majority of problem instances.

Note that this framework is actually not far from the framework being used for clusterizing human emotional states, which is spanned by the three axes: coping, valence and novelty. Both frames are constituted of axes representing fuzzy variables to be suitably handled in view of a final goal rather than used to produce a definite label to emotions – there – or set of data in our case. Thus it is not surprising to realize that, once the tight cage of axiomatic theories has been abandoned, we are driven in our computations by complex thought paradigms that are not far from those ruling our emotions. Hence, on the one hand, it is not shocking to say that our computational framework may be viewed as a particular specification of the Affective Computing framework. On the other hand, let us reassure the reader that the key subject of our discourse will be the information granule, whose location in the above three dimensional framework specializes at level of the cluster of observed data items. In this setting, rules represent a suitable processing of granules.

Our underlying didactic strategy is to isolate fundamental conceptual bricks of Granular Computing in terms of some key problems and underlying methods, and discuss their rigorous implementation. Namely we will focus on the following problems:

1. classifying data through
    - decision rules, and
    - Support Vector Machines
2. extracting features denoting Relevant Components from data (ICA, kernels, etc.),
3. learning functions regressing data in
    - symbolic form, and
    - subsymbolic way.

What we have said so far is the logic description of the matter. In the book, however, we will proceed along the chapters with groups of homogeneous topics that should be better suitable to the reader thanks to the compression of the results we may deal with, that is a general goal for any learning system – human brain included. Therefore, we formerly devote one chapter to introduce the information granule as atomic unit of our procedures and declining its modalities. Then, we will devote a collection of chapters to the fundamentals. Namely, Part 1 is focused on the statistical ways of processing data, with chapters dedicated to: i) understanding what a sample really represents, ii) devising tools for exploiting it in terms of algorithmic inference, and iii) the most typical implementations of these tools, learning algorithms included. A second part is focused on the essentials of fuzzy sets, with chapters on : i) transferring information from the user to granules around data, and ii) refining granules from data evidence. Then, in order to prepare the reader to move to real applications, we toss the bases on three fundamental slots. Thus in Part III we expound the problem of classifying data through the two main sections on feature extraction and their classification, where the boundary between these two subjects is quite often delineated in a not clearly defined manner. Assuming that after these chapters the reader is able to isolate the atoms of his reasoning (*what she/he is speaking about*), we next introduce tools for combining these atoms/granules into structured sentences representing the rules of the ontologies he is setting up. They represent the logical and operational tools we may consider for interacting with the environment. It is commonly recognized that most rules are inherently subsymbolic: connectionist according to the eighties cognitive paradigms, or the product of social computations to a wider extent, within the recent threads. This does not mean that are meaningless rules. On the contrary, they spring from a rigorously cultivated thinking attitude finding a synthesis level that is wider in respect to strictly formal theories. With this perspective, we discuss in the fourth Part some key paradigms of social computing. We include a chapter on evolutionary computing and neural networks with some successful training algorithms, and a chapter on swarm intelligence strongholds. A compromise way between algorithms mainly aimed at the dry successful results and human brain need of understanding methods to get success is treated in fifth Part. It runs in terms structured information granules leading to fuzzy rules as a synthesis of deductions from user experience and inductions from data observations. We devote a chapter to the conceptual aspects and another to computational methods. Finally in the last chapter of this book we envision the role of knowledge engineers engaged to use all what we can of the tools assessed in the previous chapters in order to solve complex problems. The core is represented by some general fusion strategies. In particular we will deepen two case studies on support vector machines specially featured to deal with data quality, and collaborative clustering as a strategy to generate consistent classification rules.

Contents

The General Framework

1 Granule Formation Around Data 3
   1.1 The Roots of the Statistical Inference 5
        1.1.1 From Granules to Distributions 8
        1.1.2 Learning Distribution Parameters 9
   1.2 The Borders of the Unawareness.............. 15
   1.3 From Data to Functions 19
   1.4 Functions of Granules 22
   1.5 The Benchmarks of This Book 25
        1.5.1 Synthetic Datasets 25
        1.5.2 Leukemia........................... 25
        1.5.3 Cox 26
        1.5.4 Guinea Pig 26
        1.5.5 SMSA 27
        1.5.6 Swiss 28
        1.5.7 Iris 28
        1.5.8 MNIST Digits Data Base............. 28
        1.5.9 Vote 29
        1.5.10 Sonar Signals 29
   1.6 Exercises 30

Further Reading 33

References 37

Part I: Algorithmic Inference

Introduction 43

2 Modeling Samples ... 45
- 2.1 Samples and Properties about Them ... 47
  - 2.1.1 Samples ... 47
  - 2.1.2 Properties about Samples ... 52
- 2.2 Organizing Granules into Populations ... 55
- 2.3 Blowing up Items into Populations ... 59
- 2.4 Conclusions ... 62
- 2.5 Exercises ... 63

3 Inferring from Samples ... 65
- 3.1 Learning from Samples ... 65
  - 3.1.1 Statistic Identification ... 69
  - 3.1.2 Parameter Distribution Law ... 71
  - 3.1.3 Drawing Decisions ... 78
- 3.2 Learning from Examples ... 84
  - 3.2.1 Confidence Regions ... 85
  - 3.2.2 Numerical Examples ... 90
  - 3.2.3 PAC Learning Theory Revisited ... 96
  - 3.2.4 A General Error Distribution Law ... 103
- 3.3 Conclusions ... 113
- 3.4 Exercises ... 114

Further Reading ... 117

References ... 121

Part II: The Development of Fuzzy Sets

Introduction ... 127

4 Construction of Information Granules: From Data and Perceptions to Information Granules ... 129
- 4.1 Semantics of Fuzzy Sets: Some Insights ... 130
- 4.2 Transferring Meaning into Fuzzy Sets ... 131
  - 4.2.1 Fuzzy Set as a Descriptor of Feasible Solutions ... 131
  - 4.2.2 Fuzzy Set as a Descriptor of the Notion of *Typicality* ... 133
  - 4.2.3 Membership Functions for Visualizing Solution Preferences ... 135
  - 4.2.4 Nonlinear Transformation of Templates ... 136
  - 4.2.5 Shadowed Sets as a Three-Valued Logic Characterization of Fuzzy Sets ... 138

		4.2.6	Interval-Valued Fuzzy Sets	143
	4.3	Fuzzy Sets of Higher Order		144
		4.3.1	Second Order Fuzzy Sets	144
		4.3.2	Linguistic Approximation	146
		4.3.3	Rough Fuzzy Sets and Fuzzy Rough Sets	147
	4.4	Type-2 Fuzzy Sets		149
	4.5	Conclusions		152
	4.6	Exercises		152
5	**Estimating Fuzzy Sets**			155
	5.1	Vertical and Horizontal Schemes of Membership Estimation		156
	5.2	Maximum Likelihood Estimate		160
		5.2.1	In Search of the Maximum Compliance between Pivotal Granules and a Model	162
	5.3	Saaty's Priority Method of Pairwise Membership Function Estimation		165
	5.4	Fuzzy Sets as Granular Representatives of Numeric Data		168
	5.5	Fuzzy Sets That Reduce the Descriptional Length of a Formula		175
	5.6	Fuzzy Equalization		179
	5.7	Conclusions		181
	5.8	Exercises		182

Further Reading 185

References 187

Part III: Expanding Granula into Boolean Functions

Introduction 191

6	**The Clustering Problem**			193
	6.1	The Roots of Cluster Analysis		193
	6.2	Genotype Value of Metrics		201
	6.3	Clustering Algorithms		204
		6.3.1	A Template Procedure	205
		6.3.2	Initial Numbers and Values of Centroids	210
		6.3.3	Aggregation vs. Agglomeration	212
		6.3.4	Implementation Mode and Stopping Rule	219
	6.4	The Weighted Sum Option		220

		6.4.1 The Fuzzy C-Means Algorithm	222
	6.5	Discriminant Analysis	227
	6.6	Conclusions	230
	6.7	Exercises	231
7	**Suitably Representing Data**		233
	7.1	Principal Components Analysis	235
	7.2	Independent Components	237
	7.3	Sparse Representations	248
		7.3.1 Incremental Clustering	249
		7.3.2 Analytical Solution	255
	7.4	Nonlinear Mappings	256
	7.5	Conclusions	261
	7.6	Exercises	262

Further Reading 265

References 269

Part IV: Directing Populations

Introduction 275

8	**The Main Paradigms of Social Computation**		277
	8.1	Extending Sampling Mechanisms to Generate Tracks	278
	8.2	Knowledge Representation: From Phenotype to Genotype Space	281
	8.3	Starting from an Artificial Population	284
		8.3.1 A Very Schematic Genetic Algorithm	285
	8.4	The Seeds of Unawareness	291
		8.4.1 The Neural Networks Paradigm	292
		8.4.2 Training a Boltzmann Machine	296
		8.4.3 Multilayer Perceptron	300
		8.4.4 From a Population of Genes to a Population of Experts	307
	8.5	Conclusions	313
	8.6	Exercises	314
9	**If also Ants Are Able...**		317
	9.1	Swarm Intelligence	318
		9.1.1 Ant Colony	319
		9.1.2 Swarm Optimization	322
	9.2	Immune Systems	325
	9.3	Aging the Organisms	327

	9.3.1 The Neuron's Life	328
9.4	Some Practical Design and Implementation	332
9.5	Conclusions	334

Further Reading 335

References 337

Part V: Granular Constructs

Introduction 343

10 Granular Constructs 345
   10.1 The Architectural Blueprint of Fuzzy
       Models 345
   10.2 The Rule System 348
       10.2.1 Designing a Rule System Layout 349
       10.2.2 Evaluating a Rule System 352
   10.3 The Master Scheme 353
   10.4 The Clustering Part 355
       10.4.1 The Cluster-Based Representation of
             the Input-Output Mappings 356
       10.4.2 The Context-Based Clustering in the
             Development of Granular Models 363
   10.5 The Connectionist Part 371
       10.5.1 Granular Neuron 371
   10.6 The Neural Architecture 374
       10.6.1 Defuzzification at the Last 375
       10.6.2 Use Numbers as Soon as Possible 378
   10.7 Conclusions 380
   10.8 Exercises 381

11 Identifying Fuzzy Rules 385
   11.1 Training of a Granular Neuron 385
   11.2 Partially Supervised Identification of Fuzzy
       Clusters 388
       11.2.1 Using Data Labels When They Are
             Known 389
       11.2.2 The Development of Human-Centric
             Clusters 392
   11.3 The Overall Training of a Neuro-fuzzy
       System 396
       11.3.1 Learning Algorithm for Neuro-fuzzy
             Systems 398
   11.4 Conclusions 403
   11.5 Exercises 404

Further Reading 405

References 407

A Conceptual Synthesis

12 Knowledge Engineering 411
  12.1 Separating Points through SVM 411
    12.1.1 Binary Linear Classification 412
    12.1.2 Regression 422
  12.2 Distributed Architectures for Granular
       Computing 431
    12.2.1 Collaborative Clustering 433
    12.2.2 The General Flow of Collaborative
           Processing 435
    12.2.3 Evaluation of the Quality of
           Collaboration 437
    12.2.4 Experience-Consistent Fuzzy Models 439
  12.3 Conclusions 446

References 449

Appendices 451

A Norms 453
  A.1 Norms 453
  A.2 General Norms 453
  A.3 Fuzzy Norms 460

B Some Statistical Distribution 463
  B.1 Some One-Dimensional Variables 463
  B.2 Multidimensional Gaussian Variable 466

C Some Membership Functions 469

D Some Algebra 471
  D.1 Eigenvalues and Eigenvectors 472
  D.2 Kernels 472

E List of Symbols 477
  E.1 Variables 477
  E.2 Functions 479
  E.3 Operations on Sets 479
  E.4 Operators 480

Index 481

The General Framework

1 Granule Formation Around Data

And when the full moon had risen, Hansel took his little sister by the hand, and followed the pebbles which shone like newly-coined silver pieces, and showed them the way.
(*Grimm's Fairy Tales – Hansel & Gretel*)

Hansel was a very smart kid. Unable to bring with him a ton of pebbles to mark the track from house into the woods, he used them as granules of information suggesting the way. Even better, he exploited their luminescence to give them an order, hence dealing with them as a sample of the road.

Don't be hurt by the fact that we start this book with a child. The essence of soft computing is exactly to aim at computations not made heavy by hard theoretical constructs and *a priori* hypotheses. Rather, we start from available data and draw from them all reasonable consequences, where such consequences are a result of having stated some order relations between data. Hansel pebbles are a true paradigmatic example of this human capability. Since they are luminescent you may order them in a sequence. You do not always see the end of the road but you know the direction. Imagine substituting each pebble with a bold mass that borders on the analogous substitutes of the previous and subsequent pebbles as shown in option (a) of Fig. 1.1. After, you have no further information in the night, so each interval between one pebble and the next is equivalent to any other of these intervals. Actually, if you know that in your way forward you drew m pebbles, each interval is a block of information – call it *statistical granule* – having a weight $1/m$ in any computation you may do in the dark of the night. This is the essence of statistics theory as a wonderful way to manage the information of ordered data adopting probability measure as the descriptive tool. We will revisit some fundamentals of the theory in this chapter, just for the extension that is necessary to establish the required

Data by first:

possibly ordered,

4   Granule Formation Around Data

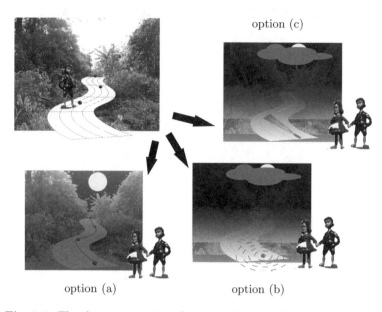

Fig. 1.1. The three perspectives for organizing granules from data.

possibly isolated, framework. We must also consider the possibility of Hansel needing to come back at dusk, when the moon does not help him to line up the pebbles. In this case each pebble works alone. It is an information granule, hence again a mass in place of the pebble. But since it has no relation with the others, since you do not know which ones come before and after, you must decide on your own on the shape of the mass, that for this reason we come up with the concept of a *fuzzy granule*. For instance, a bell shape as in the option (b) in Fig. 1.1 tells you that you cannot expect a point too far from the actual pebble position (bell center) to belong to the path home, apart from the ones close to another bell center. This is our starting point for the granular computing theory.

possibly structured, As a third option we may imagine Hansel to have been already trained in an orienteering course, so that he uses oblong stones denoting a direction like arrows. In this case, each time he examines a pebble he knows an angular region, say 30 degrees wide, where to confine the direction toward the next pebble as assumed in option (c) in Fig. 1.1. We will speak of a *logical granule* since the information we draw from it is connected to a formal rule we may apply: *the true direction is between* 15 *degrees left and* 15 *degrees right with respect to the direction shown by the pebble.* An early discussion will be devoted to this kind of granules in a section of this chapter as well.

We may see that these three perspectives are not conceptually different. Initially we have data and position ourselves in any intermediate place between the three views depending on the knowledge we may use to draw information from the data. A typical instance could occur when at semi-dusk Hansel understands, seeing at a given pebble, that within the 30 degrees indicated by the pebble, some directions are privileged by the local landscape and he sees at distance a pebble that is not sure to be exactly the next one in the track. Thus we adopt a hybrid approach *per se*, let's say between symbolic and sub-symbolic paradigms, between deductive and heuristic methods, declarative and imperative programming, etc., where the frontiers between the boundaries of the various dichotomies become more and more vague and undefined.

<small>possibly sharing the three conditions.</small>

Summing up, we have introduced three kinds of information granules: statistic, fuzzy and logic. In the rest of this chapter we will show the basic ways through which we can manipulate them in order to draw useful information for our everyday operational problems. Moreover, since the data are the main actors of our approach, we will introduce the benchmarks benchmark on which we will toss the granular computing procedures of this book.

1.1 The Roots of the Statistical Inference

In the everyday life we are used to meet probabilistic sentences, like those concerning stock market profits, weather forecasting, examination successes and so on. Aim of this section is to connect the meaning of these sentences to the statistical granules information within a paradigm fundamentally considering the sentences as a smart way of organizing the information supplied by the sample data rather than a fallout of the amusement of Nature tossing dices.

According to the *formal as it needs* strategy declared in the prolusion, we start with a *non definition* of probability P of an event e as the tendency to occur of the conditions denoting e. The reader may find a lot of philosophical questions and perspectives connected to give a rigorous meaning to P. We just limit ourselves to the operational connotations given by the following fact:

<small>Probability as a tool for accounting frequencies</small>

Fact 1.1. *In a record reporting the unpredictable values of a discrete variable X connected with a phenomenon we are observing, the asymptotic frequency $\phi(x)$ with which we meet the value x when the length of the record goes to infinity equals $P(X = x)$, i.e. the probability of the event $e = (X = x)$.*

Notation: we denote variables like X in the previous fact as *random variables*, where capital letters (X, U) denote the variables and small letters (x, u) describe the values they assume, that represent *realizations* of the variables; the sets the realizations belong to will be denoted by capital gothic letters $(\mathfrak{U}, \mathfrak{X})$.

<small>Random variables for detailing frequencies</small>

For the sake of simplicity we will consider in principle random variables X having discrete realizations, where the behavior of continuous random variables is deduced from an asymptotic decrease of the discretization grain. Namely, we will assume each variable taking values from $-\infty$ to $+\infty$ with a probability that is specified by a *distribution law* that, inherently discrete, may be smoothed into a continuous function. In case of discrete variables, only at most countably infinite values will be taken with a probability different from 0; in case of continuous variables, we will consider a probability density over uncountably infinite values. Of the functions characterizing X distribution law, we will consider in this chapter only the following two:

<small>through standard functions,</small>

Definition 1.1. *A random variable X is a variable taking values with probability* P *according to a probability function f_X (p.f.) such that for each $x \in \mathfrak{X}$*

$$P(X = x) = f_X(x) \qquad (1.1)$$

or, equivalently, according to a cumulative distribution function *F_X (c.d.f.) such that for each $x \in \mathfrak{X}$*

$$P(X \leq x) = F_X(x). \qquad (1.2)$$

By definition f_X values belong to the interval $(0, \infty)$ and F_X monotonically ranges from 0 to 1.

Of the most elementary functionals on the X distribution law, we will consider the *mean* $E[X]$ here:

<small>and standard parameters.</small>

$$E[X] = \sum_{x \in \mathfrak{X}} x f_X(x). \qquad (1.3)$$

As mentioned in the beginning of this section, the statistical granules may be built around observed data if we have an ordering criterion on them. In this case a granule identifies with the more technical concept of *statistically equivalent block*, as follows. Consider a set of data you have available and call it a sample if they satisfy a simple *fairness* requisite: they represent a sequence of observations of a same phenomenon without any reciprocal conditioning. In this case, they may be regarded in mathematical terms as realizations $\{x_1, \ldots, x_m\}$ of a set of m random variables $\{X_1, \ldots, X_m\}$ following identical distribution law and being mutually independent (i.i.d.), where m is the cardinality of the dataset. We denote the dataset to be a *sample of*

<small>From i.i.d. samples</small>

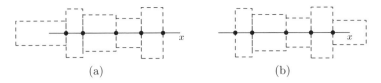

Fig. 1.2. Statistically equivalent blocks: (a) starting from $-\infty$, (b) ending at $+\infty$. Bullets: observed items; dashed lines: contours of probability masses.

the common random variable X and m the *sample size*. Willing to get more information from the data than simply their values, we may consider any possible additive measure μ_X on them, described through the cumulative function on the interval $(-\infty, x)$, i.e. $M_X(x) = \mu_X((-\infty, x))$. It may be, for instance, the number of bits through which each value is represented, or a suitable coding of the value into natural numbers and so on. With the additional normalization constraint $M_X(\infty) = 1$ we are coming to the set of all possible c.d.f.s on X. This gains the measure the meaning of probability and we may distinguish between distribution laws more or less compatible with the sample we have observed. For instance, you may judge less compatible with a sample of 10 defeating matches played by a given team a distribution law giving this team a winning probability close to 1 rather than close to 0.

We may deal with the compatibility measure as a further probability measure and get some quantitative appreciation of it. In particular, if we order the data as a vector $(x_{(1)}, \ldots, x_{(m)})$ such that $x_{(i)} \leq x_{(i+1)}, \forall i$ we discover that each interval between two points has a probability measure whose mean value is equal to the analogous intervals'. More formally we may say that:

Definition 1.2. *Given the ordered values $(x_{(1)}, \ldots, x_{(m)})$ of an i.i.d. sample specification, we define statistical blocks either the segments $(x_{(i)}, x_{(i+1)}]$ with $x_{(0)} \equiv -\infty$ or the segments $[x_{(i)}, x_{(i+1)})$ with $x_{(m+1)} \equiv +\infty$ (see Fig. 1.2).*

<small>we get statistically equivalent blocks</small>

The statistical blocks are *equivalent* under many respects [14]. In particular:

Fact 1.2. *Given a sample $\{X_1, \ldots, X_m\}$, denoting with U_k the probability measure of the union of any k statistical blocks, we have that:*

$$\frac{k}{m+1} \leq \mathrm{E}[U_k] \leq \frac{k+1}{m+1}. \tag{1.4}$$

The value that people generally assume satisfying the above inequalities is k/m. This corresponds to the rearrangement of the

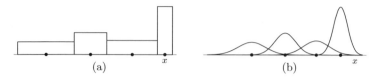

Fig. 1.3. (a) Equivalent blocks each having a measure $1/m$; and (b) information granules associated with the observed points by assigning them a bell shape. Bullets: observed items.

equal block mean measures $E[U_1]$ around the single points, by cutting the measure of one block, essentially the left half of the first block and the right half of the $(m+1)$th block and rearranging the values of the others to the common value $1/m$, with the nice feature that their sum is 1 (see Fig. 1.3(a)). Moreover, the frontiers between blocks are free to range between consecutive points, with the constraint of containing the one on which the block is closed and excluding the one on which the block is open.

The benefit of this definition is to deal with uniformly weighted points. *Vice versa*, if the probability measure π of a region is given and you question on how many points of the sample will fall inside it, the number of these points is such that, if you give equal probability $1/m$ to each sample point, the sum of probabilities of those falling in the above region tends to be exactly equal to π asymptotically with m. This closes the loop between properties inferred from points and operational use of these properties. You give a same probability mass to the sample points and interpret it as the measure of a cluster of points surrounding each sample item according to the same metric through which you ordered the sample. You don't care about the single element in the cluster, hence they uniformly share the property of the cluster that ultimately identifies with the coordinates of the sample point and a uniform probability $1/m$. All the rest of the statistics we will use comes exactly from this granulation.

Each observed value must be considered with same weight,

the rest is a consequence.

1.1.1 From Granules to Distributions

Describe data through a distribution function

Concentrating the weights on the sample points you may represent the sample as shown in Fig. 1.4(a). The ordinates are equally spatiated by $1/m$ while the abscissas are spread according to the observed phenomenon.

We denote the function in the graph an *empirical c.d.f.*

Definition 1.3. *Given a set of data* $\{x_1, \ldots, x_m\}$, *its empirical cumulative distribution function* $\widehat{F}_X$ *(e.c.d.f.) is defined as:*

$$\widehat{F}_X(x) = \frac{1}{m} \sum_{i=1}^{m} I_{(-\infty, x]}(x_i), \quad (1.5)$$

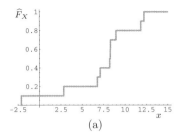 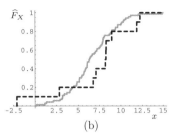

Fig. 1.4. (a) Representation of a sample of size $m = 10$ through the graph of the cumulative function $\widehat{F}_X$; (b) a smooth version of (a) obtained using a larger sample ($m = 70$).

where $I_{(a,b]}(y)$ is the indicator function *of the segment* $(a, b]$ outputting 1 *if* $a < y \leq b$, 0 *elsewhere*.

It is empirical since is based on limited amount of data, a drawback that is paid by the fact that it reports on the sampled points the expected value of F_X and is not accurate on the other points. We may rely on this way of organizing the data information if at the end of the day we may hit the specific distribution law the data belong to. The following fact denotes that we may tend to this distribution by increasing the size of the sample.

Fact 1.3. *If no other relation than the used order relation exists between the x_is in Definition 1.3, then for each i:*

$$|\widehat{F}_X(x_i) - \mathrm{E}[F_X(x_i)]| \leq \frac{1}{m+1}. \tag{1.6}$$

This fact proposes the empirical c.d.f. as representative of the expected c.d.f. with m increasing. Hence a reasonable strategy is to associate the data distribution to the empirical curve on a very large sample – what is called a *population* of data.

> Describe a distribution function either by using many data,

Example 1.1. In Fig. 1.4(b) we see the convergence of $\widehat{F}_X$ with the growth of the sample size to a smooth distribution that we may consider the *true* distribution of X.

1.1.2 Learning Distribution Parameters

In the previous section we are left with a smoothing of the e.c.d.f. into the asymptotical c.d.f. resuming the information about all possible X determinations we will meet in the future. We are allowed to smooth the staircase diagram in different ways, however, also if we rely on a few sample data, provided some essential constraints are satisfied. We start from a diagram denoting in the y axis the probability of finding a point before a given abscissa, in the preliminary hypothesis that only points from the sample will be observed in the future, rather, that all the other points are confused with (assimilated to) them. We are allowed to smear the $1/m$ probability masses along the intervals between consecutive points, according to smoothing functions that we assume suitable (maybe being easy to compute, enjoying special properties of symmetry or satisfying some logical requisites, etc.). This means that the resulting graph must still be monotone and ranging between 0 and 1, as shown in Fig. 1.5.

> or by introducing further knowledge,

Moreover, in the light of the fact that $1/m$ represents a mean value of the surrounding points probability, we are not obliged to make the curve passing through the corners of the original graph. Rather we may assume that some properties of the observed points must be taken into account, while the rest is irrelevant as irrelevant are the shifts of surrounding points from the sampled ones.

> with the effect of focusing on a few free parameters and on a few statistics of the sampled data.

This reverberates in the fact that if we already know the family of compatible distribution laws – possibly the result of having observed a very large sample in the past – it proves more efficient to focus on the parameters of the family and relate them to relevant properties of the sample, as mentioned before and as it will be done in the next section. As a conclusion, we move from distributions compatible with the extensions of the various blocks, to the ones compatible with relevant sample properties – call them *statistics*. The general scheme is the following:

> Sample is a matter of algorithms.

1. Sampling mechanism. We move from a sample $\{z_1, \ldots, z_m\}$ of a completely known random variable Z to a sample $\{x_1, \ldots, x_m\}$ of our random variable X through a mapping g_θ (called *explaining function*) such that $x_i = g_\theta(z_i)$ having θ as a free parameter. We call $\mathscr{M}_X = (Z, g_\theta)$ a *sampling mechanism* of X, and focus on learning θ from samples. We are ensured by the literature that there exist *universal* sampling mechanisms through which it is possible to generate samples belonging to any computable distribution law [38].

> The mather of all samples

Fact 1.4. *A universal sampling mechanism is represented by* $\mathscr{M}_X^* = (U, \widetilde{F}_{X_\theta}^{-1})$, *where* U *is a* $[0,1]$-*uniform variable and* $\widetilde{F}_{X_\theta}^{-1}$ *is a generalized inverse function of the c.d.f* F_{X_θ} *of the random variable* X *we have selected, having a free parameter* θ. *Namely* $\widetilde{F}_{X_\theta}^{-1}(u) = \min\{x | F_{X_\theta}(x) \geq u\}$. *As it emerges from Fig. 1.6, the above definition reads* $\widetilde{F}_{X_\theta}^{-1}(u) = (x | F_{X_\theta}(x) = u)$ *for* X *continuous, with an obvious extension for* X *discrete.*

The following examples show the implementation of this mechanism in common inference instances.

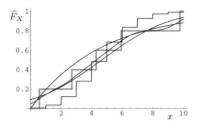

Fig. 1.5. Smooth versions of the graph in Fig. 1.4(a) (gray plot) obtained using further knowledge.

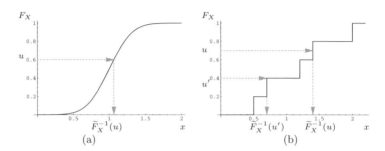

Fig. 1.6. A sampling mechanism for continuous (a) and discrete (b) random variables.

Example 1.2. We may produce a sample $\{x_1, \ldots, x_m\}$ of a Bernoulli variable X with parameter p from a sample $\{u_1, \ldots, u_m\}$ through the sampling mechanism $\mathcal{M}_X^* = (U, g_p)$ with the explaining function of the form:

$$g_p(u) = \begin{cases} 1 \text{ if } u \leq p \\ 0 \text{ otherwise.} \end{cases} \quad (1.7)$$

For X negative exponential random variable with parameter λ the explaining function is:

$$g_\lambda(u) = \frac{-\log u}{\lambda}. \quad (1.8)$$

For X uniform random variable in $[0, a]$ the explaining function has the form:

$$g_a(u) = au. \quad (1.9)$$

Remark 1.1. The blocks coming from U samples enjoy Fact 1.2 as well. Thus, blocks of uniform samples are at the basis of any variable sample given the universality of $\mathcal{M}_X^*$. The algorithm mapping from these to our actual blocks is the matter of our learning (this is why this approach is called algorithmic inference [3]). We may figure a backward trip, where starting from a sample $\{x_1, \ldots, x_m\}$ we grasp any functional commonalities between data till reducing them to a uniform sample. Put it differently, we terminate this back mapping when we have nothing else to understand about the values of the data, so that they behave as the purely random output of a uniform number generator within a standard range that we conventionally assume to be $[0, 1]$. In the reverse way, we put at the basis of m information granules a *seed* constituted by m uniform samples $\{u_1, \ldots, u_m\}$ and try to understand why the x_is displace from this sample.

> *Grasping structure from data*

2. Compatible populations. Let Z distribution and g to be given in the sampling mechanism, while θ is the free parameter to be learnt. Assume that the statistic s_Θ is the sample property to be explainable through θ; we are interested in how to move from $\widehat{F}_X$ to F_X, hence how to fix Θ values with the commitment of preserving s_Θ. Said in other words, you may invent any excuse for the fact that your soccer team defeated in the last 10 matches, but you cannot scratch this data item in whatever ability profiles you may imagine for the players. Summing up, we are interested in populations compatible with s_Θ, i.e. populations that, according to the sampling mechanism $\mathcal{M}_X$, could have generated $\{x_1, \ldots, x_m\}$ or *equivalent* samples. As nobody knows the sample seeds, we must face this unawareness with a reasoning that is uniformly true w.r.t. the seed population. Namely, assuming $s_\Theta = \rho(x_1, \ldots, x_m)$ and $x_i = g_\theta(z_i)$, we have the *master equation*:

> *Which mechanism could have generated the sample?*

> *All boils up around a logical pivot,*

$$s_\Theta = \rho(g_\theta(z_1), \ldots, g_\theta(z_m)) = h(\theta, z_1, \ldots, z_m). \quad (1.10)$$

We ideally consider all possible seeds $\{z_1, \ldots, z_m\}$ and obtain the Θ specifications that would have brought to the same main property of the sample by solving (1.10) in θ in correspondence to each seed. The remaining properties of the sample are assumed irrelevant, so that we may obtain s_Θ even generating different samples that are equivalent on this respect. In this way we obtain a population of θs that reflects a population of populations compatible with s_Θ. Let us remark our notations as a non marginal tool to explain the point. The parameter θ generating our sample through the related sampling mechanism is a specification of the random variable Θ; s_Θ is the observed statistic, hence in correspondence to unknown parameter θ since the corresponding seed $\{z_1, \ldots, z_m\}$ is unknown; s_θ is the value we would observe for a given seed if $\Theta = \theta$; capital symbols S_Θ and S_θ denote

> *Parameters described by random variables*

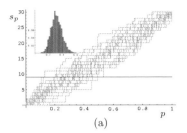
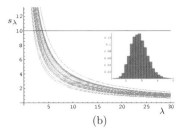

Fig. 1.7. (a) Course of s_p w.r.t. the unknown parameter p and histogram of the parameter when the statistic $s_P = 9$; and (b) course of s_λ w.r.t. λ and histogram of the parameter when $s_\Lambda = 10$. Both figures show 20 trajectories starting from a sample of size 30.

the related random variables generated by the random seed $\{Z_1, \ldots, Z_m\}$.

We follow the procedure sketched so far in the next example where: i) we randomly tossed a large amount of seeds in place of considering all possible them, and ii) we graphically solved (1.10) by drawing the trajectories of s_θ with θ ranging in a suitable interval in correspondence to each tossed seed and crossing these curves with the line $s_\theta = s_\Theta$.

Example 1.3. With reference to the sampling mechanism (1.7) and the statistic $s_P = \sum_{i=1}^{m} x_i/m$, Fig. 1.7(a) shows the course of s_p as a function of p for a fixed seed $\boldsymbol{u} = \{u_1, \ldots, u_m\}$ and the histogram of the values of P crossed by the line $s_P = 9$. Namely, the plotted function is $s_p = \sum_{i=1}^{m} I_{(0,p)}(u_i)/m$. The trajectories refer to seeds of size $m = 30$ and the statistic refers to a sample of bits where nine of them are set to 1. Note that equations like $s_P = \sum_{i=1}^{m} I_{(0,p)}(u_i)/m$ have no unique solution. Instead there exists an interval of p values solving it, which gives rise to a slight indeterminacy that we ignore for the moment.

> as a function of sample features.

Analogously, in Fig. 1.7(b) with reference to the sampling mechanism (1.8) and the statistic $s_\Lambda = \sum_{i=1}^{m} x_i/m$ we report the course of s_λ with λ and the histogram of λs crossed by the line $s_\Lambda = 10$. The sample size is still 30 while the plotted curve is now $s_\lambda = \sum_{i=1}^{m} -\log u_i/\lambda$.

3. Relevant properties. In spite of the persistency of the statistic $\sum_{i=1}^{m} x_i$ in the above example and its similarity to the most common evaluation we are used to do about data, i.e. their arithmetic mean $\sum_{i=1}^{m} x_i/m$, the identification of the data properties pivoting our computation is neither trivial in general nor arbitrary. The key requirement is to have a unique and statistically correct solution of (1.10) in θ. To be the identified θ statistically correct: i) it must be feasible in

> Which feature of the sample in particular?

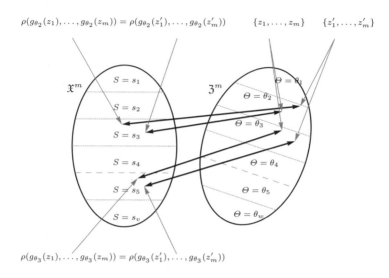

Fig. 1.8. Sufficient statistics and partition induced in $\mathfrak{X}^m$ and $\mathfrak{Z}^m$.

the sense that the observed sample can be generated with this value of the parameter, and ii) a collection of θs must be a true sample of Θ, hence must represent realizations of i.i.d. random variables.

Example 1.4. The statistic satisfying the above requirements in case of the family of uniform distribution in $[0, A]$, with $A > 0$ is $s_A = x_{\max} = \max\{x_i; i = 1, \ldots, m\}$, that in force of sampling mechanism (1.9) reads:

$$s_a = a u_{\max}, \tag{1.11}$$

for $u_{\max} = \max\{u_i; i = 1, \ldots, m\}$. Vice versa, the statistic $s'_A = \sum_{i=1}^m x_i$ may give rise to a solution a of (1.11) that is not feasible since smaller than $x_{\max}$.

In a next chapter, these considerations will be extended given that these properties are related to the ordering capability of the relevant statistic s. Consider the left picture in Fig. 1.8. The oval represents the space $\mathfrak{X}^m$ where each sample $\{x_1, \ldots, x_m\}$ is a point that is ordered according to the value of the S statistic computed on it. The relevance is reflected in the companion right figure. Both produce a partitioning of the space $\mathfrak{Z}^m$ where each point is a seed $\{z_1, \ldots, z_m\}$ of a sample. On the left, given a value of θ each $\mathfrak{Z}^m$ contour maps into a $\mathfrak{X}^m$ contour. On the right, for a given value of s the same contour is affected by a single value of Θ so that the sampling mechanism (Z, g_θ) produces an X sample with exactly the s value of the statistic. Thus on any pair (s, θ) affecting the actually observed sample, a correspondence exists between S and Θ contours. Hence, to draw the Θ distribution law we rely on the ordering functionality of the relevant statistics that in a relative (in place of absolute) framework brings to the notion of distance, possibly in its specification as a norm, as defined in Appendix A.2. In turn, the distance relation between data, also called metric, is a key feature through which to synthesize the information boiling up around them that we will consider at the basis of many inference procedures along this book.

In the next chapter we will see a pair of computational tools to efficiently implement the procedure producing the parameter distribution law.

1.2 The Borders of the Unawareness

"I know I do not know" (Socrtates, Apologia 21-23): a so good as illustrious sentence could prove a bit pretentious in opening of this section. It associates the young Hansel with the old Socrates in facing the unknown. In any case, Hansel has the operational target of coming back home. Hence he must set up a strategy that is suitable even when no moon is in the night. Out of the metaphors, our problem with fuzzy granules is still to exploit data information, but with some slot between them.

We may characterize the less informative framework with a lack of some quantitative links between the data. Image you have some local information about the distribution law of a variable you are interested in, for instance the fuel consumption of a car in various driving conditions, say in highway, in crowded traffic, in *normal* conditions, and assume, for the sake of simplicity, that these cover all your driving scenarios. Willing to know the probability of consuming more than ten liters for one hundred kilometers you must combine the probabilities, say $p_h(>), p_c(>)$ and $p_n(>)$ that this happens in the three mentioned driving

Combine information globally, if you know how,

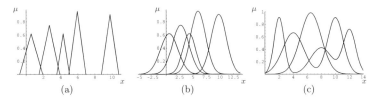

Fig. 1.9. Possible shapes of information granules around sample points: (a) triangular shape, (b) bell's shape, and (c) multi-modal granules having a bell shape.

conditions with the probabilities $p(h), p(c)$ and $p(n)$ that you meet these conditions during your driving life. Namely:

$$p(>) = p_h(>)p(h) + p_c(>)p(c) + p_n(>)p(n). \quad (1.12)$$

otherwise work locally. With fuzzy granules we are essentially in the situation where someone hid the weights $p(h), p(c)$ and $p(n)$, so that we are left with the commitment of a convex combination and a request of suitable weights.

With this scenario the three operations described in the previous section translate into:

1. Granule generation. Let us start with information granules *Granule shape up to you* around sampled points. Their shape, for instance of those considered in Fig. 1.3(b) in contrast with the statistical granules, comes from our past experience on the current phenomenon or similar ones, hence from past statistics on past samples that we forgot. Possibly we are left with some free parameters, such as the slopes in Fig. 1.9(a), or the scale factors in Fig. 1.9(b), and so on. Note that in the cases considered in the figure we assume that a broad order relation exists along the x-axis. Multi-modal graphs interleaving each other would denote the absence of even this feeble relation (see Fig. 1.9(c)).

2. From granules to fuzzy sets. As you cannot rely on a global *Work locally, but coherently.* vision, you must promise a local coherence as a direction for fixing weights. To this aim, let us shade the notion of set by first, in order to obtain more logically permissive, though still coherent, contexts. Given a set A, instead of defining the indicator function of elements with respect to A giving strict yes/no answers, let us allow a more variegate indication of the elements' membership to A. We obtain what is called a *fuzzy set* A which is described through its *membership functions* μ_A. Formally

Definition 1.4. *Given a set* $\mathfrak{X}$ *that we call the universe of* *A set plus a membership function* *discourse, a* fuzzy set A *is described through a* membership function $\mu_A : \mathfrak{X} \to [0,1]$. *The value* $\mu_A(x)$ *quantifies the grade of membership of the element* x *to the set* A.

Fuzzy set: a granule of granules, A granule is a fuzzy set. We may imagine granules Gs around each point of a set; some of them are around observed points, all of them constitute a granule population that we want to describe in a synthetic way through a fuzzy set A. We may define A as the union of all Gs, where the intersection between two granules is generally not empty. To give an operational meaning to this framework we are used to compute union and intersection between granules through the

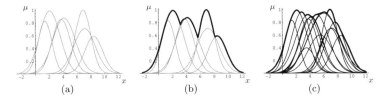

Fig. 1.10. (a) Granules having a bell shape; (b) membership function of the fuzzy set synthesizing the seven granules (black curve); (c) a smooth version of (b) obtained using a larger number of observed granules.

max and min operators according to the Gödel norm. In this chapter we adopt this option with the following definition.

Definition 1.5. *We compute the union* $A \cup B$ *of two fuzzy sets* A *and* B *on the universe of discourse* $\mathfrak{X}$ *through a fuzzy set on the same universe whose membership function in each point* x *is the maximum of the original membership functions in the same point. In formulas:*

$$\mu_{A \cup B}(x) = \max\{\mu_A(x), \mu_B(x)\}. \qquad (1.13)$$

The intersection $A \cap B$ *of two fuzzy sets* A *and* B *on* $\mathfrak{X}$ *is computed through a fuzzy set on* $\mathfrak{X}$ *whose membership function is the minimum of the original membership functions in each point:*

$$\mu_{A \cap B}(x) = \min\{\mu_A(x), \mu_B(x)\}. \qquad (1.14)$$

With these operators, a synthesis analogous (though with different meaning) to the e.c.d.f. is the union of observed granules, as in Fig. 1.10(a-b). It has analogous convergency properties. Its membership function, indeed, converges to an even smoother curve with the growth of the number of observed granules, asymptotically representing the membership function of the fuzzy set A that resumes the granule population (see Fig. 1.10(c)). As we will see in Appendix A.3, there exist many other ways of defining union and intersection of two fuzzy sets. They all satisfy the logical constraint to be based on norm operations and become the classical union and intersection operators for degenerate *fuzzy sets* A and B having membership funcions constituting indicator functions, hence taking values in $\{0, 1\}$. With fuzzy sets we are not able to give a clear ordering of the data. This reflects on the relaxing of the 0 definition of the related norm: not yet $\nu(\boldsymbol{w}) = 0$ if and only if $\boldsymbol{w}$ is the null vector (see point N3 in

either synthesized through norm operators,

– a fuzzy norm in any way at the basis of fuzzy set management –

Definition A.2), hence $\mathsf{T}(a,b) = 0$ if and only if $a = b$. Rather $\mathsf{T}(a,0) = 0$ and $\mathsf{T}(a,1) = a$ (see point T4 and T5 Definition A.4). This is the root of the fuzziness data structure, set contours included, that we manage in a specific way.

With a few granules you may try to infer the fuzzy set membership function in a very specular way we do with distribution laws, but centered on local properties. The key idea is that with expanding a sample point into a granule we are stating an equivalence relation between the points within a granule. It says that with observing a sampled point we are virtually observing any of the points of the equivalence set, though with different membership function. Thus contrasting the membership functions of the observed granules with the membership function of the fuzzy set A resuming the entire population, we are looking for the representative points in each granule, i.e. the points that really did contribute with their membership function to fix the values of μ_A. Given the idempotence of max operator we identify them in the crosses of granule and fuzzy set membership functions. Indeed, let us resume the maximum of all granules membership functions exactly in μ_A. Then, by definition, on each such x, $\mu_A(x)$ is a fixed point w.r.t. $\max\{\mu_A(x), \mu_G(x)\}$. We are logically saying that these points belong to both G and A, where the membership function to the intersection of the two sets in x is given by $\min\{\mu_A(x), \mu_G(x)\}$, according to Definition 1.5, and this quantity still equals $\mu_A(x)$, thanks to the idempotence of the min operator as well (see Fig. 1.11(a)). If we have more than one cross, then we have many representatives of the granule. Hence, we take the maximum of their membership function in order to compute the membership of the entire granule to the fuzzy set.

Do the same with all granules and take the min of the granule membership functions, so that you may associate, at the end of the day, a membership function to μ_A. In case we have a family M_A of membership functions with some free parameters, we may interpret the above membership function as a compatibility measure of any element μ_A of the family with the granules.

Example 1.5. Let us come back to the set of 7 granules already considered in Fig. 1.10. Now we have the further assumption that the fuzzy set A resuming the granule population as a membership function belongs to the Gaussian family $\mathscr{F}_A$ (see Appendix C for commonly used membership function families), where each element is characterized by the parameters h, ν and σ, denoting respectively the height,

[margin note: or fitted across fixed points.]

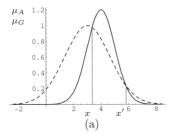 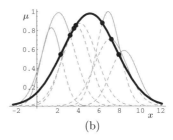

Fig. 1.11. (a) Membership functions of a granule G (plain curve) and a fuzzy set A (dashed curve). The granule points x and x' are candidates for contributing to the A membership function; (b) the most compatible fuzzy set with the observed granules. Bullets denote the granule coordinates and their degrees concurring to determine the fuzzy set membership function (thick curve).

central position and width of the Gaussian bell. With the above procedure we may give a compatibility measure to each triplet (h, ν, σ). In Fig. 1.11(b) we represent the membership function with maximum compatibility.

1.3 From Data to Functions

Using oblong stones fixing the cone width containing the correct direction is another way Hansel pursues to granulate points. In this case we have the benefit of sharp contours of the granule but the drawback of no indications inside it. This is a typical instance of interval computation that we generalize to boolean functions. Said in other words, given an observed data we consider the equivalence set of points with respect to a given boolean function. In the case of the arrow-like pebble, the function answers 1 if the direction chosen by Hansel is not farther than 15 degrees on the left or on the right of the arrow direction. In the case that the observed data are sets of boolean variables, in the book we will refer to monomial and clauses as favorite granules. Before discussing them, let us introduce the following notation.

Flat granules with sharp frontiers, i.e.

Definition 1.6. *On the assignments space $\mathfrak{X} = \{0,1\}^n$ we define a boolean function $c : \mathfrak{X} \to \{0,1\}$ through boolean operations on the propositional variables $\{\mathsf{v}_1, \ldots, \mathsf{v}_n\}$ such that v_i is satisfied if $x_i = 1$, not satisfied if $x_1 = 0$. In particular, for literal ℓ_i an affirmed (v_i) or negated propositional variable ($\overline{\mathsf{v}}_i$), a monomial is a conjunction ($\ell_{i_1} \wedge \ldots \wedge \ell_{i_k}$) of literals and a clause is a disjunction ($\ell_{i_1} \vee \ldots \vee \ell_{i_h}$) of literals. A boolean formula is monotone if is expressed as a function of sole affirmed variables. The support of a boolean formula c is the set of all assignments satisfying c. By*

the support of boolean functions

abuse, we will often confuse a formula with its support so that, for instance, $x \in c$ and $c \subseteq c'$ means respectively that x belongs to the support of c and that the support of c is included in the support of c'.

The preference for the above kinds of granules comes from the following fact.

Fact 1.5. *Each boolean formula* $\{0,1\}^n \to \{0,1\}$ *may be expressed*

- *in a Disjunctive Normal Form (DNF), i.e. as the disjunction of monomials;*
- *in a Conjunctive Normal Form (CNF), i.e. as the conjunction of clauses.*

Hence our granules belong to the set of building blocks of any boolean formula. Moreover denoting positive points w.r.t. a formula c those belonging to the support of c and negative points the remaining points in a given assignment space $\mathfrak{X}$, if we observe a set of m^+ positive points $\{x_1, \ldots, x_{m^+}\}$, the smallest formula containing them, hence consistent with the data and included in any other formula with this property, is a DNF made up of the disjunction of m^+ monomials. Each monomial corresponds to an observed point: it is the conjunction of n literals, where each boolean variable is affirmed if in the point is assigned by 1, negated otherwise. Analogously the largest formula excluding a set of m^- negative points is a CNF disjoining m^- clauses, one for each point. Namely, the clause is made of the disjunction of n literals, where a boolean variable is affirmed if in the point is assigned by 0, negated otherwise.

<small>Favorite granules: monomials and clauses, if the data have a label,</small>

Example 1.6. The minimal boolean formula on $\{0,1\}^5$ including assignments $\{00110, 10011, 00001\}$ is the DNF: $(\bar{v}_1 \wedge \bar{v}_2 \wedge v_3 \wedge v_4 \wedge \bar{v}_5) \vee (v_1 \wedge \bar{v}_2 \wedge \bar{v}_3 \wedge v_4 \wedge v_5) \vee (\bar{v}_1 \wedge \bar{v}_2 \wedge \bar{v}_3 \wedge \bar{v}_4 \wedge v_5)$. The maximal boolean formula excluding assignments $\{11000, 00011\}$ is the CNF: $(\bar{v}_1 \vee \bar{v}_2 \vee v_3 \vee v_4 \vee v_5) \wedge (v_1 \vee v_2 \vee v_3 \vee \bar{v}_4 \vee \bar{v}_5)$.

A key feature of these granules is that, if negative and positive points refer to a same formula, then the union of monomial is included in the intersection of clauses. This is an *elegant* property that proves very ineffective, however, since each monomial contains exactly one point – the union m^+ of the 2^n points – and each clause excludes one point – the intersection m^- of the points. Things are different if we have some *a priori* hypothesis on c. The most general and suitable one is that c is monotone. Now, on the one hand we may re-enunciate Fact 1.5 saying that each monotone boolean formula can be expressed either in terms of monotone DNF or in terms of monotone CNF. On the other hand the dummy granules around points, constituted by monomials whose supports identify exactly with the points, and analogously for clauses, now become real and generally very wide granules that we call canonical monomials and canonical clauses, respectively. We may realize it thanks to the following property introduced by the monotonicity hypothesis.

Fact 1.6. *i) if an assignment* s *(say, $(0, 1, 0, \ldots, 1, 0)$) satisfies c, then any s' with one or more 0 flipped into 1 (say, $(0, 1, 1, \ldots, 1, 1)$) satisfies c as well. The set of all such s' constitutes the support of the following monomial:*

in particular,

Definition 1.7. *Given a set E^+ of points denoting positive examples in $\{0, 1\}^n$, a monotone monomial* m *(conjunction of affirmed variables) is a* canonical monomial *if $x \in E^+$ exists such that*

canonical monomial, and

$$\text{for each } i \in \{1, \ldots, n\} \begin{cases} \mathsf{v}_i \in \text{set}(\mathsf{m}) & \text{if } x_i = 1 \\ \mathsf{v}_i \notin \text{set}(\mathsf{m}) & \text{otherwise;} \end{cases} \quad (1.15)$$

ii) if an assignment s *(say, $(0, 1, 0, \ldots, 1, 0)$) does not satisfy c, then any s' with one or more 1 flipped into 0 (say, $(0, 0, 0, \ldots, 1, 0)$) does not satisfy c as well. The set of all such s' constitutes the complement of the support of the following monomial:*

Definition 1.8. *Given a set E^- of points denoting negative examples in $\{0, 1\}^n$, a monotone clause* c *(disjunction of affirmed variables) is a* canonical clause *if $x \in E^-$ exists such that:*

canonical clauses

$$\text{for each } i \in \{1, \ldots, n\} \begin{cases} \mathsf{v}_i \in \text{set}(\mathsf{c}) & \text{if } x_i = 0 \\ \mathsf{v}_i \notin \text{set}(\mathsf{c}) & \text{otherwise.} \end{cases} \quad (1.16)$$

Example 1.7. With reference to Example 1.6 the canonical monomials corresponding to the positive examples $\{00110, 10011, 00001\}$ are $x_3 x_4$, $x_1 x_4 x_5$ and x_5 respectively. These are represented in Fig. 1.12 as the inner gray granules together with their generating points.

Example 1.8. Still referring to Example 1.6, the canonical clauses corresponding to the negative examples $\{11000, 00011\}$ are $x_3 + x_4 + x_5$ and $x_1 + x_2 + x_3$ respectively. The big circles in Fig. 1.12 excluding negative points (crosses in the figure) are a visual representation of canonical clauses.

Now, a canonical monomial m is a minimal formula, in the sense that no monotone monomial with shorter support could contain

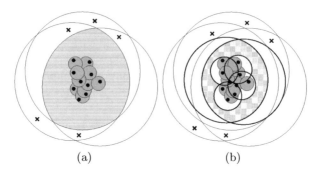

Fig. 1.12. Monomials (inner granules) and clauses (outer granules) consistent with the observed sample and identifying a region (gap) within which to find the contour of the target concept c. (a) and (b) show the granules at two different levels of abstraction.

the example generating m. Hence, the union of canonical monomials constitutes a DNF representation of the minimal formula consistent with (containing) the positive examples. Similarly, a canonical clause is a maximal formula and the intersection of canonical clauses represents a maximal (possible) consistent hypothesis on c containing c. Visualizing these formulas through the Venn diagrams in Fig. 1.12, we realize that the expansion of points into granules leaves a gap within which to find the contour of any formula c discriminating positive from negative granules. This formula may represent the overall synthetic description of the granules like fuzzy sets do with local granules. The tighter is the gap the more precise is the identification of c.

<small>delimiting the divide between positive and negative records.</small>

Not always the monotonicity hypothesis may be assumed. However we may in principle bypassing it by using a duplication trick: we may duplicate the propositional variables, setting $v_{n+k} = \bar{v}_k$ for each $k \in \{1, ..., n\}$, so that the same example which assigns value 1 to v_k assigns value 0 to v_{n+k} and *vice versa*. In this way each canonical monomial and clause has all n literals (affirmed or negated) in its argument, with the drawback of having a support again composed of only one point for the monomial and of all but one point for the clause. Granulating will occur as soon as we are able to remove one of the opposite variables of a given literal. Analogously, additional hypotheses could still enlarge the granule size as in Fig. 1.12(b).

1.4 Functions of Granules

At the end of the journey Hansel would learn how to come back home in any moment, without renewing the pebbles' track any

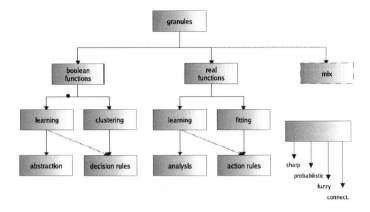

Fig. 1.13. A synopsis of the puzzle pieces.

time. He must associate some landscape features to the sites marked by pebbles (positive examples) and to analogous sites he found distracting him from the right road. In this manner he will understand from these features the direction to follow from any place he is located. This may represent a template of the final use we want of the granules: choosing an action as a reaction to the observed data and current stimulus. We have variegate tools for managing information to reach this goal. They fall in the realm of learning activities in late sense. As mentioned in the title of the book, these methods look like puzzle pieces to the eyes of the user, to which we try to offer an interpretation key through the graph in Fig. 1.13 as a dual perspective of Fig. 0.1, of course without any exhaustiveness pretense. Namely, we make a first distinction based on the nature of the investigated function: a boolean function having a decision rule as operational counterparts, and a real valued function bringing to action rules. In both cases we further distinguish between the scope of our computation: i) true learning of a function in order to have a meaningful representation of it, or ii) its efficient use, formalized into a clustering problem with boolean functions and into a fitting problem with non boolean functions. The operational results of the computation split consequently. On the one hand you obtain results that may constitute intermediate elements to be reprocessed within the above taxonomy. Hence you have properties of observed points as a result of learning a boolean function that may be reused as *metapoints* within an abstraction process. Similarly you have a more analytic view of the data if you learn functions from them, a view that may be refined through further learning procedures. On the other hand, clustering and fitting are more directly finalized to get either decision or action rules. Nothing is definitely rigid. So you may translate learnt functions

In search of the links connecting data,

by tossing a variety of ways,

24   Granule Formation Around Data

explored with a variety of perspectives

into rules, or you may have a mix of these situations as sketched in the box "mix". In the book we will cross this graph with four different perspectives: sharp, probabilistic, fuzzy and connectionist, loosely ordered from left to right in function of the quantity of information they are able to put on the desk. We color the boxes of the diagram in Fig. 1.13 (admittedly, with no distinction between them) with levels of shading to qualitatively indicate the relevance of perspectives on tasks. In greater details:

Chops of formal knowledge,

- **sharp.** *Dear Glaucon, you will not be able to follow me here, though I would do my best, you should behold not an image only but the absolute truth* [32]. On the part of Glaucon, we are not looking for the absolute truth, but we need some formal statement (actually the current synthesis of past data), the generally mentioned inductive bias [26], to start our inferences. Rather than *universal laws* such as $f = m \cdot a$ we look for pieces of properties we have already discovered about data (for instance their metrics). These fragments of sharp knowledge affect all the boxes in the diagram. For instance, we may make a logical assumption about the formulas monotony as in Section 1.3. Analogously, in a regression model like:

$$y_i = g_\theta(x_i) + \epsilon_i, \qquad (1.17)$$

where Y specifications are linked to X, the noise term ϵ_i gives room to a large family of regression functions since it defines equivalence regions around the sampled point (x_i, y_i), so that a fitting curve $g_\theta : \theta \in \Theta$ is variously satisfactory if it crosses any point in this region. Also in this case *a priori* assumptions are done on g, leaving θ as free parameter.

glued with hints from the framing of the data

- **probabilistic.** The scenario introduced by the inductive biases may be described by pieces of functions in a functional space to be properly completed and glued together. This may be read as the task of learning a function within a class of functions. If we may devise a suitable ordering on them, hence if we have a metrics for the related sampled data, we may deduce from the latter (jointly with emerging features) a distribution law of the functions, and solve the learning problem in terms of optimization of some cost function.

and from specific experience,

- **fuzzy.** If we do not have a global metric, but we may order data (and functions) only locally, we necessarily broaden the cost function and use specific defuzzifying criteria to isolate the goal function.

or just by numerical fitting in the worst case.

- **connectionist.** If you have no information about granules then you have no pretenses about function classes, metrics and so on. You just ask for rules working well enough, don't care why. This is realm of *subsymbolic devices* such as neural networks or genetic algorithms. Here we just give some smooth functions to expand granules, then leave their actual configuration as the output of a general fitting algorithm. The basic criterion is exactly granular: if you meet an input *similar* to something you have already observed you may expect results *similar* to those you have correspondingly observed. Hence you learn to reproduce the input/output observed pairs and expect the function you have learnt to work well on similar data as well. Actually we don't limit ourselves to cross the fingers. We work to ensure the learnt function some general properties.

1.5 The Benchmarks of This Book

Finally, we do not forget the main claim of this book: *data by first, rules just as it needs*. Hence in this section we detail on the various datasets from which we will squeeze information with the methods and algorithms proposed in the book. We will first describe the structure of the data base. Then, for the largest bulk of data we will refer the reader to the website where they are reposed. The small sets are directly reported on the book. We will toss our methods on these datasets for mere didactical reasons, hence to asses their rationale rather than in search of outperforming results. As for a notational remark, in the book we will quote these benchmark exactly with the titles of the sections where they are described, omitting mentioning the sections as well.

No good idea if not tossed on benchmarks

1.5.1 Synthetic Datasets

When willing to describe some general features or peculiarities of a given method, we will often resort to *synthetic* datasets. These benchmarks will be generated within a two-stage procedure, consisting of: i) drawing a set of points from standard distributions (typically, extracting vectors uniformly in a given region), and ii) mapping each point through suitable explaining functions (see Section 1.1.2) to vectors of variables constituting the instances of the problem we want to solve with our methods.

Artificial data for specific checks

1.5.2 Leukemia

We will use two sets of data reporting critical times in the cancer treatment. Namely, Table 1.1 shows the illness re-occurence time (in years) of leukemia after surgery [15], while Table 1.2 the remission times (in weeks) of patients with acute myelogenous leukemia undergoing maintenance chemotherapy [7].

Clinical data/cancer

Table 1.1. Leukemia re-occurrence times after surgery.

| 1 1 2 2 3 4 4 5 5 8 8 8 8 11 11 12 12 15 17 22 23 |

Table 1.2. Censored (marked with >) and non censored remission times (in weeks) of patients with acute myelogenous leukemia undergoing maintenance chemotherapy.

| 9 13 >13 18 12 23 31 34 >45 48 >161 |

Typical distinction with these records is made between censored and non censored data. The former refer to subjects that are not observed for full duration of the phenomenon (reoccurrence/dismission) we are interested in – maybe due to loss, early termination, or death by other causes, etc. – representing in any case incomplete information. The latter reports times that are completely observed.

Data in Table 1.1 are non censored. Those in Table 1.2 are partly censored partly not – a typical experimental record – where the formers are marked with symbol ">" before the data.

1.5.3 Cox

Clinical data/genetics

This biomedical dataset describes a set of 209 observations related to a rare genetic disorder; it was used as a benchmark in 1982 at the annual meeting of the American Statistical Association [8]. Each observation consists of one value for the patient's age and four values for blood measurements. As usual in clinical investigations these data come both from subjects not affected by the disease, that are called *normal* individuals, and from subjects bearing the disease, that are denoted *carriers*. As the investigated disease is rare, the latter class has a smaller number of representatives in the whole dataset (namely, 75 points out of 209). Moreover, in the sample there are 15 points deceptive in regard to some measure. The dataset can be downloaded from [8] by clicking on the biomedical data base `biomed`.

1.5.4 Guinea Pig

Biomedical data

This dataset, shown in Table 1.3, concerns the uptake of B-methyl-glucoside in the guinea pig intestine as a function of its concentration ρ.

The data represent the averages of 5 measurements in small pieces of the intestine of 8 pigs (in the column heads) at 10 different B-metyl-glucoside concentrations (in the row heads).

Table 1.3. Guinea pigs dataset: $\rho \rightarrow$ B-methyl-glucoside concentration, guinea pig $\rightarrow$ pig identifier, cells $\rightarrow$ B-methyl-glucoside uptake.

		\multicolumn{8}{c}{guinea pig}
		1	2	3	4	5	6	7	8
	.5	332	312	309	347	324	334	339	335
	1	263	268	264	289	263	278	284	281
	2	226	223	216	229	236	246	233	254
	4	189	187	200	201	198	193	206	215
ρ	7	177	171	161	177	156	158	183	178
	10	139	149	150	168	135	166	162	166
	15	139	125	132	146	132	144	154	143
	20	151	101	128	139	114	164	139	145
	35	98	100	97	120	86	124	113	122
	50	47	87	84	92	82	118	106	103

Formerly studied by Johansen [18], the Guinea Pig dataset has been subsequently modeled by other researchers [25].

1.5.5 SMSA

The SMSA dataset [35] describes air pollution and mortality of 60 Standard Metropolitan Statistical Areas (a standard Census Bureau designation of the region around a city) in the United States, collected from a variety of sources. The data include information on the social and economic conditions in these areas, on

Ecology concern

Table 1.4. Structure of the SMSA dataset.

Name	Description
City	City name
JanTemp	Mean January temperature (degrees Farenheit)
JulyTemp	Mean July temperature (degrees Farenheit)
RelHum	Relative Humidity
Rain	Annual rainfall (inches)
Mortality	Age adjusted mortality
Education	Median education
PopDensity	Population density
%NonWhite	Percentage of non whites
%WC	Percentage of white collar workers
Pop	Population
Pop/house	Population per household
Income	Median income
HCPot	HC pollution potential
NOxPot	Nitrous Oxide pollution potential
SO2Pot	Sulfur Dioxide pollution potential
NOx	Nitrous Oxide

their climate, and some indices of air pollution potentials. Each record is made up of 17 values as described in Table 1.4. The dataset is downloadable from [35]. The typical use of this benchmark is in multiple regression analysis to determine whether air pollution and other socio-economic conditions are significantly related to mortality.

1.5.6 Swiss

Socio-economic concern

This dataset reports a set of socio-economic indicators, as well as a standardized fertility measure, for 47 French-speaking provinces of Switzerland in 1888 [29, 6]. The benchmark contains one record per province and each record is made up of 6 values ranging in [0, 100], as shown in Table 1.5. The dataset is downloadable from [29].

Table 1.5. Structure of the Swiss dataset.

Name	Description
Fertility	Standardized fertility measure
Agriculture	% of males involved in agriculture as occupation
Examination	% of people receiving highest mark on army examination
Education	% of people with education level beyond primary school
Catholic	% of catholic people
Infant mortality	% of live births who live less than 1 year

1.5.7 Iris

Plant features

The Iris Plants data base created by R. A. Fisher [13] is one of the best known datasets to be found in the pattern recognition literature. It is constituted by a set of records describing 3 classes of plants: Iris Setosa, Versicolor and Virginica. For each class 50 instances are reported in terms of four canonical features: sepal and petal length and width. It is a very simple domain: one class (Iris Setosa) is linearly separable from the other 2; nevertheless the latter are not linearly separable from each other. The original version of the IRIS dataset may be downloaded from [16].

1.5.8 MNIST Digits Data Base

Handwritten digits

This handwritten digits benchmark [23] was obtained as the result of some modifications (size-normalization and centering) on the original NIST data base [31] which extends the former by providing a huge number of images representing handwritten characters. It consists of a training and a test set of 60000 and 10000 examples, respectively. Each example contains a 28×28 grid of 256-valued gray shades (between 0 and 255, with 0 meaning white and 255 black). Fig. 1.14 reports a sample of these

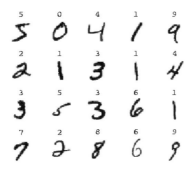

Fig. 1.14. A sample of 20 digits written by different people as collected in the MNIST dataset.

data. Actually their number is so huge that we often worked with a subset of them, while the processing of the entire data base may prove prohibitive for non-optimized algorithms. The dataset (together with the NIST data base) is downloadable for free from [27]. To obtain it, is sufficent to reach the Data section which lists benchmarks typically used in the machine learning community. In particular, for the MNIST dataset a list of classification algorithms may be found together with a description of the preprocessing operations, classification errors and reference to the corresponding articles.

1.5.9 Vote

This benchmark consists in the votes for each of 300 U.S. House of the Representatives Congress people on 16 key votes. The goal is to recognize, on the basis of the votes, whether the Congress-people is a Democratic or a Republican. The records consist of 16 three-valued variables, reporting a favorable (y), unfavorable (n) or unexpressed vote (u), plus a binary variable declaring the party of the voter. This dataset is included in the C4.5 distribution available on the web [33].

True democratic voters

1.5.10 Sonar Signals

The dataset [4] contains a total of 208 patterns: i) 111 items are obtained by bouncing sonar signals off metal cylinders, and ii) 97 items off rocks. The transmitted sonar signal is a frequency-modulated chirp, rising in frequency. The dataset contains signals obtained from a variety of different aspect angles, spanning 90 degrees for the cylinder and 180 degrees for the rock. Each pattern is a set of 60 numbers in the range 0.0 to 1.0. Each number represents the time integrated energy within a particular

Sonar inspections

frequency band. The label associated with each record contains the letter R if the object is a rock and M if it is a metal cylinder.

1.6 Exercises

1. Given $\mathfrak{X} = \{2, 3, 5, 8, 9\}$, consider the random variable X associated to the following probability function f_X:

x_i	2	3	5	8	9
$f_X(x_i)$	0.10	0.54	0.23	0.02	0.11

   compute $P(X = 5)$, $P(X = 6)$, $F_X(0)$, $F_X(5)$, and the mean of X.
2. Draw the graph of the c.d.f. of the random variable X described in Exercise 1.
3. Let $\{3, 8, 9, 3, 3, 2, 5\}$ be a sample drawn from X as in Exercise 1. Compute its empirical c.d.f. $\widehat{F}_X$ and compare its graph with that of F_X obtained in Exercise 2. Then check for which values of the sample the inequality (1.6) is satisfied with $\mathrm{E}[F_X] = F_X$.
4. Generate a sample of 20 specifications of a negative exponential random variable of parameter $\lambda = 0.3$ through the explaining function (1.8), then compute their empirical cumulative distribution with the related c.d.f., which assumes the form $F_X(x) = 1 - \mathrm{e}^{-\lambda x}$. Repeat the experiment using a uniform random variable described by the explaining function (1.9) with $a = 7$.
5. Rewrite the equation (1.10) for the particular case of a negative exponential distribution, $s_\Lambda = \sum_{i=1}^m x_i$, and uniform random seeds.
6. Given the universe of discourse $\mathfrak{X} = [0, 1]$, let A and B be two fuzzy sets on it having as membership functions, respectively, $\mu_\mathrm{A}(x) = x^2$ and $\mu_\mathrm{B}(x) = 1 - x^2$. Compute the membership functions of $\mathrm{A} \cup \mathrm{B}$ and $\mathrm{A} \cap \mathrm{B}$, then draw their graphs.
7. Consider a set of granules $\{\mathrm{G}_1, \ldots, \mathrm{G}_6\}$, each with a triangular membership function with extremes in the points 0 and 16 on the real line, and vertices $b_i = i^2/2.5$. Infer a triangular membership function of a fuzzy set with same extremes resuming the granules.
8. Express the boolean formula $((\mathsf{v}_1 \vee (\mathsf{v}_2 \wedge \bar{\mathsf{v}}_3)) \wedge \mathsf{v}_4) \vee \bar{\mathsf{v}}_2$ in disjunctive normal form and in conjunctive normal form.
9. Convert the boolean formula $(\mathsf{v}_1 \vee \bar{\mathsf{v}}_3) \wedge (\mathsf{v}_2 \vee \mathsf{v}_3) \wedge (\bar{\mathsf{v}}_2 \vee \mathsf{v}_4)$ from the conjunctive to the disjunctive normal form.
10. Given the assignments $\{0101, 0001, 1011, 0000, 1101\}$, compute the minimal monotone DNF satisfying them.

11. Compute the empirical c.d.f. of the values in Table 1.1.
12. Broaden the values in Table 1.2 into granules with membership functions belonging to one of the families in Appendix C. How may you take into account that some values are censored?
13. Design an appropriate representation for the digits contained in the MNIST dataset allowing a metric. Then compute the empirical c.d.f. based on a sample of these digits.

Further Reading

This chapter refers to a quite wide spectrum of philosophical aspects that we preferred to avoid, however, in favor of a very pragmatic approach: *introduce concepts not as pieces of truth but just as tools for suitable computing*. The interested reader may deepen the bases of probability for instance in [3, 28], of fuzzy set in [24] and of granular computing in [5]. Fact 1.2 comes from a revisiting of Tukey's results on statistically equivalent blocks [37]. Therein the blocks have random delimiters $\{X_{(1)}, \ldots, X_{(m)}\}$. In our approach the blocks have fixed delimiters $\{x_{(1)}, \ldots, x_{(m)}\}$, while the randomness of their probability measure depends on the randomness of the probability distributions compatible with the observed sample. The basic theorem obtained by paraphrasing results of [2] is the following:

Theorem 1.1. *Given a sample* $\{x_1, \ldots, x_m\}$, *denoting with* U_k *the probability measure of the union of* k *statistical blocks, we have that:*

$$I_\alpha(k, m-k+1) \geq P(U_k \leq \alpha) \geq I_\alpha(k+1, m-k) \qquad (1.18)$$

where $I_\alpha(h_1, h_2)$ *is the Incomplete Beta function [38] with parameters* h_1, h_2.

With Incomplete Beta function is meant the c.d.f. of a special random variable denoted Beta variable (see Section 3.1.3 and Appendix B.1). The bounds on $E[U_k]$ in Fact 1.2 derive from the fact that the expected value of a Beta variable with parameters $(k, m-k+1)$ is $k/(m+1)$.

The notion of compatibility throws a bridge way of inference between Agnostic Learning [20] and Prior Knowledge [21] based on an inference goal represented not by the attainment of truth but simply by a suitable organization of the knowledge we have accumulated on the observed data. In a framework where this knowledge is not definite, we smear it across a series of possible models that we characterize through a probability measure effectively explaining the observed data which denotes their *compatibility* with the data. According to Laplace [22], Fisher [12], De Finetti [9] and others, we assume the sample data as fixed point and aim at synthesizing all what we know about them in terms of distribution over the functions that my have generated them. The operational value is that if we maintain the physic of the phenomenon unchanged (whatever it will be), hence the generating function as well, then an analogous distribution applies to the operational synthesis (such as mean, variance, and so on) we will use in the future. Our starting point is a sampling mechanism where we concentrate both all prior knowledge and the part that still remains unknown. The framework of parallel universes (multiverses) has been considered in many disciplines such as cosmology, physics, philosophy, theology, and in fictions as well, as a key for solving the indeterminacy of various phenomena [17] – a true trigger of the agnosticism. With a less ambitious goal we move from universe to our surrounding world and aim to describe a reasonable set of its instantiations that we realize to be *compatible* with our experience. Compatibility plays an analogous role of likelihood in classical statistics. The latter appreciate the compliance of the observed data with a model; the former the compliance of a model with the data. As a matter of fact compliance is not a symmetric relation, and our perspective enables a more clear and operational understanding of computational learning mechanisms.

The relation we state between granules and fuzzy sets goes ideally in the opposite direction than the refinement operations on fuzzy sets that recently are deeply studied for calibrating fuzzy rules [30, 10]. Therein we move from synthetic to detailed descriptions of fuzzy sets at the basis of decision rules. Here we move from the most detailed description, the granules, to a synthesizing set. Also the way of inferring the membership function of the latter is new. In Chapter 5 it will be extended to more complex instances.

Finally the gauging of boolean functions through maximal and minimal hypothesis made up of logical granules as in Fig. 1.12 is a logical counterpart of the statistical concept of confidence interval [3] that meets similar concepts devised in other theories such as interval-valued fuzzy sets [34], interval computation [19] or possibility theory [11].

With these conceptual implementations we approach the wide field of computational intelligence with an own perspective. As inference results in a computational task and we do not lack

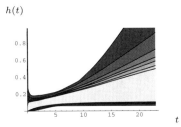

Fig. 1.15. Nested confidence regions containing hazard functions for the re-occurrence of leukemia after surgery described in Table 1.1. Different gray levels denote different probability α that hazard functions fall in the nested domains: from light to dark gray $\alpha = 0.70, 0.75, 0.80, 0.85, 0.90, 0.95, 0.99$.

computational resources, we abandon the old *divide et impera* scheme where data uncertainties are solved by first and then strict deterministic rules are implemented [1]. Rather we look for solving the uncertainty of the data exactly in view of their final use. This means that we move from granulated data to granulated functions and we solve indeterminacy exactly on the latter. To give an early example, in Fig. 1.15 we report a series of nested regions containing with the attached probability the curve representing the reoccurrence rate of leukemia as described in Section 1.5.2.

As for datasets, web facilities made available large benchmarks of data with which to stress our algorithms, see for instance [4, 36]. Most of them reports really granular data, so that the values inside must be considered as indicators of a phenomenon, rather than its univocally exact expression. Thus they will be used in order to stress the peculiarities of the methods we show in the book, rather than targets to be identically reproduced by the methods. In other words, w.r.t. the sampling mechanism (1.7) our goal will be to identify the explaining function abandoning the idea of identifying the random seed. While increasing the power of the former is an objective worth pursuing, the success in the latter task could be necessarily non repeatable, when it depends on specifications of variables not forecastable by definition.

References

1. Aho, A., Hopcroft, J., Ullman, I.: The Design and Analysis of Algorithms. Addison-Wesley, Reading (1974)
2. Apolloni, B., Esposito, A., Malchiodi, D., Orovas, C., Palmas, G., Taylor, J.G.: A general framework for learning rules from data. IEEE Transactions on Neural Networks 15, 1333–1349 (2004)
3. Apolloni, B., Malchiodi, D., Gaito, S.: Algorithmic Inference in Machine Learning, 2nd edn. International Series on Advanced Intelligence, Advanced Knowledge International, Magill, Adelaide, vol. 5 (2006)
4. Asuncion, A., Newman, D.J.: UCI machine learning repository (2007),
   http://www.ics.uci.edu/~mlearn/MLRepository.html
5. Bargiela, A., Pedrycz, W.: Granular Computing, An Introduction. Kluwer Academic Publishers, Boston (2003)
6. Becker, R.A., Chambers, J.M., Wilks, A.R.: The New S Language. Wadsworth & Brooks/Cole (1988)
7. Carpenter, J., Bithell, J.: Bootstrap confidence intervals: when, which, what? A practical guide for medical statisticians. Statistics in Medicine 19, 1141–1164 (2000)
8. Cox, L.H., Johnson, M.M., Kafadar, K.: Exposition of statistical graphics technology. In: ASA Proc. Stat. Comp. Section, pp. 55–56 (1982),
   http://lib.stat.cmu.edu/datasets
9. De Finetti, B.: Theory of Probability. A Critical Introductory Treatment, vol. 2. John Wiley & Sons, New York (1975)
10. Driankov, D., Hellendoorn, H., Reinfrank, M.: An introduction to fuzzy control, 2nd edn. Springer, New York (1996)
11. Dubois, D., Prade, H.: Possibility Theory. Plenum Press, London (1976)
12. Fisher, M.A.: The fiducial argument in statistical inference. Annals of Eugenics 6, 391–398 (1935)
13. Fisher, R.A.: The use of multiple measurements in taxonomic problems. Annual Eugenics, Part II 7, 179–188 (1936); Also in Contributions to Mathematical Statistics. John Wiley, NY (1950)

14. Fraser, D.A.S.: Nonparametric Methods in Statistics. John Wiley, New York (1957)
15. Freireich, E.J., Gehan, E., Schroeder, L.R., Wolman, I.J., Anbari, R., Burgert, E.O., Mills, S.D., Pinkel, D., Selawry, J.H., Moon, B.R.G., Spurr, C.L., Storrs, R., Haurani, F., Hoogstraten, B., Lee, S.: The effect of 6-mercaptopurine on the duration of steroid-induces remissions in acute leukemia: a model for evaluation of other potentially useful therapy. Blood 21, 699–716 (1963)
16. IRIS dataset, ftp://ftp.ics.uci.edu/pub/machine-learning-databases/iris
17. William, J.: Pragmatism's conception of truth. Lecture 6 in Pragmatism: A New Name for Some Old Ways of Thinking 83 (1907)
18. Johansen, S.: Functional relations, random coefficients, and nonlinear regression with application to kinetic data. In: Brillinger, D., Fienberg, S., Gani, J., Hartigan, J., Krickeberg, K. (eds.) Lecture Notes in Statistics, vol. 22, pp. 1–126. Springer, New York (1984)
19. Kearfott, R.B., Kreinovich, V.: Applications of Interval Computations. Kluwer, Dordrecht (1996)
20. Kearns, M.J., Schapire, R.E., Sellie, L.: Toward efficient agnostic learning. In: Computational Learing Theory, pp. 341–352 (1992)
21. Krupka, E., Tishby, N.: Incorporating prior knowledge on features into learning. In: Eleventh International Conference on Artificial Intelligence and Statistics (AISTATS 2007) (2007)
22. de Laplace, P.S.M.: Essai philosophique sur les probabilités. Veuve Courcier, Paris (1814)
23. LeCun, Y., Bottou, L., Bengio, Y., Haffner, P.: Gradient-based learning applied to document recognition. Proceedings of the IEEE 86(11), 2278–2324 (1998)
24. Lee, K.H.: First Course on Fuzzy Theory and Applications. Springer, Heidelberg (2005)
25. Lindstrom, M.J., Bates, D.M.: Nonlinear mixed effects models for repeated measures data. Biometrics 46, 673–687 (1990)
26. Mitchell, T.M.: Machine Learning. McGraw-Hill series in computer science. McGraw-Hill, New York (1997)
27. MNIST dataset, http://yann.lecun.com/exdb/mnist/
28. Mood, A.M., Graybill, F.A., Boes, D.C.: Introduction to the Theory of Statistics. McGraw-Hill, New York (1974)
29. Mosteller, F., Tukey, J.W.: Data Analysis and Regression: A Second Course in Statistics. Addison-Wesley, Reading (1977), Files for all 182 districts in 1888 and other years have been available at http://opr.princeton.edu/archive/eufert/switz.html or http://opr.princeton.edu/archive/pefp/switz.asp
30. Nauck, D., Klawonn, F., Kruse, R.: Foundations of neuro-fuzzy systems. John Wiley, New York (1997)
31. National institute of standards and technology, nist scientific and technical databases (July 2006), http://www.nist.gov/srd/optical.htm
32. Plato: The Republic. The Colonial Press, P.F. Collier & Son, New York (1901)

33. Quinlan, R.: C4.5 Release 8 (1992),
    http://www.rulequest.com/Personal/c4.5r8.tar.gz
34. Sambuc, R.: Fonctions Phi-floues, Application l'aide an diagnostic en pathologie thyroidienne. PhD thesis, University of Marseille, France (1975)
35. SMSA dataset,
    http://lib.stat.cmu.edu/DASL/Datafiles/SMSA.html
36. StatLib. Data, software and news from the statistics community (2007),
    http://lib.stat.cmu.edu/
37. Tukey, J.W.: Nonparametric estimation, ii. statistically equivalent blocks and multivariate tolerance regions. the continuous case. Annals of Mathematical Statistics 18, 529–539 (1947)
38. Wilks, S.S.: Mathematical Statistics. Wiley Publications in Statistics. John Wiley, New York (1962)

Part I
Algorithmic Inference

Introduction

We have been left in the previous chapter with a set of pebbles that here we assume aligned, so that we may affect them with a coordinate x_i denoting their distance from Hansel's home. We also realized that, in absence of further information, the best to do is to assign equal weight in our reasoning to each pebble. Finally, we agreed on the possibility of enriching our vision on the phenomenon we are facing with (being more specific, the track back home) and are open to embark on new knowledge either in terms of further observed points, or of properties about the family of points. We devote this part of the book to carry out these two tasks, by proceeding along the parallel rails of elaborating both on probability models as a tool for efficiently synthesizing observed data, and on statistics as tools for improving the synthesis efficiency. Chapter 2 essentially answers the question: which kind of structure may I find within data to be reused in the following via a probability model? Chapter 3 will toss the primary tools for identifying this structure in the light of the data we are observing, which stand for the properties that are worth to be observed about them. Then, it will move to operations, thus answering the fundamental question: what can I do with these data? i.e. how may they help the operational choices I'm called to take about the phenomenon generating them? At the end of this chapter you should be enabled to become a "laic" diviner: no way of forecasting next number in a lottery, but if some indication may emerge from data about next positions of Hansel pebbles, you are able to capture it.

Giving equal weight to information granules,

with possible correctives coming from the experience:

the ultimate diviner cocktail.

2 Modeling Samples

We will assume the mechanism shown in Fig. 2.1 as the *mother of all samples*. You may see the contrivance in its center as a spring generating standard random numbers for free which are transmuted into numbers following a given distribution law after having passed through the gears of the mechanism. By default, we assume that the standard numbers are uniform in $[0, 1]$ like in this figure. Thus, we have an universal sampling mechanism $\mathcal{M}_X^*(U, g_\theta)$, and you are ensured that, like from a cornucopia, you may draw any kind of random variable X, provided you may device the gears in order to compute the related $\widetilde{F}_X^{-1}$ defined in Fact 1.4. Two considerations are in order to remove the wrong idea that we are univocally, hence *a priori*, describing the world.

1. The above source of randomness does not exist but may be suitably approximated. Indeed:
   - On the one hand we may interpret randomness as a lack of information. Thus any computer operating system includes a routine for recursive generation of a specifications' sequence of uniform random variable U over the $[0, 1]$ interval with a decided accuracy at an entry level (see Algorithm 2.1 for a typical Rand routine). These specifications are used for standard simulation tasks, i.e. tasks: i) not employing a so huge number of specifications that the routine is obliged to pass again through the first item of the sequence and its successors – a constraint that is easy to satisfy as this number is a power of the number of bits coding U, and ii) not biased by the structure induced on the sequence by the routine instructions – a structure that is hard to identify by direct computation.

Random numbers in a computer output

46   Modeling Samples

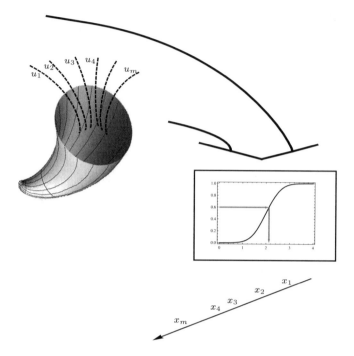

Fig. 2.1. The mother of all samples.

Vice versa, we expect to be able to generate stronger U specifications for any sophisticated application violating the above constraints – this is why you may manage electronically your bank account without relevant risks.
- As a routine, your random number generator does not come for free. At least it consumes electricity to run. Hence you may maintain this assumption in order to define a random variable X as an infinite collection of its specifications distributed according to a given f_X. As a matter of fact you will always deal with finite sequences of xs that may be more or less adequately resumed through X.

Algorithm 2.1. Linear congruential generator

Given a, b and m generator-specific integer constants, a linear congruential generator is defined by the recurrence relation:

$$u_{j+1} = \frac{(amu_j + b) \bmod m}{m} \tag{2.1}$$

where $\{u_j\}$ is the sequence of random values, u_0 the initial seed and mod is the modulo operation. Typical choice of the parameters is $a = 1664525, b = 1013904223, m = 2^{32}$.

---

2. The seeds produced by the randomness source may be different. Normal variable Z specifications are commonly used [35]. To result universal seeds, hence in order to be transmuted into any random variable specifications, it is enough (and necessary as well) that they may be mapped into U specifications.

At the end of the day, a sampling mechanism is a way of studying random variables. It is our favorite way for two reasons:

1. It deals directly with finite sets of random variable specifications, normally formalized as *samples*, identifying the variable with the function mapping from seeds to specific realizations, the explaining function indeed.
2. It clearly identifies the structure we are interested in when analyzing data, identified with the explaining function, while confines the rest in the seed distribution law.

We are interested in the features of this computer

In this chapter, we proceed with a few formal definitions of what we may be interested in when we observe data (Section 2.1), and we discuss some basic relations between data structure and elaborate their distribution law in Section 2.2.

2.1 Samples and Properties about Them

In the case that we observe so huge a number of points to be confident of having exhaustively explored the phenomenon at hand, if we may assume that Hansel will maintain the same dropping habit in a next trip from home to the forest, the cumulative distribution function we introduced in Definition 1.1 achieves the meaning of *probability* of finding a pebble before the coordinate in argument to the function. This sentence requires a set of definitions to be made unambiguous. We will do this by adopting the risky strategy of tailoring definitions that fit exactly our framework, though agreeing with the general statement the reader can find in the literature.

2.1.1 Samples

Definition 2.1. *Given an ordered set* $\mathfrak{X}$ *a random variable* X *is a pair* $(\mathfrak{X}, \mathrm{P})$, *where* P *is a function associating a probability measure to each subset of* $\mathfrak{X}$. *The elements* $x \in \mathfrak{X}$ *are denoted as* X specifications.

What we observe are random variables,

We refer the reader to any probability textbook regarding the definition of probability measure. Here we simply report some facts about X that will prove useful in the book.

48   Modeling Samples

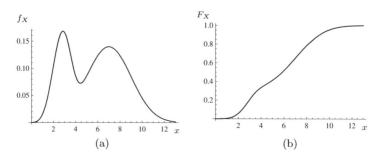

Fig. 2.2. (a) d.f., and (b) c.d.f. of a bimodal continuous random variable.

Fact 2.1. • *Moving from the record of observed data to the infinite population of the data we could ideally observe in an unlimited future, if these data may assume infinitely many values in own turn, it may turn useful to consider these values gathered into a continuous range, so that functions like c.d.f. become continuous (see Fig. 2.2), de facto smoothing steps of disregardable heights. This allows in general quicker and easier computations drawing from the infinitesimal calculus, but requires a refinement of definitions like the d.f.'s. Namely, Definition 1.1 is augmented in the following:*

<small>that may be useful to idealize as continuous.</small>

Definition 2.2. *The c.d.f. of a random variable X is defined as:*
$$F_X(x) = \mathrm{P}(X \leq x). \tag{2.2}$$

The probability function *(p.f.) of a discrete random variable X is defined as:*
$$f_X(x) = \begin{cases} \mathrm{P}(X = x) & \text{if } x \in \mathfrak{X} \\ 0 & \text{otherwise.} \end{cases} \tag{2.3}$$

The density function *(d.f.) of a continuous random variable X is the derivative of its c.d.f.*
$$f_X(x) = \frac{\mathrm{d} F_X(x)}{\mathrm{d} x}. \tag{2.4}$$

Appendix B.1 reports the analytical expressions and graphs of six template distribution laws (uniform, Bernoulli, negative exponential, Pareto, Gaussian and Beta) that are at the basis of the most used ways of modeling probabilistic granules.

Remark 2.1. With this more complete definition we must split many formulas where the non cumulative distribution is

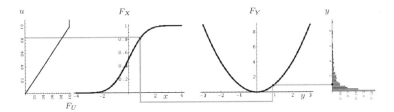

Fig. 2.3. Mapping from uniform U to Gaussian distribution X and from Gaussian to chi square distribution Y with 1 degree of freedom.

involved. For instance the definition of $E[X]$ that reads as in (1.3) for discrete X becomes:

$$E[X] = \int_{-\infty}^{+\infty} x f_X(x) \mathrm{d}x \qquad (2.5)$$

for X continuous. In general, moving from discrete to continuous variables the sum is substituted by an integral, while the rest remains the same. Thus we are used to denote with f_X a density function also in case of discrete variables and use sum or integral in the formulas on our convenience.

- Taking a function $h : \mathbb{R} \to \mathbb{R}$ and substituting its argument with a random variable X, we can interpret the whole as a generator of a new random variable Y assuming a specification $h(x)$ each time X assumes specification x. The c.d.f. F_Y of Y may be computed either analytically, in the simplest cases, or numerically using X or directly U as a primary source of randomness and drawing as many seeds are needed to get the desired accuracy. Fig. 2.3 shows the c.d.f. change when we pass from the c.d.f. of X distributed according to a Gaussian distribution law to the c.d.f. of $Y = X^2$. The basic logic when $Y = h(X)$ and the function h is monotonically growing with x – so that $h^{-1}(y) = x$ each time $h(x) = y$ – is:

  Using a random variable as seed you obtain other random variables.

$$\begin{aligned} F_Y(y) &= \mathrm{P}(Y \leq y) = F_X(h^{-1}(y)) \\ f_Y(y) &= \mathrm{P}(Y = y) = f_X(h^{-1}(y)) && \text{if } X \text{ is discrete} \\ f_Y(y) &= \frac{\partial F_Y(y)}{\partial y} = f_X(h^{-1}(y)) \frac{\partial h^{-1}(y)}{\partial y} && \text{if } X \text{ and } Y \text{ are continuous,} \end{aligned}$$

$$(2.6)$$

and obvious extensions with different h.

- A widely employed function of random variable is $(X - E[X])^2$, whose expected value is the variance $V[X]$ of the variable, also identified with the square of the standard deviation σ_X of the variable. In formulas:

50   Modeling Samples

$$V[X] \equiv \sigma_X^2 = E\left[(X - E[X])^2\right]. \qquad (2.7)$$

Joining random variables you obtain new random variables.

- Observations described by more than one variable may be synthesized by joint distribution laws, whose c.d.f. $F_{X_1,\ldots,X_\nu}$ $(X_1,\ldots,X_\nu)$ has the obvious meaning of probability of the joint inequalities $(X_1 \leq x_1,\ldots,X_\nu \leq x_\nu)$ which is usually factorized as:

$$F_{X_1,\ldots,X_\nu}(X_1,\ldots,X_\nu) = \\ F_{X_1}(x_1) F_{X_2|X_1 \leq x_1}(x_2) \ldots F_{X_\nu|X_1 \leq x_1,\ldots,X_{\nu-1} \leq x_{\nu-1}}(x_\nu), \qquad (2.8)$$

where $F_{X_2|X_1 \leq x_1}(x_2)$ has the meaning of probability of the event $(X_2 \leq x_2)$ conditioned to the fact that $(X_1 \leq x_1)$ and the operational definition of asymptotic ratio of the number of observations with $X_1 \leq x_1$ and $X_2 \leq x_2$ over the number of those with $X_1 \leq x_1$. Special simplifications of sampling mechanism and related inferences occur when the above ratio equals the analogous ratio of the number of all observations

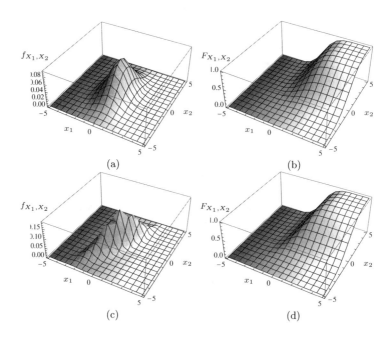

Fig. 2.4. Bidimensional Gaussian distribution. (a-b): d.f. and c.d.f. in the independent case; (c-d): same functions in the dependent case.

Table 2.1. Distribution of the sum of ν independent random variables. Notation: $\Gamma(\nu)$ denotes the Gamma function, $s_\mu = \sum_{i=1}^\nu \mu_i$ and $s_\sigma^2 = \sum_{i=1}^\nu \sigma_i^2$.

Sum of independent random variables $Y = \sum_{i=1}^\nu X_i$	
$X_i \sim$ Bernoulli	$Y \sim$ Binomial
$f_{X_i}(x;p) = p^x(1-p)^{1-x} I_{\{0,1\}}(x)$	$f_Y(y;p,\nu) = \binom{\nu}{y} p^y(1-p)^{\nu-y} I_{\{0,\ldots,\nu\}}(y)$
$X_i \sim$ Poisson	$Y \sim$ Poisson
$f_{X_i}(x;\mu) = \dfrac{e^{-\mu}\mu^x}{x!} I_{\mathbb{N}\cup\{0\}}(x)$	$f_Y(y;\nu\mu) = \dfrac{e^{-\nu\mu}(\nu\mu)^y}{y!} I_{\mathbb{N}\cup\{0\}}(y)$
$X_i \sim$ Exponential	$Y \sim$ Gamma
$f_{X_i}(x;\lambda) = \lambda e^{-\lambda x} I_{[0,+\infty)}(x)$	$f_Y(y;\nu,\lambda) = \dfrac{\lambda}{\Gamma(\nu)}(\lambda y)^{\nu-1} e^{-\lambda y} I_{[0,+\infty)}(y)$
$X_i \sim$ Gaussian	$Y \sim$ Gaussian
$f_{X_i}(x;\mu_i,\sigma_i^2) = \dfrac{1}{\sqrt{2\pi}\sigma_i} e^{-\frac{(x-\mu_i)^2}{2\sigma_i^2}}$	$f_Y(y;s_\mu,s_\sigma^2) = \dfrac{1}{\sqrt{2\pi s_\sigma^2}} e^{-\frac{(y-s_\mu)^2}{2s_\sigma^2}}$

with $X_2 \leq x_2$ over the total number of observations, a condition saying that the additional condition $X_1 \leq x_1$ does not change the information we cumulate on X_2. In this case we say that X_1 and X_2 are independent variables and express this property in terms of c.d.f. by stating that

$$F_{X_1,X_2}(x_1,x_2) = F_{X_1}(x_1)F_{X_2}(x_2), \qquad (2.9)$$

with obvious extensions to more than two variables. The independence between X_1 and X_2 is equivalently stated by the equality $f_{X_1,X_2}(x_1,x_2) = f_{X_1}(x_1)f_{X_2}(x_2)$.

Fig. 2.4 reports d.f.s and c.d.f.s of a bidimensional random variable in both cases that its components are either independent or not.

- Combining the previous points we can manage functions of more than one random variable. In particular, many statistics of our interest are based on the sum of random variables and for a wide class of probability distributions these sums have a simple analytical form. For instance, with reference to Table 2.1 the sum of ν independent random variables, each following either i) a Bernoulli distribution of parameter p, or ii) a Poisson distribution of parameter μ, or iii) a negative exponential distribution of parameter λ, follow respectively a Binomial distribution of parameters ν and p, a Poisson distribution of parameter $\nu\mu$ and a Gamma distribution of parameters ν and λ. Similarly, the sum of ν independent Gaussian random variables of parameter μ_i and σ_i^2 is a Gaussian random variable of parameters

> You may interleave the two steps.

$\sum_{i=1}^{\nu} \mu_i$ and $\sum_{i=1}^{\nu} \sigma_i^2$ *(see [6] for the definition of these new distributions)*.

Given this notation it is suitable to frame our observations into a sample specification defined as follows:

<small>Gathering information granules you obtain a sample.</small>

Definition 2.3. *Given a string of data* $\boldsymbol{x} = \{x_1, x_2, ..., x_m\}$, *we call it an i.i.d. sample specification if it is constituted by a specification of a set of independent identically distributed random variables* $\boldsymbol{X}_m = \{X_1, X_2, ..., X_m\}$. *We call* sample *the set of i.i.d. variables, and* statistic *any random variable that is a known and computable function of the sample or the function itself by abuse of notation. We call* likelihood *of a sample the joint probability (density) of its elements, namely*

$$L(x_1, \ldots, x_m) = \prod_{i=1}^{m} f_X(x_i). \qquad (2.10)$$

Often we will simply denote sample a sample specification as well for the sake of simplicity.

2.1.2 Properties about Samples

We call *statistic* any property we may actually compute of sample, i.e. the output of any algorithm having sample $\{x_1, x_2, ..., x_m\}$ and possible ancillary data in input. For instance $\sum_{i=1}^{m}(x_i - \mu)^2$ is a statistic if μ is known, otherwise it is still a property of the sample but not a statistic.

<small>Statistic is what you are able to compute on data.</small>

Sampling mechanism $\mathcal{M}_X$ has great value of highlighting what is deprived of any information to us – the seed – since its distribution law is perfectly known, and what deserves innovation – the explaining function parameters – since we do not know them but we try to argue from the observations of X specifications. In order to better realize this split, let us sketch why and how this mechanism works with a uniform seed in case of continuous random variables, with obvious extensions for the other ones.

<small>Also with random input, double inverting a function leads to the identity function.</small>

Fact 2.2. *Consider the random variable* Y *that is a function of* X *through the mapping* $Y = F_X(X)$. *For whatever continuous* X, $F_Y(y) = y$ *for* $y \in [0, 1]$, *i.e.* Y *follows a uniform distribution law over the continuous set* $[0, 1]$. *Indeed for each* $u \in [0, 1]$:

$$F_Y(y) = \mathrm{P}(F_X(X) \leq y) = \mathrm{P}(X \leq F_X^{-1}(y)) = F_X(F_X^{-1}(y)) = y, \qquad (2.11)$$

where by h^{-1} *we denote the inverse function of* h, *i.e. a function such that* $h^{-1}(h(x)) = x$ *for each* $x \in \mathfrak{X}$. *Such a function always* exists for $h = F_X$ thanks to the monotonicity and continuity of F_X.

This statement is commonly known as the *probability integral transformation theorem* [44]. Thus, given a device producing specifications $\{u_i\}$ of a $[0,1]$-uniform random variable U, we obtain a device like the one shown in Fig. 2.1 producing specifications $\{x_i\}$ of a continuous random variable X via the simple algorithm:

$$x_i = F_X^{-1}(u_i). \qquad (2.12)$$

Indeed,

Fact 2.3. *Denoting Z the variable $F_X^{-1}(U)$, whose specifications we are producing through (2.12), we have:*

$$F_Z(z) = P(F_X^{-1}(U) \leq z) = P(U \leq F_X(z)) = F_X(z). \qquad (2.13)$$

While for entertainment sake we may enjoy of the variety of X specifications primed by the u_is, to get innovation about θ in the explaining function, we want to capture properties that are independent of u_is, hence owned by all X specifications. In their place we come to properties owned at least by the x_is we observe, with the understatement that they are so more interesting as less biased by the u_is. As functions computed from the sample, *statistics* are exactly the local properties we may access to. Since sample $\{x_1, \ldots, x_m\}$ is a specification of the set of random variables $\{X_1, \ldots, X_m\}$, a statistic s we may compute from the former is a specification of a random variable S as a function of the above set of variables. A study of its distribution law lets us understand the bias degree coming from the seeds. Typical properties whose variance decreases with the increase of the sample size are:

Still looking for noiseless results,

you cannot avoid their randomness.

$$\overline{X} = \frac{1}{m}\sum_{i=1}^m X_i; \quad X_{\max} = \max\{X_1, \ldots, X_m\}; \quad X_{\min} = \min\{X_1, \ldots, X_m\}.$$
$$(2.14)$$

As for the former – commonly denoted *sample mean* – you may exploit Table 2.1 to compute its distribution law in special cases. For instance, since for $Y = aX$ and continuous X you have $f_Y(y) = f_X(y/a)/a$ and $F_Y(y) = F_X(y/a)$ (see (2.6), the distribution law of $\overline{X}$ when X is distributed according to an exponential distribution law with parameter λ is described by:

$$f_{\overline{X}}(x) = \frac{m\lambda}{\Gamma(m)} x^{m-1} e^{-\lambda x} I_{[0,\infty)}(x). \qquad (2.15)$$

In the general case, for the values of m large enough you may exploit the approximation of $\overline{X}$ with a Gaussian random variable so that:

With huge samples life is easy,

$$\lim_{m \to \infty} f_{\overline{X}}(x) = \frac{1}{\sqrt{2\pi}\sigma_X/\sqrt{m}} e^{-\frac{1}{2}\left(\frac{x - \mu_X}{\sigma_X/\sqrt{m}}\right)^2}, \qquad (2.16)$$

provided X has finite mean μ_X and finite variance σ_X^2. In any case we may exploit the Glivenco-Cantelli theorem [23], to rely on the following fact:

Fact 2.4. *For a non pathological random variable X (i.e. for almost all distribution laws we will meet in operational frameworks),*

thanks to a few general-purpose statistics.

- *for any parameter θ that can be expressed as the expected value $\mathrm{E}[h(X)]$ for a suitable h, the statistic $1/m \sum_{i=1}^{m} h(x_i)$ will be so close to θ as you want with almost all samples provided their size m is large enough;*
- *for any delimiter θ so that $(h(X) \leq \theta)$ (or so that $(h(X) \geq \theta)$) for a suitable h, the statistic $\max\{h(x_1), \ldots, h(x_m)\}$ (or $\min\{h(x_1), \ldots, h(x_m)\}$, respectively) will be so close to θ as you want with almost all samples provided their size m is large enough.*

While we refer to basic textbook such as [38] to identify the X pathologies, we characterize the convergence of the estimators to the statistic with the weaker condition:

$$\lim_{m \to \infty} \mathrm{P}\left(\theta - \frac{1}{m}\sum_{i=1}^{m} h(X_i)\right) = 0 \qquad (2.17)$$

for the former statistic, idem for the latter.

Once (2.17) has been stated, we could consider ourselves done with almost all inference problems, adopting the operational rule: observe $h(X)$ for a long time, than either average the observed values, or compute their extremes. The drawback is that the required sample size m often exceeds our capacities. The reason is either in the large amount of this value or in the high cost of collecting observations. The former condition is typical of modern observations. We often deal with complex phenomena characterized by a plenty of either free parameters or hidden variables. With reference to the SMSA dataset, consider for instance the age adjusted mortality M as shown in Fig. 2.5(a), where on the x axis we report the percentage of non white in the population (%NW) and on the y axis the sulfur dioxide concentration (SO_2). Being interested only in regional areas where the %NW lies between 20 and 23% of the 60 data items only 4 can still be used in our inference procedure (see Fig. 2.5(b)).

But, how big is a large sample?

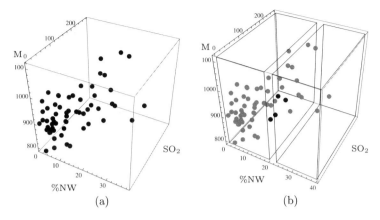

Fig. 2.5. Age adjusted mortality index M vs. percentage of non white in the population (%NW) and sulfur dioxide concentration (SO_2), as reported in the SMSA dataset, when: (a) all data items are available; and (b) only 4 items with $20\% \leq \%NW \leq 23\%$. Black points: available data; gray points: shadow data.

2.2 Organizing Granules into Populations

The cumulative distribution function is a very efficient tool for synthesizing our observations about a phenomenon. Its efficiency comes from the scalability of the representation and the on-line updatability on the basis of new observations. Indeed, the following relations are satisfied:

Probability as a robust mirror of what we have seen

- For whatever discretization grid G' and $\widehat{F}_{X_{G'}}$, $\widehat{F}_{X_G}(x) = \widehat{F}_{X_{G'}}(x)$ for any x discretized according to G, where G' is a grid finer than G. Thus in Fig. 2.6 we see that $\widehat{F}_X(10)$ remains the same either if we differentiate $x = 9.51$ from $x = 10$ or we group the two points into $x = 10$ because we decide working with integer numbers.
- Upon a new observation x_{m+1}, $\widehat{F}_X$ moves to $\widehat{F}_X^+$ with

$$\widehat{F}_X^+(x) = \widehat{F}_X(x)\frac{m}{m+1} + \frac{1}{m+1}I_{[x_{m+1},+\infty)}(x). \qquad (2.18)$$

Once an order relation has been established on $\mathfrak{X}$, $\widehat{F}_X$ denotes how the data are structured in respect to this relation. The definitions in the previous section introduce tools to exploit this information in order to describe how the associated random variable behaves under various perspectives. Using a biological analogy, until now we were talking about the phenotypical properties of the observed data. It is the mental attitude with which ancient naturalists were answering to the question why opium puts

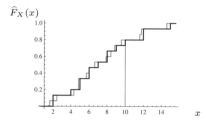

Fig. 2.6. Course of the empirical c.d.f. using two different discretization grids: working with integer numbers (black lines) and with 1 decimal digit (gray lines).

From phenotypic to genotypic expressions of data

people to sleep: because opium owns *the vis dormitiva* (power to induce sleep of a sleeping pill). Modern science, and modern statistics as well, are drawn by the true causes at the basis of the phenomena they observe, i.e., still in the biological analogy, with their genotype. In pure computer science terminology they are interested in the algorithms, maybe implemented in a computer or using another device in nature as well, which produces an output. Hence if you want to be confident of coming back home, you must elaborate why Hansel dropped pebbles with the pattern you observed, i.e., which relation exists with this pattern and neighboring conditions such as track slope, vegetation etc. If you succeed you may hope to gain the way back also when your trip has been to the fountain rather than to the forest. This deeper way of understanding data comes not *ex abrupto*. However in the last years it is resumed in the target of learning functions underlying examples.

Genotype is a concise computer program, i.e.,

But which kind of function could you extract from the Hansel pebbles having access to a PC in his lonely cottage? A very fundamental answer has been supplied by Kolmogorov. In order to assess how a data item is complex, he considers the shortest program π that, in input to a general purpose computing machine like yours, outputs exactly the item [29]. He identifies the complexity of the item with the length of π and assumes random those items that cannot be produced by short programs. Said in other words, one of the first conceptual task requested in the primary schools is to synthesize a story, like the Hansel and Gretel's tale. The best strategy is to rephrase the story in terms of the essential facts it is composed of. Rather, if the student just repeats the same sentences used by Andersen, then the teacher may guess that he was not able to understand the meaning of the tale, i.e. the text looks like a random sequence of words. The following definitions provide a slight improvement of this notion of complexity based on a more precise description of the computing machine. First of all let us identify with $h : x \to y$ the notions of function, algorithm and (computing) machine. All are intended to univocally mapping an input x into an output y. The first term is drawn by the analytical aspects of h, the second by its operational description, while the third one focuses on the physical implementation of the computation.

Definition 2.4. *Let $\mathfrak{X}$ be the set of all binary strings and $|x|$ the length of the string x. Denote with $h(x) < +\infty$ the fact that a function h may be evaluated on x. A computing machine $\psi : \mathfrak{X} \mapsto \mathfrak{X}$ is called* prefix *if $\psi(x) < +\infty$ and $\psi(y) < +\infty$ implies that x is not a proper prefix of y. A prefix machine $\mathscr{U}$ is called* universal *if it is capable of computing any computable function (according to the Church-Turing thesis [43]).*

<small>the most concise description of a string,</small>

Remark 2.2. For short, minding ψ a function for coding words, in order for ψ to be a prefix you cannot have two codewords having a same prefix. As a matter of fact, every programmer is used to deal with prefix functions, as he delimits a code section with particular instructions (such as `Begin` and `End`, or a balanced pair of curly brackets, and so forth).

<small>without ambiguity on the end.</small>

Definition 2.5. *[32] For a universal prefix machine $\mathscr{U}$ denote $\mathscr{U}(\pi, y)$ the output of machine $\mathscr{U}$ fed by program π having y as input. The* conditional prefix (or Levin's) complexity $K(x|y)$ *of x given y is defined as:*

$$K(x|y) = \min_{\pi \in \mathfrak{X}}\{|\pi| \text{ such that } \mathscr{U}(\pi, y) = x\} \tag{2.19}$$

and the unconditional prefix complexity $K(x)$ *of x as $K(x) = K(x|\epsilon)$, where ϵ is the empty string.*

<small>A data item measure that:</small>

Below we list some facts of life of prefix complexity in order to familiarize the reader with this notion and show its relevance to probability.

1. The prefix complexity of all data items is independent of the special universal machine employed to implement the related programs, apart from a constant that depends on the machine and not on the items. Said in formal terms, for each pair of universal prefix machines $\mathscr{U}$ and $\mathscr{U}'$ we have:

   <small>is independent of its use,</small>

   $$K_{\mathscr{U}}(x) \leq K_{\mathscr{U}'}(x) + c, \tag{2.20}$$

   for each x in any set $\mathfrak{X}$, where c is the above constant. This is why we omit the K subscript in (2.19). Following similar reasonings we may claim that for a fixed conditioning information y, $K(x|y)$ and $K(x)$ differ by a constant, so that all what we say in the next points, mentioning the unconditional complexity, holds for any conditioning y different from the empty string.

2. The complexity of any item x is less than the length of its representation through a given set of characters plus a constant. Actually x will be represented through a string of characters $(c_1, c_2, \ldots, c_\nu)$ where c_is belong to the set of characters handled by the computer. A very elementary program for computing x is represented through the instruction: "write the following string: $c_1, c_2, \ldots, c_\nu$", where the head of the instruction needs a number of characters to be transmitted to the computer that does not depend on x. We may expect that a smarter program may possibly be coded in a shorter number of characters.

[margin: with sharp upper bounds]

3. At least one half of the data items represented by all strings of characters up to a given length have a complexity higher than their own length. This comes from the fact that from one side we may use all different strings to represent different data items (for instance we may represent $\sum_{i=0}^{n} 2^\nu = 2^{n+1}-1$ items using at most $n+1$ bits). *Vice versa* you need different programs to compute different items, but you cannot use all different strings to have different programs, otherwise the computer you use is not a prefix machine. For instance, the string representing the instruction **End** put at the end of any program cannot be confused with the sequences of characters that you want to process within your program. Hence you must freeze at least one character to encode **End**. In particular, in case of binary strings you need at least two bits (only one bit would mean that "1" for instance means **End** and you have only a sequence of zeroes to code the program, that is impossible) which means that you have only $n - 2$ bits for really encoding the different programs, which makes $2^{n-1} - 1$ different programs coding in any case less than half of the 2^n strings of length n. Things get still worse for a richer set of characters. You may decree these items random not allowing any compression of their representation. In general as more random is any item as longer is its most compressed representation. Note that when the shortest program is longer than the original representation you are *essentialy* referring to the elementary program "write what you have read" discussed in the previous point.

[margin: and lower bounds,]

4. There is no a computer that in input any given item outputs the shortest program computing it. This comes by evident circular reasoning. Hence we may look for *suboptimal*

[margin: is non computable]

programs having the shortest length up to our computing framework.

5. For any data item x we have [33]:

$$f_X(x) \leq 2^{-(K(x)-c)}, \qquad (2.21)$$

but very meaningful.

where c is a constant depending on f_X but not on x.

The above facts delineate a sufficiently complete set of relationships between structure of data and probability of meeting them in the future. Referring to a binary representation for simplicity, you pass from totally unconnected bits, declared so by the assumption of their independence, to highly compressible sequences, possibly represented by a string of constant length when they are the output of a function whose inputs are given as conditioning information in (2.19). The right side of (2.21) forms an upper bound to the probability $f_X(x)$. On the one hand, it tells that you may decide to attribute many different distribution laws to your data, but if they have not sufficient structure you cannot tightly peak the distribution around some values. On the other hand, deducing the distribution from data, i.e. using it to describe them, the gap in probability is due to the gap in the structure you find in the data with respect to the optimal one. It is in part a matter of non computability of the shortest program, as mentioned in point 4, but this drawback is weakened in the gap introduced by the computational context the data are handled with, which reduces the domain where to optimize the program. For instance, it does not make sense to find a structure in the string "0000000111" if we declare in advance that the bits are independent. This information represents a conditioning on the programs, i.e. shifts our operational scenario from a universal to a specific computing machine. As this machine is universal again with respect to the reduced scenario, we are pushed to search for the optimal program coding our items with the constraint induced by the scenario. The reward is to identify the asymptotic frequency with which we will meet the items in the future.

The complexity of a string is a matter of the context it belongs to.

2.3 Blowing up Items into Populations

In the previous section we have focused on single item phenotype and genotype, a task that proves useful for instance when we are looking for a possibly approximate solution to a problem. In this case we look for a single x owning a good probability of being a good solution to the given problem; this probability strictly depends on the tight analysis we did for x structure in relation to the problem. It could be the task of Hansel, which reflects any time where is the best site for the next pebble drop. Here, to the contrary, we are interested in the characteristics of the problem itself, as a generator of random numbers, for instance representing approximate solutions. Hence we are considering pebble distribution law properties as a manifestation of Hansel dropping strategy. Starting from a set of observed data each affected with equal weight $1/m$ we have no other chance than adding their values or any singularly evaluated functions of them to obtain a statistically correct properties. We obtain in this way, for a function h, the empirical expected value:

$$\overline{h(X)} = \frac{1}{m} \sum_{i=1}^{m} h(x_i) \qquad (2.22)$$

Phenotypic as a sample version of the expected value operator:

$$\mathrm{E}[h(X)] = \sum_{x \in \mathcal{X}} h(x) f_X(x) \mathrm{d}x, \qquad (2.23)$$

with analogous extension for continuous variable. Applying it to either X or a generic function $h(X)$ of it we obtain a phenotypic synthesis of the random variable. It turns in a genotypic syn-
and genotypic thesis when h returns the length of the program π as in (2.19). According to our previous discussion, a good approximation of
synthesis, the latter is the function denoted *entropy* $H[X]$ of the random variable and computed as follows

$$H[X] = -\sum_{x \in \mathcal{X}} f_X(x) \log f_X(x) \approx \sum_{x \in \mathcal{X}} f_X(x) |\pi(x)|. \qquad (2.24)$$

with $|a|$ denoting the length of a. Let us shift the X origin to the minimum of its specifications, so that the new X has only positive values. Forget the ideal cases where this minimum does not exist. Then, in the approximation of $\mathrm{E}[\pi(X)]$ with $H[X]$ you have:

$$H[X] \leq \mathrm{E}[\chi(X)] \leq \mathrm{E}[X] \qquad (2.25)$$

by definition, where $\chi(x)$ is the length of any computable approximation of π related to our actual computational framework. Still
and the real more, for any biunivocal function h,
use of the data.

$$H[X] \leq \mathrm{E}[\chi(h(X))] \leq \mathrm{E}[h(X)]. \qquad (2.26)$$

$\mathrm{E}[h(X)] - \mathrm{E}[\chi(h(X))]$ is the benefit of going to the root of the things. What about the gap $E[\chi(h(X))] - H[X]$? This is the room occupied by the granules around the data point. From a

Table 2.2. Sample specifications drawn from: 1) a Gaussian, and 2) a negative exponential distribution law.

	Sample specifications
1)	$-0.72, -0.77, -0.32, -1.92, 0.86, 0.05, -0.49, 1.44, 1.49$
	$-0.02, -0.77, -0.01, 0.04, -0.81, 0.31, -1.12, 0.39, 0.01, -2.53$
2)	$2.6, 3.61, 7.07, 1.18, 1.75, 8.57, 22.88, 6.34, 16.98$
	$0.62, 0.36, 0.04, 9.64, 2.98, 4.6, 3.51, 9.89, 6.03, 2.29$

strict information point of view it is a waste. Referring the pebble dislocation on the ground to your advisor, the first question he raises is "what does it mean", and the best answer is the program computing the allotment. Still preliminarily, your advisor will be very happy if you are able to describe the allotment in a very succinct way. To this aim you may use standard codings such as Huffman code, representing with shorter codewords the word having high probabilities, hence smaller complexity, so having the minimal entropy of the codeword distribution. In many instances we will deal with in this book genotype length of a single data will identify with its norm according to a metric (see Appendix A.2) decided by the context, hence the length of a different coding. In light of (2.21) a data is essentially coded through its probability. This is not the most efficient nor univocal encoding but is feasible in light of the overall knowledge we have of the context. Entropy in turn will by replaced by the mean norm, in some sense a fuzzy radius of the questioned set of data (see Definition 5.3 later on).

All that to realize that shortest is the best. However waste is not always a drawback and granules have not to be always disclaimed. Rather we abandon the pure compression of the data when we look for their actual use. Consider the following information loop. Starting, for instance, from samples as in Table 2.2, we go to their distribution law as the preliminary step for their exploitation by adding elementary structure from abroad. Namely, we start with probability masses spread into blocks like in Fig. 1.3(a) and subdivide them according to a suitable discretization of the x axis. Then, we obtain a Gaussian distribution from the former sample after shaking the probability masses associated to the smaller blocks as shown in the sampled frames in Fig. 2.7(a-d) of a long trip where iteratively we force both the symmetry around the median and a monotone thread in: i) the blocks heights (increasing before the median, decreasing after it), and ii) the slope between two consecutive blocks (depending on their location w.r.t. the distribution tails). Analogously (see Fig. 2.7(e-h)) we obtain an exponential distribution

Short is useful,

but don't blame the aesthetics of phenotypes!

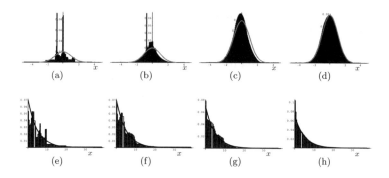

Fig. 2.7. Sampled frames of strips generated by a suitable iterated shaking of the probability masses associated to sample 1) (a-d) and sample 2) (e-h) in Table 2.2 and bringing a Gaussian and a negative exponential distribution, respectively.

from the second sample after forcing both a decreasing monotonicity in the blocks height and an increasing monotonicity in their slope. Inference procedures, such as those which we will discuss on, for instance, in the next chapter, rely on intermediate expectations concerning structures, and hence programs computing properties of the sample neither so specific as the values of the single variables nor so deep to relate to the prefix complexity, but sufficiently relevant to reproduce common characteristics to be employed to take decisions.

Note that we have not imposed any limitations on X. It may be a scalar, a vector, but also a function or a graph. In synthesis X may be any set of items on which an order relation may be stated. Using this relation we may discuss on and infer the structure with which the items take relative position. This may be assumed a step ahead in the way of considering statistics, an attitude that is generally denoted as *learning*. While in the past both observations and parameters have concerned the phenotypical world, with modern paradigms such as hidden Markov models, function learning, or neural networks, we are strongly discriminating the phenotypic realm, from which we take observations, from the genotypic realm, within which we search a deep explanation of the observed phenomenon constituting the true target of the inference.

<small>Learning is to pass from phenotypical observations to genotypical conclusions.</small>

2.4 Conclusions

It is not too hazardous to claim that the most of computations we do on data represent statistics on them. This is because, apart from the accuracy with which they are collected, granulation comes from the fact that the common use we do of the computation output refers to a wider operational framework. It may be that we will use them to program next activities, to generalize conclusion to other instances, and so on. It is not astonishing, therefore, that with these goals we focus on computational features of the data itself, such as the shortest program π replicating the data and its variants.

What we learn from this chapter are some ways of relating phenotypical aspects of the granules – that we deal with in a much sophisticated way in terms of distribution laws – with the mentioned computational features representing their genotypical counterpart. Focusing on the single granules a computable version of π is the explaining function g_θ, where the non computability of the former is bartered with the non identifiability of the input in the latter, namely the random seed. The shape of g_θ is very eloquent, however, on the kind of phenomenon – and distribution law, consequently – we are facing with. Whereas the value of statistics on granules constitutes more or less focused observations on θ.

Moving to the entire population of granules as the true target of our operational decisions, we like to formalize a decision effect as the expected value of a function h of a random variable X. On the one hand, the expected value may be (relatively, see Section 2.1.2) easily approximated by the corresponding sample mean with obvious operational benefits. On the other hand, X and h represent a breakthrough between what we may assume to be the interface signal with the phenomenon (its phenotype) and the inverse mapping from phenotype to genotype, in its turn the appearance of the utility function in our decision framework. We range from *primordial* cases where h is the identity function – hence we are just interested in the mean value of what we observe – to the *extreme* case where h coincides with π – hence the inner structure of data reflecting all what we know about them – through a large set of variants denoting how large part of this structure is suitable to meet our current goal.

2.5 Exercises

1. The Java function Math.random() returns random values in the interval $[0, 1]$, with the approximation mentioned in the beginning of this chapter. Generate 100 such values, compute the corresponding empirical c.d.f. and draw its graph. Repeat the experiment subsequently increasing the number of values and check how the obtained graph changes. Repeat the experiment using the rand function of the C programming language (you should include the stdlib library in your

program in order to be able to access this function) and implementing the Algorithm 2.1. Compare the graphs obtained.
2. The c.d.f. of a negative exponential random variable X having $\lambda > 0$ as parameter is $F_X(x) = 1 - e^{-\lambda x}$. Compute the corresponding d.f. and the mean of X.
3. Compute mean and variance of a Bernoulli variable X identified through the parameter p as in Appendix B.1.
4. Let X be the random variable whose specifications are outputted by a routine in your favorite programming language. Write a program that outputs a sample of the random variable $Y = X^2$ and use this output in order to draw the graph of the corresponding empirical c.d.f.
5. Let (X_1, X_2) be two joint random variables, and let $F_{X_1,X_2}(x_1, x_2) = x_1 x_2 I_{(0,1)}(x_1) I_{(0,1)}(x_2)$. Show that X_1 and X_2 are independent.
6. Using (2.6) derive the analytical forms of the distributions of the sample mean $\overline{X}$ of the random variables considered in Table 2.1.
7. Express the analogous convergence condition of (2.17) for the statistic $\max\{x_1, \ldots, x_m\}$ computed on a sample drawn from a X uniform in $[0, a]$.
8. Express the likelihood of a sample drawn from a Gaussian random variable.
9. Compute the entropy of a random variable X distributed according to a Bernoulli law of parameter p. Check for which value of p it reaches its maximum.
10. Compute the entropy of a random variable X distributed according to a Gaussian law of mean μ and variance σ^2.

3 Inferring from Samples

A traditional way of introducing the inference facility in operational contexts is through the match box metaphor. You buy your box and wonder how many matches will fire, how many not. You cannot check all them otherwise you will be satisfied with your knowledge but cannot use the obtained information on the current box because it became empty. Thus you are challenged to understand the quality of the box by firing a small number of its matches. We link the metaphor to statistical framework we are building up through the noodles' picture in Fig. 3.1. For a fixed string of bits recording the sample of a random variable X you have a plenty of suffixes along which the population of subsequent X observations may develop. The compatibility requirement stresses the prefix-suffix relation between the strings as observations of the same phenomenon. With the increase of the complexity of the phenomena we study, and of computational power as well, the inference methods augmented their scope. Not only we want to forecast something of the future of X specifications, but we want to understand better what we have really observed. *Per se* the second goal complies with the noodles picture as well, since, in any case, what you will have understood will be employed to frame future data, as the current one belongs to the past. However the new objective calls for drilling relations between data, an attitude that is particular relevant in learning tasks, as we mentioned above, but is now going up till the most elementary parametric inference problems, when they are afforded in terms of genotypical features (see Section 3.2).

One past, many possible futures.

3.1 Learning from Samples

Mechanisms like the one shown in Fig. 2.1 have their further appeal that the way of implementing it is up to you, provided

Sampling mechanism is a tool for understanding,

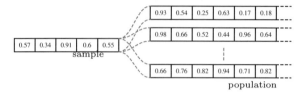

Fig. 3.1. Observed sample and its possible continuations.

Algorithm 3.1. Sampling Mechanism for a Binomial variable of parameters n and p: direct algorithm

set $f_Y = (1-p)^n$;
$u = \mathtt{rand}()$;
$x = 0$;
while $u \geq f_Y$ **do**
  $f_Y = f_Y \frac{p}{1-p} \frac{n-i}{i+1}$;
  $x = x + 1$;
end while
return x;

Algorithm 3.2. Sampling Mechanism for a Binomial variable of parameters n and p: using sum of Bernoulli variables

$x = 0$
for $i = 1$ to n **do**
  $u_i = \mathtt{rand}()$;
  $x = x + \mathrm{bernoulli}(u_i, p)$;
end for
return x;

---

any i.i.d. sample for the seed variable U, or Z in general, maps into an i.i.d. sample from the goal variable X. This means that:

1. the mechanism has not univocal implementation. For instance, in order to generate a Binomial sample we may use directly Fact 1.4, with the assignment statement implemented through Algorithm 3.1. Otherwise you may simply consider the Binomal variable as a sum of Bernoulli variables, so that a possible sampling mechanism is the one described in Algorithm 3.2.
2. an implementation of the same mechanism may be selected in order to put in evidence specific properties of X, provided that this property is *observable* by the sample.

The first option is mainly a matter of computational efficiency of the mechanism, a very appreciated property when we generate large samples like with Monte Carlo methods. The second option concerns the use you will make of the data, i.e. the properties of X, say its parameters, you are interested in. Thus, in the hypothesis that the pebble throws cover a distance that follows a Gaussian distribution law as in Example 1.5, you may be interested in their mean value and in the mean of their fourth power, and may consider an explaining function that explicitly depends on these parameters. But you cannot expect to *observe* from the sample the maximum of X. In light of the general inference framework outlined in Section 1.1.2, the notion of observability takes in this framework a clear definition.

but you may understand only what you observe.

Definition 3.1. *Given a random variable X and a property Θ on it, Θ is* observable *if a sampling mechanism and a master equation exist so that for each seed $\{z_1, \ldots, z_m\}$ the master equation admits a solution in θ that is univocal and feasible.*

Observability requires

In own turn,

Definition 3.2. *A statistic s is* well behaving *if Θ is observable through it, i.e. if it satisfies the following properties:*

having a statistic behaving well.

monotonicity. *A uniformly monotone relation exists between s and θ for any fixed seeds $\{z_1, \ldots, z_m\}$ – so as to have a unique solution of (1.10) in θ with $s_\Theta = s$;*

well definition. *Denote the range D_X^θ as the set of values the random variable X may assume with probability/density > 0 for a given parameter specification θ, and the support $\mathfrak{X}$ of X as the union of the D_X^θ corresponding to all Θ specifications. On each observed s the statistic is well defined for every value of θ, i.e. any sample specification $\{x_1, \ldots, x_m\} \in \mathfrak{X}^m$ such that $\rho(x_1, \ldots, x_m) = s$ has a probability density different from 0 – so as to avoid considering non surjective mapping from $\mathfrak{X}^m$ to $\mathfrak{S}$, i.e. associating via s to a sample $\{x_1, \ldots, x_m\}$ a θ that could not generate the sample itself;*

local sufficiency. *The solutions $\{\theta_1, \ldots, \theta_N\}$ of (1.10), for suitable N, constitute a true sample of Θ compatible exactly with the observed s. This means that Θ distribution must not depend on the special $\{x_1, \ldots, x_m\}$ we have observed but only on s. Consider the sheaf of trajectories in Fig. 1.7 crossing the line $s_P = s$ at a surrounding of a given p. For a same neighborhood width it will be constituted by a different number of trajectories depending on p. What we ask is that their distribution is the same independently of p. More in general, mapping z_is into x_is, we require that the joint distribution of the X_is for a fixed value s assumed by their statistic S is independent of θ. This insures that, denoting $\boldsymbol{x} = \{x_1, \ldots, x_m\}$*

68   Inferring from Samples

$$f_{\Theta|X=x}(\theta) = \frac{f_{X|S=s,\Theta=\theta}(x) f_{S|\Theta=\theta}(s)}{f_{X|S=s}(x)} \frac{f_\Theta(\theta)}{f_S(s)} \qquad (3.1)$$

is a sole function of θ and s when the above independence is satisfied, i.e. $f_{X|S=s,\Theta=\theta}(x) = f_{X|S=s}(x)$.

Finally,

In this case you may think of compatible worlds.

Definition 3.3. *For a random variable and a sample as in Definition 3.1 a* compatible distribution *is a distribution having the same sampling mechanism $\mathscr{M}_X = (Z, g_\theta)$ of X with a value $\breve{\theta}$ of the parameter θ derived from a master equation (1.10) rooted on a well behaving statistic s.*

For instance, considering Example 1.3 for both the negative exponential variable with parameter λ and the Bernoulli variable with parameter p the statistic $\sum_{i=1}^m x_i$ is well behaving. The same holds for the power law random variable when using $\sum_{i=1}^m \log x_i$ as statistic. Vice versa, in the case of $[0, A]$ uniform variable both statistics do not meet the second requirement in the definition. For instance, for $A = a$ the observed sample $\{a/2, a/2, a/2\}$ gives $s'_A = 3a/2$, but sample specification $\{3a/2, 0, 0\}$ giving rise to the same statistic has a probability density equal to 0. The statistic $\max\{x_1, \ldots, x_m\}$ is well behaving in this case. As a general statement, which we will refer to in the book, under weak conditions it is possible to prove that jointly sufficient statistics [50] well behave with respect to the related parameters [6]. For a wide family of distribution laws the well behaving property coincides with sufficiency property defined as follows.

Features of a sufficient statistic

Definition 3.4. *For any family $\mathscr{X}$ of random variables generated by a family g_Θ of explaining functions, the following statements are equivalent:*

- *a statistic S is* sufficient *w.r.t. Θ;*
- *denoted by X the random variable corresponding to an observable sample, a statistic S induces a partition $\mathfrak{U}(S)$ on D_X such that the ratio $f_X(x^1; \theta)/f_X(x^2; \theta)$ does not depend on θ when x^1 and x^2 belong to a same element of $\mathfrak{U}(S)$;*
- *a statistic S is such that the joint density/probability function $f_{X|s}$ of a sample given the value of the statistic is independent of θ;*

The factorization criterion

- *a statistic S is such that the likelihood $L(x_1, \ldots, x_m; \theta)$ of a sample $x = \{x_1, \ldots, x_m\}$ can split in the product of a function independent of θ and another one dependent on x only through s (factorization criterion) [37]. Namely:*

$$L(x_1,\ldots,x_m;\theta) = \prod_{i=1}^{m} f_X(x_i;\theta) = \eta(x_1,\ldots,x_m)\gamma(s,\theta).$$
(3.2)

The last statement in Definition 3.4 provides a powerful tool for discovering sufficient statistics. The second point refers to the contours outlined in Fig. 1.8. We realize that samples $\{x_1,\ldots,x_m\}$ in a same s contour in the left picture have same probabilities, apart from constant factors, because the corresponding seeds are affected in the right picture by a same θ parameter. This is the root of the ordering/distance correspondence, hence of probability exchange, between statistics and parameters we will discuss in the next section.

Example 3.1. Given an exponential random variable X with unknown parameter Λ having specification λ, it is easy to see that starting from a sample $\boldsymbol{x} = \{x_1,\ldots,x_m\}$ the statistic $s_\Lambda = \sum_{i=1}^{m} x_i$ is sufficient. In fact, according to Definition 3.4, if $\boldsymbol{x}$ and $\boldsymbol{x}'$ belong to the same element of the partition $\mathfrak{U}(S)$, i.e. $\sum_i x_i = \sum_i x'_i$, then $\lambda^m e^{-\lambda \sum x_i}/\lambda^m e^{-\lambda \sum x'_i} = 1$ does not depend on λ.

Example 3.2. The statistic $s'_A = \sum_{i=1}^{m} x_i$, where x_i is a specification of a random variable X uniformly distributed in $[0, A]$ is not sufficient, as can be proved by direct inspection of the likelihood according to the factorization criterion. Rather, the statistic $s_A = \max\{x_1,\ldots,x_m\}$ is such that the distribution of $(X_1,\ldots,X_m|S_A = s_A)$ does not depend on a, property that (following the third point in Definition 3.4) assures S_A to be a sufficient statistic.

In Table 3.1 we report a list of sufficient statistics s_Λ with regard to the parameters Λ of some common distribution laws.

Having clarified the rationale underlying the inference of a parameter Θ in our framework, we may reconsider the previous examples as specifications of a general estimation procedure that runs in the following three steps:

- identification of a well behaving statistic,
- description of the parameter distribution law, and
- drawing of the operational decision.

Winning strategy in a few moves

3.1.1 Statistic Identification

As a general rule, we first look for a sufficient statistic. If we do not find it, we look for well behaving-ness minded like a *local* sufficiency. Otherwise we look for *well done* statistics, hence statistics satisfying asymptotically the above properties or at

If not well behaving, at least well done statistics.

Table 3.1. Common distribution laws together with related sufficient statistics.

Distribution	Definition of density function	Sufficient Statistic
Uniform discrete	$f(x; n) = 1/n I_{\{1,2,\ldots,n\}}(x)$	$s_n = \max_i x_i$
Bernoulli	$f(x; p) = p^x (1-p)^{1-x} I_{\{0,1\}}(x)$	$s_P = \sum_{i=1}^m x_i$
Binomial	$f(x; n, p) = \binom{n}{x} p^x (1-p)^{n-x} I_{0,1,\ldots,n}(x)$	$s_P = \sum_{i=1}^m x_i$
Geometric	$f(x; p) = p(1-p)^x I_{\{0,1,\ldots\}}(x)$	$s_P = \sum_{i=1}^m x_i$
Poisson	$f(x; \mu) = e^{-\mu x} \mu^x / x! I_{\{0,1,\ldots\}}(x)$	$s_M = \sum_{i=1}^m x_i$
Uniform continuous	$f(x; a, b) = 1/(b-a) I_{[a,b]}(x)$	$s_A = \min_i x_i$ $s_B = \max_i x_i$
Negative exponential	$f(x; \lambda) = \lambda e^{-\lambda x} I_{[0,\infty]}(x)$	$s_A = \sum_{i=1}^m x_i$
Gaussian	$f(x, \mu, \sigma) = 1/(\sqrt{2\pi}\sigma) e^{-(x-\mu)^2/(2\sigma^2)}$	$s_M = \sum_{i=1}^m x_i$ $s_\Sigma = \sqrt{\sum_{i=1}^m (x_i - \bar{x})^2}$
Gamma	$f(x; r, \lambda) = \lambda/\Gamma(r) (\lambda x)^{r-1} e^{-\lambda x} I_{[0,\infty]}(x)$	$s_A = \sum_{i=1}^m x_i$

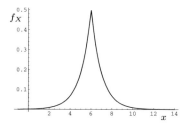

Fig. 3.2. Plot of the d.f. for the symmetric exponential distribution with parameter $\lambda = 6$.

least satisfying some of them. For instance, with the *double exponential distribution* characterized by a d.f. whose analytical expression reads:

$$f(x; \lambda) = \frac{1}{2} e^{-|x-\lambda|}, \tag{3.3}$$

defined for every $x \in \mathbb{R}$ and $\lambda \in \mathbb{R}^+$, and whose graph looks like in Fig. 3.2, we have no sufficient statistic for Λ. Rather, the median τ of the sampled values is approximately a well behaving statistic, where we define median value within the sample $\{x_1, \ldots, x_m\}$ to be the $(m+1)/2$-th value of the sorted sample $\{x_{(1)}, \ldots, x_{(m)}\}$ if m is odd, and any value between $x_{(m/2)}$ and $x_{(m/2+1)}$ if m is even (a typical choice is in this case the arithmetic mean of these values). Indeed, a sampling mechanism for this variable reads:

$$g_\lambda(u) = \begin{cases} \lambda + \log(2u) & \text{if } u < 0.5 \\ \lambda - \log(2(1-u)) & \text{if } u \geq 0.5. \end{cases} \tag{3.4}$$

Consequently we have the following master equation:

$$\tau = \begin{cases} \lambda + \log(2\breve{u}) & \text{if } \breve{u} < 0.5 \\ \lambda - \log(2(1-\breve{u})) & \text{if } \breve{u} \geq 0.5 \end{cases} \quad (3.5)$$

where $\breve{u}$ is the median of the seed sample $\{u_1, \ldots, u_m\}$. From this equation we see that τ satisfies the first two properties of a well behaving statistic; the third one is only approximately satisfied given the analogy of $|a|$ and a^2 functions and the coincidence of mean with median of a symmetric population.

3.1.2 Parameter Distribution Law

As pointed out in Fig. 1.7, as a function of the seed, Θ is a random variable whose distribution law comes from the iterated solution of master equations with different values of the sample seeds. This may bring to either numerical or analytical procedures.

Engineering the procedures

Bootstrapping populations

The first option looks as the simplest one, just by throwing seeds and computing the c.d.f. of Θ as:

The numerical option,

$$F_\Theta(\theta) = \lim_{N \to +\infty} \sum_{j=1}^{N} \frac{1}{N} I_{(-\infty, \theta]}(\breve{\theta}_j), \quad (3.6)$$

denoting by $\breve{\theta}_j$ the generic solution of the above master equation (1.10). Some indeterminacies remain with discrete X that we will shortly consider. The whole procedure may be summed up in the form of Algorithm 3.3.

through a standard procedure.

Algorithm 3.3. Generating parameter populations through bootstrap.

Given a sample $\{x_1, \ldots, x_m\}$ from a random variable with parameter θ unknown,

1. Identify a well behaving statistic S for the parameter θ;
2. compute a specification s_Θ from the sample;
3. repeat for a satisfactory number N of iterations:
    a) draw a sample z_i of m seed random variables;
    b) get $\breve{\theta}_i = \text{Inv}(s_\Theta, z_i)$ as a solution of (1.10) in θ with $z_i = \{z_1, \ldots, z_m\}$;
    c) add $\breve{\theta}_i$ to Θ population.

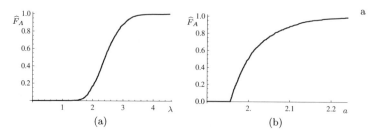

Fig. 3.3. Compatible parameter empirical c.d.f. generated from: (a) parameter Λ of a negative exponential distribution rooted in $s_\Lambda = 12.39$; (b) parameter A of a uniform distribution in $[0, A]$ with $s_A = 1.96$. Sample size $m = 30$.

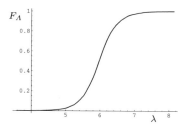

Fig. 3.4. Λ c.d.f. computed by approximating (3.6) with 5000 bootstrap replicas.

You may easily realize from Table 3.1, (1.8) and (1.9) that we obtain the curve in Fig. 3.3(a) computing the empirical distribution (3.6) on the population obtained through Algorithm 3.3 when $\text{Inv}(s_\Lambda, \boldsymbol{u}_i) = \sum_j(-\log u_{ij})/s_\Lambda$, and the curve in Fig. 3.3(b) when $\text{Inv}(s_A, \boldsymbol{u}_i) = s_A/\max_j\{u_{ij}\}$.

Remark 3.1. Note that the accuracy with which a parameter distribution law of populations compatible with a sample is obtained is not a function of the sample size. Rather it is a function of the number of seeds we draw. In own turn this number is just a matter of purely computational time but does not require any extension of the observed data.

Example 3.3. Consider again the d.f. (3.3) of a double exponential distribution. As already shown in Section 3.1.1, we may rely on τ as an approximately well behaving statistic. Hence, from a set of N seeds we sample N values of $\check{\lambda}$ that are used to approximate the Λ c.d.f. via (3.6). In particular from the sample $\{5.71, 7.13, 5.98, 5.89, 5.92, 9.66, 7.15\}$ we obtain the c.d.f. as in Fig. 3.4.

Deducing populations

The second option aims to identify directly the analytical form of F_Θ. It may require more technicalities, but it will prove the sole feasible in some instances that will be discussed in the next section. The key idea is that, in case sample properties are sufficient statistics – and, with some approximation, with poorer statistics – we may deduce an ordering on the parameter specifications from an ordering on the statistics so that we may translate S c.d.f. into inference of Θ c.d.f. A template reasoning is the following.

The logical option

We deduce an ordering on the parameter from the ordering on the observations.

We generate a sample $\{x_1, \ldots, x_m\}$ of a Bernoulli variable X with parameter p through sampling mechanism (1.7). Forget about the problem of how to draw $\{u_1, \ldots, u_m\}$ (tossing a sequence of dice, using a pseudo-random number generator). As a matter of fact you do not know the us. We only observe $\{x_1, \ldots, x_m\}$, and maintain their sum $s_P = \sum_{i=1}^m x_i$ – coinciding with the number k of 1 in the sample – as relevant statistic. For a given seed we obtain different binary strings with different values of k depending on the value p of the threshold. However, for each sample $\{x_1, \ldots, x_m\}$, hence for each fixed – though unknown – p and $\{u_1, \ldots, u_m\}$ in (1.7), and for whatever number k of 1 observed in the sample and any other number $\widetilde{p} \in [0, 1]$ we can easily derive the following implication chain:

Raise p, raise k

For any fixed seed,

$$(k_{\widetilde{p}} \geq k) \Leftarrow (p < \widetilde{p}) \Leftarrow (k_{\widetilde{p}} \geq k + 1), \qquad (3.7)$$

where $k_{\widetilde{p}}$ counts the analogous number of 1 if the threshold in the explaining function (1.7) switches to $\widetilde{p}$ *for the same specifications* $\{u_1, \ldots, u_m\}$ of U. The asymmetry, that disappears if we are working with a continuous random variable, derives from the fact that: i) raising the threshold parameter p in the explaining function g_p cannot decrease the number of 1 in the observed sample, but ii) we can recognize that such a raising occurred only if we really see a number of 1 in the sample greater than k.

In order to perceive the randomness of parameter P, we must consider that each specification p of P accounts for the asymptotic frequency with which 1 will appear in the prosecution of the observation log, hence with which uniform seeds will trespass p itself. The indeterminacy of this fixed point problem is broken by the fact that the same p must explain also the actually observed values of the sample. We may interpret (3.7) in the light of Fig. 3.5 saying that a raising of the threshold line which increments the number of 1 in the population cannot decrease the number of 1 in the sample and *vice versa*. Extending this sentence to any string of $m + M$ specifications of U, we may read the three conditions in (3.7) in terms of events of the $[0, 1]^{m+M}$ sample space jointly spanned by the $m + M$ uniform Us for fixed

and summing up over all seeds

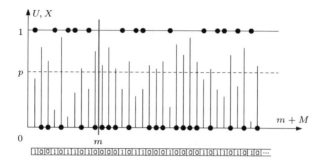

Fig. 3.5. Generating a Bernoullian sample. Horizontal axis: index of the U realizations; vertical axis: both U (lines) and X (bullets) values. The threshold line $y = p$ (dashed line) realizes a mapping from U to X through (1.7). The higher is p, the greater is the number of 1 in the binary string.

k and $\widetilde{p}$. This amounts to considering analogous set relations between the three events $(K_{\widetilde{p}} \geq k)$, $(P < \widetilde{p})$ and $(K_{\widetilde{p}} \geq k+1)$, namely:

$$(K_{\widetilde{p}} \geq k) \supseteq (P < \widetilde{p}) \supseteq (K_{\widetilde{p}} \geq k+1). \qquad (3.8)$$

It is a simple geometrical argument

For instance, let us focus on the cross formed by the vertical line $\phi = 0.3$ and the horizontal line $\psi = 0.4$ (the plain black lines in Fig. 3.6) as the pivot $(k, \widetilde{p})$ of our implications. It is clear that any trajectory crossing the line $\phi = 0.3$ at any ordinate $\psi = p < 0.4$, say $\psi = 0.2$ (the dashed line in the figure), will meet the line $\psi = \widetilde{p} = 0.4$ at an abscissa $\phi = k_{\widetilde{p}}/30 \geq k/30 = 0.3$. These random trajectories are exactly what commutes the randomness of the abscissa with the randomness of the ordinate. In other words, trajectories having a p less than $\widetilde{p}$ are all and only those having a $\widetilde{k} \geq k$, apart from some uncertainty about those having $\widetilde{k}$ exactly equal to k.

allowing us to identify the c.d.f. of P.

Hence, passing to the probabilities P, i.e. to the asymptotic frequencies of the above trajectories, we have the consequent bounds:

$$\mathrm{P}\,(K_{\widetilde{p}} \geq k) \geq \mathrm{P}\,(P < \widetilde{p}) = F_P\,(\widetilde{p}) \geq \mathrm{P}\,(K_{\widetilde{p}} \geq k+1), \qquad (3.9)$$

which characterize the c.d.f. F_P of the parameter P when the statistic $s_P = \sum_{i=1}^{m} x_i$ equals k, as no assumption has been introduced against the continuity of P [1]. In Fig. 3.7 we plot these bounds for $k = 9$ and $m = 30$, using formula (3.22) which will be introduced later on. The graph of F_P lies between the two

---

[1] Summing up: P integrates probabilities in $\mathfrak{U}^{m+M}$, with $\mathfrak{U} = [0,1]$; P asymptotically accounts frequencies over M bits with $M \to +\infty$.

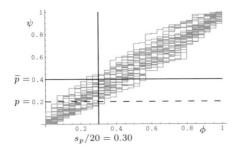

Fig. 3.6. Twisting sample with population frequencies along sampling mechanism trajectories. Population and sample sizes: $M = 200$ and $m = 30$ elements, respectively. $\phi = s_p/m =$ frequency of 1 in the sample; $\psi = h_p/M =$ frequency of 1 in the population. Stepwise gray lines: trajectories described by the number of 1 in sample and population when p ranges from 0 to 1, for different sets of initial uniform random variables. Black lines mark the coordinates of a possible twisting implication pivot.

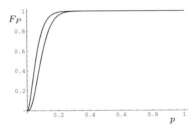

Fig. 3.7. Bounds on the c.d.f. of the parameter P of a Bernoulli distribution for $k = 9$ and $m = 30$.

curves. This gap represents the analytical counterpart of the numerical indeterminacy we found in solving $\text{Inv}(s_P, \boldsymbol{u}_i)$ with the bootstrap procedure.

We denote with *twisting arguments* any logical implication like (3.7) that is rooted on a well behaving statistic and is aimed to twist sample properties with population properties.

Twisting sample's with population's properties,

Definition 3.5. *Let X be a random variable with unknown parameter Θ and s_Θ a well behaving statistic according to Definition 3.2. Then, for any pair $(s_\Theta, \widetilde{\theta})$, the satisfaction of the relation:*

basing on a well done statistic.

$$\left(s_{\widetilde{\theta}} \geq s_\Theta\right) \Leftarrow \left(\widetilde{\theta} > \theta\right) \Leftarrow \left(s_{\widetilde{\theta}} \geq s_\Theta + \ell\right) \quad (3.10)$$

for bounded ℓ, is denoted a twisting argument for Θ.

In true operational terms, with a well behaving statistic s master equations such as (1.10) state a *clean* monotone relationship between s and the parameter θ for whatever $\{z_1, \ldots, z_m\}$ by definition. In case of both scalar statistic and parameter, this means that for any pair of scalar (s, θ) satisfying the equation, a new pair (s', θ') is such that:

$$s' > s \Rightarrow \theta' \geq \theta; \quad \theta' > \theta \Rightarrow s' \geq s$$
if the relastionship is monotonically non decreasing, (3.11)

$$s' > s \Rightarrow \theta' \leq \theta; \quad \theta' > \theta \Rightarrow s' \leq s$$
if the relastionship is monotonically non increasing. (3.12)

As pivot of our reasoning, we fix s to the observed s_Θ and θ' to a value $\widetilde{\theta}$ upon us. Then we consider the seeds z underlying the inequalities and come to the synthesis about events. Namely, they are seeds z such that $s' = h(\widetilde{\theta}, z_1, \ldots, z_m) > s_\Theta$ identifying in force of (3.11) with the ones such that $h^{-1}(s_\Theta, z_1, \ldots, z_m) \leq \widetilde{\theta}$ in case of non decreasing monotony, or according to (3.12) with $h^{-1}(s_\Theta, z_1, \ldots, z_m) \geq \widetilde{\theta}$ in case of non increasing monotony – analogously with right implications in (3.11) and (3.12). With S_θ and Θ continuous we have:

$$(S_\theta \geq s_\Theta) \Leftrightarrow (\Theta \leq \theta) \quad \text{or} \quad (3.13)$$
$$(S_\theta \leq s_\Theta) \Leftrightarrow (\Theta \leq \theta), \quad (3.14)$$

correspondingly, which leads to:

$$F_\Theta(\theta) = 1 - F_{S_\theta}(s_\Theta) \quad \text{or} \quad (3.15)$$
$$F_\Theta(\theta) = F_{S_\theta}(s_\Theta). \quad (3.16)$$

While we do not consider the case with Θ discrete, as a working hypothesis in absence of ancillary informations, if S_θ is discrete the mentioned local non unicity of the solution $s = \rho(x_1, \ldots, x_m)$ must be taken into account. This leads to (3.10) in place of (3.12) and thresholds on F_Θ as follows:

$$1 - F_{S_\theta}(s - \ell) \geq F_\Theta(\theta) \geq 1 - F_{S_\theta}(s). \quad (3.17)$$

Still with scalar statistic and alternative notation for parametric dependence: $F_{X|\Theta=\theta} = F_{X_\theta}$, a companion of Algorithm 3.3 is the Algorithm 3.4. A complete form of statement (3) in the algorithm should be $F_{\Theta|S=s}(\theta) \in \left(q_1(F_{S|\Theta=\theta}(s)), q_2(F_{S|\Theta=\theta}(s))\right)$ which highlights the symmetry of the conditions. Instead, we generally omit parametrizing Θ since $S = s$ represent a consistency condition with the environment that we assume to be satisfied by default.

Algorithm 3.4. Generating parameter distribution law through twisting argument.

Given a sample $\{x_1, \ldots, x_m\}$ from a random variable with parameter θ unknown,

---

1. Identify a well behaving statistic S for the parameter θ and its discretization granule ℓ (if any);
2. decide the monotony versus;
3. compute $F_\Theta(\theta) \in (q_1(F_{S|\Theta=\theta}(s)), q_2(F_{S|\Theta=\theta}(s)))$ where:
   - if S is continuous
   $$q_1 = q_2$$
   - if S is discrete
   $$\begin{cases} q_2(F_S(s)) = q_1(F_S(s-\ell)) & \text{if } s \text{ does not decrease with } \theta \\ q_1(F_S(s)) = q_2(F_S(s-\ell)) & \text{if } s \text{ does not increase with } \theta \end{cases}$$

   and
   $$q_i(F_S) = \begin{cases} 1 - F_S & \text{if } s \text{ does not decrease with } \theta \\ F_S & \text{if } s \text{ does not increase with } \theta \end{cases}$$

   for $i = 1, 2$.

When parameters are vectors, things are similar though a bit more complicated by the management of joint inequalities. Note that identifying the monotony versus of the statistics with the parameters often does not require an analysis of the sampling mechanism. For instance we have no doubt that the sum of sample items grows with the population mean. Moreover in some cases $q(F_S)$ has the same shape of F_Θ so that the statistic takes in the former an analogous or even equal meaning of the latter w.r.t. the variable behavior.

Example 3.4. We know from Example 1.2 that for X negative exponential random variable with parameter Λ (having specification λ), the explaining function (1.8) denotes that the statistic $s_\Lambda = \sum_{i=1}^m x_i$ is well behaving. Hence we have the twisting argument:

$$\left(\lambda \leq \widetilde{\lambda}\right) \Leftrightarrow \left(s_\lambda \geq s_{\widetilde{\lambda}}\right) \tag{3.18}$$

where $s_{\widetilde{\lambda}} = \sum_{i=1}^m \widetilde{x}_i$ and $\widetilde{x}_i$ is the value into which u_i would map substituting λ with $\widetilde{\lambda}$ in (1.8). Hence S_θ is continuous and we have a decreasing monotonic relation of s_λ with λ. This options $q_i = q_2$ and $q_i(F_S) = F_S$. In addition, S_λ follows a Gamma distribution law with parameters m and λ [50], whose analytical form commutes the roles of variable specification and parameter between s_λ and λ, so as Λ too follows a Gamma distribution of parameters m and s_Λ.

3.1.3 Drawing Decisions

The first two steps were dedicated to the analysis of the data in search of a profitable representation, not far from what happens in other frameworks, for instance with the Fourier transform of the data. Now we need synthesizing them into a decision. The most used alternatives consist in focusing on probabilities of keen events or mediating some features on the entire population.

Confidence intervals

A well grouped set of highly compatible parameters

Let us start with the formal definition of confidence interval.

Definition 3.6. *Given a random variable X with parameter Θ and a real number $0 \leq \delta \leq 1$, $[\theta_{\mathrm{dw}}, \theta_{\mathrm{up}}]$ is called a $1-\delta$ confidence interval for Θ if*

$$\mathrm{P}(\theta_{\mathrm{dw}} \leq \Theta \leq \theta_{\mathrm{up}}) = 1 - \delta. \tag{3.19}$$

The quantity δ is called the confidence level of the interval. Analogous definitions describe open and semiclosed intervals.

supporting operational decisions.

Facing a possibly infinitely wide range of θ, we look for a narrower interval where we are *almost* sure of having the parameter of the observations' population we will meet in the future. For instance:

- if we are designing the aqueduct for a village we could rely on the right extreme of the interval $(0, \theta_{\mathrm{up}})$ of the mean water consumption Θ per family of the village;
- if we are considering the critical budget of our retailing company we should rely on the left extreme of the interval $(\theta_{\mathrm{dw}}, +\infty)$ of the mean income of our shops;
- we should consider the whole interval $(\theta_{\mathrm{dw}}, \theta_{\mathrm{up}})$ if the value of Θ affects non linearly our decision on a random phenomenon.

Another way of delineating the gap between possibility and necessity

This represents a sharp form of granulation: we interpret from another perspective the *necessity* and *possibility* notion with which we gauge sets connected with granular information [16]. In Section 1.3 this gives rise to two nested regions in whose gap we locate the contour of a set of our interest. Here we identify directly a gap where to locate a parameter of our interest.

From a technical point of view, a confidence interval is a pair of quantiles of the Θ distribution, where the α-quantile x_α of a

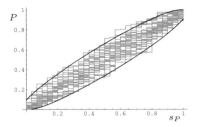

Fig. 3.8. 0.90 confidence region for the parameter P of a Bernoulli distribution (black curves) vs. the value of the statistic s_P. Gray lines: trajectories describing the course of the parameter with the statistic, based on different random seeds as in Fig. 3.6.

random variable X is such that $F_X(x_\alpha) = \alpha$. Thus, we are called to solve a pair of equations:

$$F_\Theta(\theta_{\text{up}}) = 1 - \delta_1 \qquad (3.20)$$
$$F_\Theta(\theta_{\text{dw}}) = \delta_2 \qquad (3.21)$$

Just a matter of solving an inverse problem

with $\delta_1 + \delta_2 = \delta$.

Example 3.5. Consider the Bernoulli case study described in Section 3.1.2. We can read the probabilities in (3.9) as:

$$I_{\widetilde{p}}(s_P, m - s_P + 1) \equiv 1 - \sum_{i=0}^{s_P - 1} \binom{m}{i} \widetilde{p}^i (1 - \widetilde{p})^{m-i} \geq F_P(\widetilde{p})$$

$$\geq 1 - \sum_{i=0}^{s_P} \binom{m}{i} \widetilde{p}^i (1 - \widetilde{p})^{m-i} \equiv I_{\widetilde{p}}(s_P + 1, m - s_P) \quad (3.22)$$

where $s_P = \sum_{i=1}^{m} x_i$ and I_β is the c.d.f. of a Beta variable $Z_{h,r}$ [2] of parameters h and r [21], that is

$$I_\beta(h, r) \equiv \text{P}(Z_{h,r} \leq \beta) = 1 - \sum_{i=0}^{h-1} \binom{h+r-1}{i} \beta^i (1-\beta)^{h+r-1-i}. \qquad (3.23)$$

From (3.22) we may find a two-sided symmetric confidence interval $(p_{\text{dw}}, p_{\text{up}})$ for P of level δ by solving

$$I_{p_{\text{up}}}(s_P + 1, m - s_P) = 1 - \delta/2 \qquad (3.24)$$
$$I_{p_{\text{dw}}}(s_P, m - s_P + 1) = \delta/2 \qquad (3.25)$$

in p_{dw} and p_{up}. In Fig 3.8 we draw the 0.90 confidence intervals computed through (3.24) and (3.25) with s_P varying from 0 to m trespassed only seldom by the stepwise lines as in Fig. 3.6.

---
[2] Also called *Incomplete Beta function*.

80   Inferring from Samples

The operational scheme

The solution of the above example consists again in a twisting operation. We twist quantiles of the parameter distribution with those of the statistic distribution. The general scheme is:

$$\theta_\alpha = (\theta | s_{g(\alpha),\theta} = s_\Theta), \qquad (3.26)$$

where:

- s_Θ is the observed statistic,
- $g(\alpha)$ is a suitable function of α depending on the monotonicity relation between s_θ and θ exploited by the twisting argument,
- $s_{\delta,\theta}$ is the δ quantile of the above statistic when $\Theta = \theta$.

For instance $g(\alpha) = \alpha$ in the case of a negative exponential variable and $g(\alpha) = 1 - \alpha$ in the case of a Bernoulli variable. The actual implementation of (3.26) may deserve some computational difficulty. However with a common PC, possibly with the help of specific packages[3], we may always write down a routine that has in input α and s_Θ and outputs θ_α with a satisfactory approximation.

Shortcuts are possible

Given the Gaussian distribution law's widespread use in many operational frameworks, its parameters play a special role in statistics. This interlinks with a very easy derivation of the distribution of their parameters through the above scheme. Let us start with the case where only one parameter, either the mean or the standard deviation of the random variable, is unknown.

Example 3.6. Let X be a Gaussian variable with parameters M and Σ^2 and $\{x_1, \ldots, x_m\}$ a specification of a sample drawn from it. We may quickly get the distribution laws of each one of these parameters, when the other is known, as follows.

We know [35] that under these conditions $s_M = \sum_{i=1}^{m} x_i$ and $s_{\Sigma^2} = \sum_{i=1}^{m}(x_i - \mu)^2$ are sufficient statistics w.r.t. M and Σ^2 respectively (here μ represents a specification of M); hence the distribution law of these parameters will also be a separate function of these statistics. Starting from seeds constituted by specifications of a Normal variable Z (i.e. a Gaussian variable with null mean and unitary variance) we have the sampling mechanism $\mathscr{M}_X = (Z, g_{(\mu,\sigma)})$ for any X:

$$x_i = \mu + z_i \sigma. \qquad (3.27)$$

Accordingly, we derive the two logical implications:

$$(\mu \leq \tilde{\mu}) \Leftrightarrow (s_M \leq s_{\tilde{\mu}}) \qquad (3.28)$$
$$(\sigma \leq \tilde{\sigma}) \Leftrightarrow (s_{\Sigma^2} \leq s_{\tilde{\sigma}}). \qquad (3.29)$$

---

[3] Such as Mathematica [51], Maple [10], Scilab [25] and so on.

Hence

$$F_{M|\sigma}(\mu) = 1 - F_{S_\mu}(s_M) = 1 - F_Z\left(\frac{s_M/m - \mu}{\sigma/\sqrt{m}}\right) = F_Z\left(\frac{\mu - s_M/m}{\sigma/\sqrt{m}}\right) \quad (3.30)$$

since we may pivot around a Normal distribution,

where the third element of the equation chain refers to the Normal variable Z, and the last item comes from the symmetry of this variable around 0. The simple operational rule for implementing (3.26) w.r.t. M is:

> consider the *pivotal quantity* $\pi = \left(\frac{\mu - s_M/m}{\sigma/\sqrt{m}}\right)$, take the quantile z_α of the Normal variable Z and solve the equation $\pi = z_\alpha$ in μ, which leads to: $\mu_\alpha = s_M/m + z_\alpha \sigma/\sqrt{m}$.

For instance, if we want a symmetric two-sided 0.90 confidence interval for M we fix $\mu_{\text{dw}} = s_M/m + z_{0.05}\sigma/\sqrt{m}$ and $\mu_{\text{up}} = s_M/m + z_{0.95}\sigma/\sqrt{m}$, so that $P(\mu_{\text{dw}} \leq M \leq \mu_{\text{up}}) = 0.90$.
Analogously,

or around a Chi-square distribution,

$$F_{\Sigma^2|\mu}(\sigma) = 1 - F_{S_\sigma}(s_{\Sigma^2}) = 1 - F_{\chi_m^2}\left(\frac{s_{\Sigma^2}}{\sigma^2}\right), \quad (3.31)$$

where χ_m^2 is a Chi-square random variable with parameter m, identifies the pivotal quantity $\pi = \left(\frac{s_{\Sigma^2}}{\sigma^2}\right)$.

For instance, if with $m = 20$ we want a one-sided 0.90 confidence interval for Σ^2 having 0 as the left extreme, we fix $\sigma_{\text{up}}^2 = \Sigma_{0.90}^2 = s_{\Sigma^2}/\chi_{0.10}^2 = s_{\Sigma^2}/12.44$, so that $P(0 \leq \Sigma^2 \leq s_{\Sigma^2}/12.44) = 0.90$.

Remark 3.2. This reasoning, subtending the *pivotal quantity methods*, extends to any parameter Θ for which a pivotal quantity Z can be found, i.e. a total and invertible function of θ that is monotone in a statistic S and in no other way depends on the sample, and such that the distribution law of Z does not depend on θ.

and in general with a pivotal quantity.

Point estimators

Synthesizing the parameter distribution into one value is problem-dependent; doing it correctly is a matter of computational skills and underlying requirements. A balance between the synthesis sophistication and its computational overhead is a result of some sound compromise. As usual in decision theory, synthesis comes from searching the $\widehat{\theta}$ value which minimizes a cost function. Namely, you define a cost deriving from the use of $\widehat{\theta}$ in place of the θ which will actually parametrize the future observation history, call it *loss function* $l(\widehat{\theta}, \theta)$. Since this parameter is random you will focus on the expected value of l, call it the

A keen point from the parameter population,

estimator risk $R(\widehat{\theta}) = E_\Theta[l(\widehat{\theta}, \Theta)]$. Thus the estimator is the solution to the minimization problem:

$$\min_{\widehat{\theta}} E_\Theta[l(\widehat{\theta}, \Theta)]. \qquad (3.32)$$

minimizing approximation consequences, Sometimes it may prove easier to compute the above risk in a twisted framework (the Kolmogorov framework): assume you are at the end of the universe time, so that you *really* have observed the infinite observation history related to the phenomenon at hand, so that you may say: the parameter of this history is θ. **variously understood.** Now, coming back at the beginning, assume the limited sample $\{x_1, \ldots, x_m\}$ you have observed as the output of the operation of picking at random values from the infinite history, hence as a specification of the random sample $\{X_1, \ldots, X_m\}$. In this scenario you decide $\widehat{\theta}$ to be a function of the sample specification $\{x_1, \ldots, x_m\}$ and average the loss function on the sample population $\{X_1, \ldots, X_m\}$. Thus the above minimization reads as:

$$\min_{\rho} E_{\{X_1, \ldots, X_m\}}[l(\rho(X_1, \ldots, X_m), \theta)]. \qquad (3.33)$$

Coming back to (3.32) below we list a series of general-purpose loss functions that give rise to easily computable solutions.

Unbiased estimator *Quadratic loss.* With $l(\theta, \widehat{\theta}) = (\theta - \widehat{\theta})^2$ you have that $E[\Theta]$ is the solution of the form:

$$\min_{\widehat{\theta}} E_\Theta[(\widehat{\theta} - \Theta)^2]. \qquad (3.34)$$

For this reason $E[\Theta]$ is denoted as both unbiased and Minimum Mean Square Error (MMSE) estimator. *Vice versa*, $\breve{g}(X) = 1/m \sum_{i=1}^{m} g(X_i)$ is a weakly unbiased estimator of $\theta = E[g(X)]$ since $E[\breve{g}(X)] = E[g(X)]$, but not always is a solution of the problem:

$$\min_{\rho} E_{\{X_1, \ldots, X_m\}}[(\rho(X_1, \ldots, X_m) - E[g(X)])^2]. \qquad (3.35)$$

If this is the case for any value of $E[g(X)]$, then $\breve{g}(X)$ is called Uniformly Minimum Variance Unbiased Estimator (UMVUE).

Example 3.7. For a Uniform variable X in $[0, A]$: i) the unbiased estimator of A is the statistic $\widehat{a} = m/(m-1) \max \{x_1, \ldots, x_m\}$, ii) the weakly unbiased estimator is $\breve{A} = 2/m \sum_{i=1}^{m} X_i$, iii) while the UMVUE is $A^* = (m+1)/m \max \{X_1, \ldots, X_m\}$.

Remark 3.3. The weakly unbiasedness property of $\breve{g}(X)$, together with the fact that $\lim_{m\to\infty} V[\breve{g}(X)] = 0$, is at the basis of the master equations:

$$\frac{1}{m}\sum_{i=1}^{m} x_i^k \simeq E[X^k] \simeq \widehat{E}[X^k], \qquad (3.36)$$

where the last term is computed with a X distribution law using an estimator $\breve{\theta}$ in place of the unknown parameter θ. Solving in $\breve{\theta}$ the equation of the first and last terms in (3.36) is a way of finding a θ estimator that is known as *method of moments*. We use a number of equations like (3.36) equal to the number of unknown parameters using the lowest values of $k \geq 1$. The first term in equation (3.36) is called k *order moment* of the sample, the second of the population.

<small>A weaker, general purpose method</small>

Example 3.8. Let X be a binomial random variable with unknown parameters n and p, i.e. $f_X(x;n,p) = \binom{n}{x} p^x (1-p)^{n-x} I_{\{0,\ldots,n\}}(x)$ and $\{x_1,\ldots,x_m\}$ a sample drawn from X. By applying (3.36) with the first two moments we obtain:

$$s_1 = \frac{1}{m}\sum_{i=1}^{m} x_i = E[X] = np \qquad (3.37)$$

$$s_2 = \frac{1}{m}\sum_{i=1}^{m} x_i^2 = E[X^2] = np(1-p), \qquad (3.38)$$

having as solution $\breve{n} = \frac{s1^2}{s1-s2}, \breve{p} = 1 - \frac{s2}{s1}$.

Understanding measure. Relation (2.21) states that you have a greater chance of observing the items that you better understood, i.e. whose descriptive program is shorter. *Vice versa*, you improve the understanding of data if you frame them into a distribution whose free parameters maximize the probability of observing the actually observed data. Hence you assume as a loss function exactly $-\log f_{X_1,\ldots,X_m}(x_1,\ldots,x_m) = -\sum_{i=1}^{m} \log f_X(x_i)$, as an over-evaluation of $K(x_1,\ldots,x_m)$ analogous of $K(x)$ in (2.21), so that the loss function is the entropy $H[X]$ of X as defined in (2.24). In greater detail, since you do not know the parameter θ, on one side you maintain the expectation in respect to the X distribution determined by this value of Θ. On the other, you optimize the risk in respect to an estimate $\widehat{\theta}$ affecting $-\log f_X(x_i)$. In conclusion, the solution $\widehat{\theta}$ of (3.32), which now reads:

<small>Select parameters best evoking what you have observed</small>

$$\min_{\widehat{\theta}} E_{\mathbf{X}}\left[-\sum_{i=1}^{m} \log \widehat{f}_X(x_i)\right] \qquad (3.39)$$

passes through the weakly unbiased estimator of the expected value – simply substituted by the observed value – and the maximum likelihood estimator of θ as the value that substituted in (2.10) to the unknown parameter θ maximizes the probability of observing the specific sample.

It is a matter of sufficient statistics again.

Remark 3.4. The method of estimating θ with a value maximizing the sample likelihood is much used and appreciated. Its keen property is that such estimator is a function of a sufficient statistic of the parameter.

Example 3.9. Consider a sample $\{x_1, \ldots, x_m\}$ drawn from a negative exponential random variable X with parameter λ. It is easy to check that the log-likelihood:

$$\log L(x_1, \ldots, x_m) = m \log \lambda - \lambda \sum_{i=1}^{m} x_i, \qquad (3.40)$$

having a maximum in correspondence of $\ddot{\lambda} = \sum_{i=1}^{m} x_i$, depends on the sample only through the statistic $s_\Lambda = m / \sum_{i=1}^{m} x_i$, which, in virtue of the factorization lemma (see Definition 3.4), proves to be sufficient.

3.2 Learning from Examples

An explaining function with partially known seeds

Granularity may require to move from a fuzzy appearance of data to their intimate root (their genotype). At an intermediate level we are often interested in a function underlying the observed data in whose respect the latter represent *examples* of how the function computes. Actually, parameter Θ in Fact 1.4 identifies a function (the explaining function) as well. But in this case the input of g_θ are unknown specifications of a random variable. Here we know also the input and we want to infer the function which maps input into output from observed co-occurencies of input-output specifications. The typical instance of this kind of inference is based on a *cause/effect* sample

$$z_m = \{(x_i, y_i),\ x_i \in \mathfrak{X}, y_i \in \mathfrak{Y}, i = 1, \ldots, m\}. \qquad (3.41)$$

We assume that a class $\mathfrak{C}$ of functions and a random variable E exist such that for every $M \in \mathbb{N}$ and every (cause/effect) population z_M a c exists in $\mathfrak{C}$ such that $z_{m+M} = \{(x_i, c'(x_i, \varepsilon_i)), i = 1, \ldots, m + M\}$ where ε_is are specifications of a random variable E softening the problem of fitting the sample with a curve. In conclusion, we are asked to identify a curve c as a function of the sole x so that $\{y_i = \phi(c(x_i), \varepsilon_i), i = 1, \ldots, m\}$ for known

ϕ and known E distribution law, apart from some parameters. In most cases ϕ just adds c to ε. The key point is that, having a family of parametric curves as the candidates for representing the above relation, the random distribution of their parameters compatible with the sample translates into a random distribution of the elements of the family. Hence we will consider curves c as specifications of a random curve C. A main difference with the previous section is that the usual instruments of the density function or the cumulative distribution law to describe the curves may prove useless, if no suitable ordering of their specifications is available. Rather, we will directly synthesize the C randomness in terms of special events, namely confidence intervals, or point estimators.

Twice seeds complicate the management of the unknown paramaters.

3.2.1 Confidence Regions

Let us focus on the problem of finding a region where we can confine curves compatible with the observed data with a fixed probability. They denote confidence regions as follows.

Definition 3.7. *For sets $\mathfrak{X}, \mathfrak{Y}$ and a random function $C : \mathfrak{X} \mapsto \mathfrak{Y}$, denote by abuse $c \subseteq \mathfrak{D}$ the inclusion of the set $\{(x, c(x)); \forall x \in \mathfrak{X}\}$ in $\mathfrak{D}$. A confidence region at level γ is a domain $\mathfrak{D} \subseteq \mathfrak{X} \times \mathfrak{Y}$ such that $\mathrm{P}(C \subseteq \mathfrak{D}) = 1 - \gamma$.*

Now let us imagine a family $\mathscr{D}$ of nested confidence regions with the obvious relation $\mathfrak{D} \subseteq \mathfrak{D}' \Leftrightarrow (1 - \gamma_\mathfrak{D}) \leq (1 - \gamma_{\mathfrak{D}'})$ for each $\mathfrak{D}, \mathfrak{D}' \in \mathscr{D}$. We are free to select $\mathscr{D}$ provided we are able to compute the probability measure of sets $\mathfrak{C} = \{C_\alpha\}$ of random curves inside it, where α denotes a suitable parameter.

Augmenting the notion of confidence interval.

We may synthesize the techniques shown in the previous section into a procedure composed of three steps:

The magic three:

- Derivation of the curve distribution [CD]. We obtain it by identifying a suitable master equation on the curve parameters to be used either as the root of the parameters bootstrapper or as the pivot of a twisting argument for them.

curve distribution +

- Identification of suitably nested regions [NR]. According to Definition 3.7, we are looking for a domain that totally contains the curves accounting for its probability measure. Thus rather than looking for single curves as delimiters of a confidence region, we look for envelopes of them such that all and only those curves of our interest do no trespass the envelopes in any point. In this way we identify a family of nested regions each characterized by the probability that a curve completely belongs to it (see Fig. 1.15 on page 32).

ordering +

- Selection of pivot [PV]. The sense of the above family is to state an implicit distance measure of a generic curve from

stakes.

a central one: the greater is the probability of the minimal nested region containing it, the farther is the curve. In this sense it is a matter of convenience whether to fix the regions' family and identify a central value as the limit with $\gamma_{\mathcal{D}}$ going to 1, or deciding upon the pivot by first and then centering the family around it.

Example 3.10. Dealing with regression lines, we assume the functional relation between sample coordinates to be

$$y_i = a + bx_i + \varepsilon_i, \qquad (3.42)$$

where ε_is are specifications of a null-mean random variable E. With our inference strategy, we assume that both sample and future observations on the same phenomenon are summed up by a line $\ell : y = a + bx$ plus a vertical shift E that is random, independent from the horizontal coordinate and symmetric around 0. To show our capability of dealing with non Gaussian E we will assume a null mean *Laplace* distribution for E:

$$f_E(\varepsilon, \lambda) = \frac{1}{2}\lambda e^{-\lambda|\varepsilon|}, \qquad (3.43)$$

with $\lambda > 0$. We disregard the randomness of X and model a Y sampling mechanism $\mathcal{M}_Y = (E, \ell)$ that holds for every X specification. This is the basis of a bootstrap-based procedure that we will enunciate in general terms here below.

The building blocks of our procedure are:

CD. We essentially rely on a certain kind of *method of moments* (see Section 3.1.3) where we consider moments of functions of the random variables in a number equal to the number of the parameters to be bootstrapped, with the further commitment of working with suitable statistics. To this aim we center the reference framework in $\overline{x} = \frac{1}{m}\sum_{i=1}^{m} x_i$, so that the above functional relation reads $y_i = a + b(x_i - \overline{x}) + \varepsilon_i$. Then, we adopt the following sampling mechanism $\mathcal{M}_E$ for the ε_is consisting of a uniform seed U and the explaining function $g_\lambda(u)$ defined as:

$$\varepsilon = g_\lambda(u) = \frac{1}{\lambda}g_1(u) = \begin{cases} \frac{1}{\lambda}\log(2u) & \text{if } u < \frac{1}{2} \\ -\frac{1}{\lambda}\log(2(1-u)) & \text{otherwise}. \end{cases}$$
(3.44)

This leads to pivotal statistics $s_A = \sum_{i=1}^{m} y_i$, $s_B = \sum_{i=1}^{m} y_i(x_i - \overline{x})$ and $s_\Lambda = \sum_{i=1}^{m} y_i^2$ for the random variables A, B and Λ underlying the parameters of the regression lines.

It works with linear functions,

even with odd distributions,

Indeed:

$$\sum_{i=1}^{m} y_i = ma + \sum_{i=1}^{m} g_\lambda(u_i) \tag{3.45}$$

$$\sum_{i=1}^{m} y_i(x_i - \bar{x}) = b\sum_{i=1}^{m}(x_i - \bar{x})^2 + \sum_{i=1}^{m} g_\lambda(u_i)(x_i - \bar{x}) \tag{3.46}$$

$$\sum_{i=1}^{m} y_i^2 = m\bar{y}^2 - \left(\frac{\sum_{i=1}^{m} y_i(x_i - \bar{x})}{\sum_{i=1}^{m}(x_i - \bar{x})^2}\right)^2 - \frac{1}{\lambda^2}\sum_{i=1}^{m}\left(g_1(u_i)\right.$$

$$\left. -\overline{g_1(u)} - \frac{\sum_{j=1}^{m} g_1(u_j)(x_j - \bar{x})}{\sum_{j=1}^{m}(x_j - \bar{x})^2}(x_i - \bar{x})\right)^2 \tag{3.47}$$

with $\overline{g_1(u)}$ defined in analogy with $\bar{x}$, so that the three requested properties of well behaving statistics are almost fulfilled by s_A, s_B and s_Λ under weak hypotheses. Namely, the explicit expressions of the bootstrap-generated specifications are:

$$\check{a} = \frac{1}{m}\left(s_A - \sum_{i=1}^{m} g_{\check\lambda}(\check{u}_i)\right) \tag{3.48}$$

$$\check{b} = \left(\sum_{i=1}^{m}(x_i - \bar{x})^2\right)^{-1}\left(s_B - \sum_{i=1}^{m} g_{\check\lambda}(\check{u}_i)(x_i - \bar{x})\right) \tag{3.49}$$

$$\check{\lambda} = \left(\frac{\sum_{i=1}^{m}\left(g_1(\check{u}_i) - \overline{g_1(\check{u})} - \frac{\sum_{j=1}^{m} g_1(\check{u}_j)(x_j - \bar{x})}{\sum_{j=1}^{m}(x_j - \bar{x})^2}(x_i - \bar{x})\right)^2}{s_\Lambda - \frac{s_A^2}{m} + \left(\frac{s_B}{\sum_{i=1}^{m}(x_i - \bar{x})^2}\right)^2}\right)^{\frac{1}{2}}. \tag{3.50}$$

Thus property 1, monotonicity, is satisfied by (3.50) on $\check{\lambda}$, and by (3.48) and (3.49) given $\check{\lambda}$. Properties 2 and 3 are assured in case E is distributed according to a density of the exponential subfamily, whose general element has the form:

$$f_E(\varepsilon;\theta) = \alpha(\theta)\exp(\gamma(\theta)\varepsilon^2), \tag{3.51}$$

by the fact that s_A, s_B and s_Λ prove to be jointly sufficient statistics [35] for the parameters A, B and Λ. In our case we base the satisfaction of the two properties on the analogies of the functions ε^2 and $|\varepsilon|$.

Starting from a sample $z_m = \{(x_1, y_1), \ldots, (x_m, y_m)\}$, the CD step translates into the procedure described in the following Algorithm 3.5.

88   Inferring from Samples

Algorithm 3.5. p-bootstrap procedure for a regression problem
Given a sample $z_m = \{(x_i, y_i), i = 1, \ldots, m\}$ according to (3.42) and (3.43), and denoting by $y = a + bx$ a target line ℓ,

---

compute $s_A = \sum_{i=1}^{m} y_i$, $s_B = \sum_{i=1}^{m} y_i(x_i - \bar{x})$ and $s_A = \sum_{i=1}^{m} y_i^2$;
set $L = \{\}$ and choose a suitable value for N;
for $i = 1$ to N **do**
  draw $\{\breve{u}_1, \ldots, \breve{u}_m\}$ uniformly in $[0, 1]$;
  compute $\breve{\lambda}$ from (3.50);
  use $\breve{\lambda}$ to compute $\breve{a}$ and $\breve{b}$ according to (3.48-3.49);
  denote ℓ_i the corresponding straight line;
  add ℓ_i to L;
end for
return L;

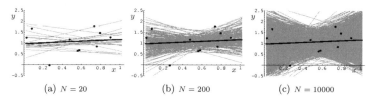

(a) $N = 20$ (b) $N = 200$ (c) $N = 10000$

Fig. 3.9. Population of regression lines (gray lines) obtained through N bootstrap replicas of an original sample of points (x_i, y_i) (bullets) where x_is are uniformly chosen in $[0, 1]$ and y_is add a laplacian drift (3.43) with $\lambda = 1$ to the $x = x_i$ intercepts of the thick line.

Fig. 3.9 shows the shape of L (the random variable corresponding to ℓ) for 20, 200 and 10000 bootstrap replicas.

NR. The shape of the envelope is up to us. The general objective is to form an optimal shape leaving out of the confidence region the less probable curves. Since the curve probability (or probability density) depends on the E distribution law [7] we may focus on a *score* which is easy to compute and is such that: i) all lines contained in the envelope have a higher score than those outside it, and ii) there is an envelope for each score ρ dividing lines with a score greater or equal to ρ from the remaining lines. With these committments, a simple score may be constituted by the product $\alpha_i(1 - \alpha_i)\beta_i(1 - \beta_i)\eta_i(1 - \eta_i)$, where η is a normalized height of the central point (with abscissa $\bar{x}$), and α and β the analogous heights of two symmetrically equidistant points w.r.t. the former (i.e. with abscissas $\bar{x} - \Delta$ and $\bar{x} + \Delta$, for a given Δ).

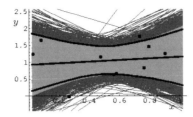

Fig. 3.10. Confidence region of level $\gamma = 0.10$ (light gray region) for the line regressing the sample in Fig. 3.9 (bullets), contrasted with the line distribution from a set of $N = 10000$ bootstrap replicas (dark gray region).

The heights are normalized so that the maxima of α and β (computed on the set of curves cleaned up from outliers) equal 1 and the minima 0, and the η scale interpolates the former scales. For short, the lines close to the center of the lines' population with a moderate vertical tilt in its respect have a high score, with a gently decreasing allowed tilt with the distance from the central point. Using this score we rank all $N = 10000$ lines in Fig. 3.9(c). Then we peel the sheaf by removing the lowest $N\gamma$ obtaining the picture in Fig. 3.10. In greater detail, we centered the region around $\bar{x}$, as said before, and selected the two extremal abscissas so that one of them coincides with one extreme of the sample x range and the latter trespasses the other extreme. The figure reports the confidence region at a level $\gamma = 0.10$.

PV. It corresponds to a point estimator of the regression curve inside the confidence region. Thus it depends on the special cost function we adopt. In our case we have that the nested regions are implicitly centered around the highest score curves that we enhance as a pivot in Fig. 3.10.

producing regions of unusual shape,

Example 3.11. In case E follows a Gaussian distribution, a suitable envelope of the lines may be defined by $a + b = k$, for $x - \bar{x} > 0$, and $a - b = k$, for $x - \bar{x} < 0$, using k as the scoring parameter. Centering the region around the MLE curve and binding singularly one of the two parameters, the nested regions may be defined by the inequalities:

but easy rendering.

$$|\Delta\alpha + \Delta\beta| \leq t_{\text{tot}}. \quad (3.52)$$

Alternatively, we establish either

$$|\Delta\alpha| \leq t_\alpha \quad (3.53)$$

like in (3.58) later on, or

$$|\Delta\beta| \leq t_\beta \quad (3.54)$$

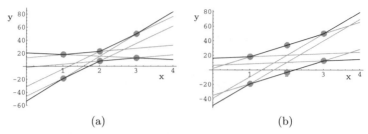

Fig. 3.11. Contour of the confidence region for a regression line as in Example 3.10 where (3.52) is coupled with: (a) (3.53), and (b) (3.54). Black piecewise lines: upper and lower contour of the confidence region; gray lines: lines summing up the contours; gray points identify the contours.

with suitable thresholds t_* in the right of the previous two inequalities giving rise to shapes similar to those shown in Fig. 3.11(a) and (b), respectively, where $\Delta \alpha$, $\Delta \beta$ denote the shifts around the central values of A and B distributions, represented by s_A/m and s_B/S_{xx}, respectively, where $S_{xx} = \sum_{i=1}^{m}(x_i - \overline{x})^2$. The choice between those or some intermediate shape is tied to developing a very concentrated confidence region and depends on the ratio between E and X variances.

3.2.2 Numerical Examples

The previous example is easily solved in all its nuances when the shift term is Gaussian or may be assimilated to it [7]. In this case typical master equations w.r.t. this mechanism are:

$$s_A = ma + \sum_{i=1}^{m} \varepsilon_i \tag{3.55}$$

$$s_B = \sum_{i=1}^{m} b(x_i - \overline{x})^2 + \sum_{i=1}^{m} \varepsilon_i (x_i - \overline{x}), \tag{3.56}$$

where s_A and s_B are the observed statistics $\sum_{i=1}^{m} y_i$ and $\sum_{i=1}^{m} y_i (x_i - \overline{x})$, respectively. The rationale behind them is the following. We have observed s_A and s_B and know neither the parameters a and b, nor the set $\{\varepsilon_1, \ldots, \varepsilon_m\}$. However we use a hypothesis regarding the random variables $\{E_1, \ldots, E_m\}$, so that to any specification $\{\breve{\varepsilon}_1, \ldots, \breve{\varepsilon}_m\}$ of them would correspond, *for the observed s_A and s_B*, the parameters:

$$\breve{a} = \frac{s_A - \sum_{i=1}^{m} \breve{\varepsilon}_i}{m}; \quad \breve{b} = \frac{s_b - \sum_{i=1}^{m} \breve{\varepsilon}_i (x_i - \overline{x})}{S_{xx}}, \tag{3.57}$$

with a consequent transfer of probability masses (densities) coming from $\{\breve{\varepsilon}_1, \ldots, \breve{\varepsilon}_m\}$ to $\breve{a}, \breve{b}$ as specifications of the random variables A, B. Since the two parameters are jointly involved in the model (3.42), this transfer is not trivial, as explained in [6, 2, 8, 7]. However, under the not heavily restrictive hypothesis of E_is being independently and identically distributed and symmetric around 0, we obtain a straightforward description of a confidence region for the line $y = A + B(x - \overline{x})$ at the basis of model (3.42). Namely, since they are random variables, A and B denote a random function C, i.e. a family of lines each affected by a probability mass (density). We may define a region exactly containing elements of this family up to a fixed probability $1 - \gamma$, as specified in Definition 3.7. We refer the reader to the references indicated above for a complete treatment of the subject. Here we come to identifying the confidence region as follows. In case E_is are Gaussian, for scalar X the region is spanned by the sheaf of lines $y = \alpha + \beta(x - \overline{x})$, with:

> We may recover the standard Gaussian models,

$$\frac{1}{m} s_A - z_{1-\gamma/2} \frac{\sigma}{\sqrt{m}} \leq \alpha \leq \frac{1}{m} s_A + z_{1-\gamma/2} \frac{\sigma}{\sqrt{m}} \quad (3.58)$$

$$\frac{s_B}{S_{xx}} - z_{1-\gamma/8} \sigma \sqrt{\frac{1}{m} + \frac{1}{S_{xx}}} + \left| \alpha - \frac{1}{m} s_A \right| \leq \beta \leq$$

$$\leq \frac{s_B}{S_{xx}} + z_{1-\gamma/8} \sigma \sqrt{\frac{1}{m} + \frac{1}{S_{xx}}} - \left| \alpha - \frac{1}{m} s_A \right|, \quad (3.59)$$

where σ is the standard deviation of E and z_η is the η quantile associated to the Normal distribution. For different distributions of the E_is, we find analogous forms when the above general hypotheses are maintained[4].

We finally report a widespread regression instance where the procedure we introduced proves to be not bound by linearity limitations on the model.

Example 3.12. The data in Table 1.2 report censored (marked with >) and non censored remission times (in weeks) of patients with acute myelogenous leukemia undergoing maintenance chemotherapy [12]. After the same therapy the 11 patients were either admitted to the hospital for a second cycle of therapy after the weeks indicated in the table – non censored data – or missed for some reason (maybe due to lost tracks, early termination, or death by other causes, etc.), so that the remission time is evaluated greater than the period marked in the table and labeled with > – censored data. From these data we want to estimate the remission risk of the specific therapy as a function of time. This is a classical prognostical problem hard to be solved. Mathematically speaking, given a random variable T (denoting the

> but also the crucial problem of hazard rate estimate,

---

[4] For instance for Laplace distribution or Zero Symmetric Pareto distribution [6].

occurring time of a certain event) and being f_T the d.f. of T and F_T its c.d.f., the *hazard function* of T is defined as

$$h(t) = \frac{f_T(t)}{1 - F_T(t)}. \tag{3.60}$$

Inferring h in terms of regression curves will result in a peculiar problem since we will infer this function from a sample $\{t_1, \ldots, t_m\}$, hence from the same variable playing the role of both argument of h and observation on h itself. Thus, in analogy with the case described in the previous example, here samples assume the form $\{(t_i, t_i), i = 1, \ldots, m\}$, where $t_i = g(h(t_i), \varepsilon)$, with g a suitable function and ε a random variable specification describing the random seed of t. To render things even more difficult, as mentioned before, sample observations refer also to censored data representing incomplete information. Let us assume a Weibull model for T as usual in health-care scenarios. Here we formulate it in the form:

$$F_T(t) = 1 - e^{-\frac{1}{\beta_0} t \beta_1^{\log t}}, \tag{3.61}$$

where $\beta_0 > 0$ and $\beta_1 > e^{-1}$ so that the hazard function and universal sampling mechanism read:

$$h(t) = \frac{1}{\beta_0} \beta_1^{\log t}(1 + \log \beta_1); \quad t = (-\beta_0 \log u)^{\frac{1}{1+\log \beta_1}}. \tag{3.62}$$

In spite of the linear relation between $\log t$ and $\log \log u$, the relation $t_i = g(h(t_i), \varepsilon)$ highlights that t_i is a fixed point of a possibly nonlinear function of t_i, and this renders the t dependence on $\log \log u$ more complex than linear. In lack of joint sufficient statistics for the parameters, with the exception of the trivial one represented by the sample itself [11], from the expression of the sample likelihood $L(t_1, \ldots, t_m)$:

$$L(t_1, \ldots, t_m) = \beta_1^{\sum \log t_i} \left(\frac{1 + \log \beta_1}{\beta_0} \right)^m e^{-\frac{\sum t_i^{1+\log \beta_1}}{\beta_0}}, \tag{3.63}$$

we see that $s_0 = \sum \log t_i$ and $s_1 = \sum (\log t_i - \overline{\log t})^2$ are relevant statistics playing the role of approximate sufficient statistics in the limit of the expansion of $e^{\sum \log t_i^{1+\log \beta_1}}$ up to the second order of the Taylor series around $\overline{\log t}$. Hence in case of non censored data we may refer to the master equations:

$$\check{\beta}_1 = \exp\left(\sqrt{\frac{\sum_{i=1}^{n} \left(g(\check{u}_i) - \overline{g(\check{u})} \right)^2}{s_1}} - 1 \right) \tag{3.64}$$

$$\check{\beta}_0 = \exp\left(s_0(1 + \log(\check{\beta}_1)) - \overline{g(\check{u})} \right), \tag{3.65}$$

where

$$g(u) = \log(-\log u); \quad \overline{g(u)} = \frac{1}{m}\sum_{i=1}^{m} g(u_i) \quad \text{and} \quad \overline{\log t} = \frac{1}{m}\sum_{i=1}^{m} \log t_i, \tag{3.66}$$

so that starting from uniform realizations $(\check{u}_1, \ldots, \check{u}_m)$, we may compute first a β_1 replica from (3.64) and use it to compute a β_0 replica from (3.65). Censored data constitute a filter that passes only those $\check{\beta}_0$ and $\check{\beta}_1$ complying with analogous master equations where inequality substitutes equality in the master equation analogous to (1.10). Actually a censored data item τ_i means that without censoring we had observed a $t_i > \tau_i$. Since we have selected statistics s_0 and s_1 as relevant properties of the sample from which to deduce the parameters, we check the effects of inequalities $t_i > \tau_i$, with i spanning the censored data, over s_0 or s_1 values.

Statistic s_0 is monotone in each t_i. Thus denoting by n the number of non censored data and summing up all m censored and non censored data we may pass those $\check{\beta}_0$ and $\check{\beta}_1$ such that:

with censored data.

$$\frac{1}{m}\sum_{i=1}^{m} \log \tau_i < \frac{\log \check{\beta}_0 + \overline{g(\check{u})}}{1 + \log \check{\beta}_1}, \tag{3.67}$$

where τ_is refer to all (censored and non censored) data in Table 1.2, and $\overline{g(\check{u})}$ is computed from a sample $\{\check{u}_1, \ldots, \check{u}_m\}$, where the last $m - n$ $\check{u}_i$s are the seeds of the unobserved T specifications. Statistic s_1 is not necessarily monotone. It depends on the location of censored $\log \tau_i$s around the sample mean. Hence we cannot use its application to non censored data for filtering the parameters[5].

Figs. 3.12 (a) and (b) show, respectively, the filtered parameters (starting from 10000 replicas represented by black points) obtained from the data in Table 1.2 and the corresponding lines. Given the role of the parameters we have a wide spectrum of their values and a non convex separation of the filtered parameters. Both drawbacks disappear in Fig. 3.12(c) where the parameters are recombined and represented through $b_0 = \log \beta_0 + \log(1 + \log \beta_1)$ in abscissa and $b_1 = \log \beta_1$ in ordinate so as to constitute the coefficients of the line connecting $\log t$ to $\log h(t)$.

Remark 3.5. The true technical difficulty arising in this section concerns the joint management of more than a single parameter. This requires enriching our sample and jointly considering many

Many parameters need many equations

---

[5] See [1] for a further discussion and possible artifacts considered for better exploiting the data.

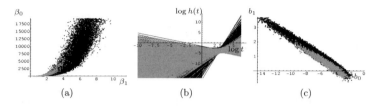

Fig. 3.12. (a) Filtered parameters replicas (gray points) computed starting from 10000 replicas (black points with truncated ordinates) obtained by applying (3.64) and (3.65); (b) hazard function curves represented in the logarithmic scale corresponding to the points shown in (a); (c) same as (a) but using as coordinate axes $b_0 = \log \beta_0 + \log(1 + \log \beta_1)$ and $b_1 = \log \beta_1$.

more master equations. If θ is a vector then also s must be a vector. Thus we are working with a system of master equations, say in $(\theta_1, \ldots, \theta_\nu)$, where we need a number of independent equations equal to ν, hence ν statistics, in order to have a definite solution:

$$\begin{aligned} s_1 &= h_1(\check{\theta}_1, \ldots, \check{\theta}_\nu, \check{u}_1, \ldots, \check{u}_m) \\ &\vdots \\ s_\nu &= h_\nu(\check{\theta}_1, \ldots, \check{\theta}_\nu, \check{u}_1, \ldots, \check{u}_m). \end{aligned} \qquad (3.68)$$

hence many statistics Typical statistics may be moments of the random variable at hand. Yet, any well behaving statistics, possibly giving rise to independent equations, are welcome. As mentioned in Section 3.1.2, if these statistics induce suitable orderings on the parameter, (in particular if they are *jointly sufficient*) then they support a twisting argument as well. Dealing with the system (3.68) we may have some lucky conditions concerning:

Better if the equations are simple
1. The separability of the equations w.r.t. the parameters. In the best case we may have equations where each relates one statistic to one parameter; otherwise, we must consider the distribution law of a parameter given other ones (for instance $F_{\Theta_1|\theta_2,\ldots,\theta_\nu}$).

and the statistics are independent.
2. The independence of the seeds' functions. Solutions $\{\check{\theta}_1, \ldots, \check{\theta}_\nu\}$ of (3.68) each depend on the whole set of us. However they may depend through functions $\{g_i(\check{u}_1, \ldots, \check{u}_m), i = 1, \ldots, \nu\}$ that are partly or totally mutually independent. In the latter case, if the equations are also separable we may partition the inference problem into ν independent inference problems. Otherwise we must start considering the joint distribution of the parameters. Then we may possibly be interested in the

marginal distributions of the parameters or in the distribution of a function of them.

Let us complete the inference of the parameters of the Gaussian distribution law as follows:

Example 3.13. Coming back to Example 3.6, logical implications (3.28) and (3.29) both have the drawback of requiring that the parameter not involved in the implication is fixed. *Vice versa*, considering that $x_i - \bar{x} = \sigma(z_i - \bar{z})$, and defining $r_{\Sigma^2} = \sum_{i=1}^{m}(x_i - \bar{x})^2$, we may substitute (3.29) with relation:

$$\sigma \leq \tilde{\sigma} \Leftrightarrow r_{\Sigma^2} \leq r_{\tilde{\sigma}}, \qquad (3.69)$$

which holds for any value of μ and still pivots around a sufficient statistic but with a *pledge point*. Namely from (3.69) we have:

The marginal Σ^2

$$F_{\Sigma^2}(\sigma) = 1 - F_{R_\sigma}(r_{\Sigma^2}) = 1 - F_{\chi^2_{m-1}}\left(\frac{r_{\Sigma^2}}{\sigma}\right). \qquad (3.70)$$

Then, using conditional distributions as in Example 3.6, the c.d.f. of the vector (M, Σ^2) is computed as:

The joint M, Σ^2

$$F_{M,\Sigma^2}(\mu, \tilde{\sigma}) = \int_0^{\tilde{\sigma}} f_{\Sigma^2}(\sigma) F_{M|\sigma}(\mu) d\sigma. \qquad (3.71)$$

By integrating the corresponding distribution density with respect to σ we obtain the marginal density function of M. In particular, defining $t = \frac{m\mu - s_M}{\sqrt{r_{\Sigma^2}}}\sqrt{\frac{m-1}{m}} = \frac{\mu - \bar{x}}{\sqrt{r_{\Sigma^2}/(m(m-1))}}$,

involving Student t distribution.

$$f_T(t) = \frac{\Gamma(m/2)}{\Gamma((m-1)/2)}\frac{1}{\sqrt{\pi(m-1)}\left(1+t^2\right)^{-m/2}}, \qquad (3.72)$$

representing the density function of the well known Student t distribution law with parameter (degrees of freedom) $m-1$ [47].

Finally, if we want a confidence region for the vector (M, Σ^2), we may follow two strategies:

1. We may look for a rectangular domain measured through the cumulative function F_{M,Σ^2}. For instance, wanting no lower bound for σ^2 except 0, and μ extremes symmetric around s_M/m (actually the mean value of M), we may manage on σ_{up} and $\mu_{\text{up}} - s_M/m = s_M/m - \mu_{\text{dw}}$ such that $F_{M,\Sigma^2}(\mu_{\text{up}}, \sigma^2_{\text{up}}) - F_{M,\Sigma^2}(\mu_{\text{dw}}, \sigma^2_{\text{up}}) = 0.90$. In this way we obtain the plain rectangle in Fig. 3.13.

A trivial confidence region

2. While easy to implement, the above solution may prove slightly informative, since the rectangle we draw contains a great portion of points with very small probability density. Thus stretching the shape of the confidence domain a

and a smarter one.

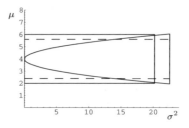

Fig. 3.13. 0.90 confidence regions for the vector (Σ^2, M) of the parameters in a Gaussian distribution. Plain rectangle: region with fixed extremes for both components. Curve: region with M extremes depending on σ^2. Dashed rectangle: region computed considering the marginal distribution of M in place of the conditional one.

bit, we may get a definitely narrower area, hence with less dispersed values among which we may expect to find the parameters with a non disregardable probability. To this aim in the measure $1 - \delta$ of the confidence region:

$$1 - \delta = \int_0^{\sigma_{\text{up}}^2} \int_{\mu_{\text{dw}}(\sigma^2)}^{\mu_{\text{up}}(\sigma^2)} f_{\Sigma^2}(\sigma^2) f_{M|\sigma^2}(\mu) \mathrm{d}\mu \mathrm{d}\sigma^2 \qquad (3.73)$$

we decide to fix

$$\mu_{\text{dw}}(\sigma^2) = (M|\sigma^2)_{(1-\sqrt{1-\delta})/2} \simeq (M|\sigma^2)_{\delta/4} \qquad (3.74)$$

$$\mu_{\text{up}}(\sigma^2) = (M|\sigma^2)_{1-(1-\sqrt{1-\delta})/2} \simeq (M|\sigma^2)_{1-\delta/4} \qquad (3.75)$$

so that $P(\mu_{\text{dw}}(\sigma^2) \leq M \leq \mu_{\text{up}}(\sigma^2)) = \sqrt{1-\delta} \simeq 1 - \delta/2$ for each σ^2. Analogously we fix $\sigma_{\text{up}}^2 = \Sigma_{\sqrt{1-\delta}}^2$ so that $P(\Sigma \leq \sigma_{\text{up}}^2) = \sqrt{1-\delta}$. Substituting these probabilities in the integral in (3.73) we find that the whole region has measure $1-\delta$. The shape of this region has the parabolic form in Fig. 3.13.

For the sake of completeness, in Fig. 3.13 we also draw with a dashed line the rectangle edges obtained by substituting the marginal distribution of M in place of its conditional distribution to compute μ_{up} and μ_{dw} with the second tool.

3.2.3 PAC Learning Theory Revisited

We move to boolean functions as a goal of our inference which definitely simplifies the terms of the inference problem. In the sampling scheme (3.41) you have $y_i \in \{0, 1\}$, which makes many of the above inference tools connected with bootstrasp strategies harmless. It generally refers to a very poor set of information granules requiring a much sophisticated rationale to be exploited.

In our leading example we may imagine ignoring Hansel habits and the pebble drawing sequence. Rather, we could be interested just to identify the Hansel habitat that in a first approximation we could identify with a circle around his home, but we do know neither the home position, hence the circle center, nor the circle radius. Hence we just randomly inspect some site of the deputed region in search of Hansel footsteps, or his pebbles. If we recognize that the inspected site is affected by them, we mark it as positive ($y_i = 1$), otherwise as negative ($y_i = 0$). In its very essential formulation the learning problem reads as follows. On the set $\mathfrak{X} \times \{0,1\}$ we have the *labeled sample* available:

Poor examples need more sophistication in the reasoning.

$$z_m = \{(x_i, b_i),\ i = 1, \ldots, m\}, \qquad (3.76)$$

whose x_is are specifications of i.i.d. random variables X_is, and b_is are expected to be output of a boolean function. Hence, from an operational perspective, we assume that for every $M \in \mathbb{N}$ and every labeled population z_M representing a possible continuation (hence a suffix) of z_m, a function c exists within a class $\mathfrak{C}$ of boolean functions (the class of circles in the Hansel example), normally called *concepts*, such that $z_{m+M} = \{(x_i, c(x_i)), i = 1, \ldots, m+M\}$. The cost function of our inference is the measure of the symmetric difference between another function we call *hypothesis h*, computed from z_m, and any of such c (see Fig. 3.14). Namely, as an operational consequence of our estimate of c through h, the latter function will be questioned about the labels of new elements of $\mathfrak{X}$ belonging to z_M, i.e. according to the same distribution law of the examples. The answer will be wrong on each point x that either belongs to c and does not belong to h (thus the answer is 0 using $h(x)$ while the *correct* answer is $c(x) = 1$) or belongs to h and not to c (answer $h(x) = 1$ in place of $c(x) = 0$). The set of these points constitutes the symmetric difference $c \div h$ between c and h. We quantify the approximating capability of h through the probability $\mathrm{P}\,(c \div h)$ of this region, usually denoted as the error probability P_{error}. This denotes that we are interested, in our example, in the probability of expecting for meeting Hansel in a site that he will never cross and *vice versa*, no matter the extension of these sites.

Ingredients are: boolean function as a goal concept,

boolean function as its approximating hypothesis, and

measure of the symmetric difference between them as a learning quality measure.

A learning algorithm is a procedure $\mathscr{A}$ that generates a family of hypotheses h_m, with their respective P_{error} converging to 0 in probability with increasing size m of the labeled sample given in input to the procedure [48].

In summary, we are interested in the value α of the probability measure of a random variable Y whose explaining function is:

A rough sampling mechanism,

$$y = \begin{cases} 1 & \text{if } x \in c \div h \\ 0 & \text{otherwise}. \end{cases} \qquad (3.77)$$

98   Inferring from Samples

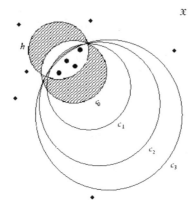

Fig. 3.14. Algorithmic inference in PAC learning. $\mathfrak{X}$: the set of points belonging to the Cartesian plane; bullets: 1-labeled (positive) sampled points; diamonds: 0-labeled (negative) sampled points. h: the circle labeling the sample; c_i: possible circles labeling possible populations. Line filled region: symmetric difference between h and c_0.

Assuming $h \subseteq c$ only for the sake of simplicity, should both the X distribution law and the c center be known, so that (3.77) reads:

$$y = \begin{cases} 1 & \text{if } (x - x_0)^T(x - x_0) \leq r^2 \\ 0 & \text{otherwise,} \end{cases} \quad (3.78)$$

where r is the radius of c and x_0 denotes its center, the inference problem would not be different from the inference of the parameter p of a Bernoulli distribution law through the guessing of height p in Fig. 3.5. But in our problem the X distribution is unknown, so that x may not play the role of random seed, and the geometry is more complex and unknown as well. In case the X distribution is known, we could remedy the second drawback by facing a peculiar regression problem having a quadratic cost function and a risk $\mathrm{E}\left[\sum_{i=1}^{m}(y_i - C(x_i))^2\right]$, whereas when the geometry is simple we would go back to usual parametric inference problems. The true original goal of this section is the inference of c with both mentioned lacks of knowledge. This means that, maintaining the number k of xs (i.e. the sum of the corresponding ys) in the $c \div h$ goal region as sufficient statistic around which to pivot the inference, twisting inequalities on α with inequalities on K may be driven both by the geometry of the sample mechanism (height p in (1.7), radius r in (3.78)) and by variations in X distribution law.

with unknown seed and unknown explaining function.

Hence we simply count variously located points.

Remark 3.6. In the previous discussion, the function c was denoted by C when its argument was regarded as a random

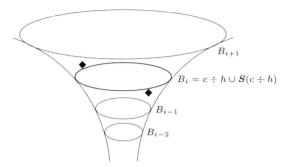

Fig. 3.15. The nested sequence $B((c \div h)^+)$ used to build a twisting argument. B_is: elements of B.

variable. For sake of readability, in the following section we will specialize the notation, using the symbol C instead of $\mathfrak{C}$ in order to denote the class $\{c\}$.

Let us move deeper in the logic that will support our inference. Hence, let us figure $c \div h$ with the domain B_i in Fig. 3.15 – no matter its shape – and consider a nested sequence $B(c \div h) = B_1 \subseteq B_2 \subseteq B_3 \subseteq \ldots$ of subsets of $\mathfrak{X}$, having B_i in the set of its elements. We arrange the figures so that B_h is under B_k if and only if $B_h \subseteq B_k$. Now, since $A \subseteq B \Rightarrow U(A) \leq U(B)$, where $U(D) = P(X \in D)$, an analogous ordering can be stated in the companion sequence $U(c \div h)$ of the probability measures of the sets in $B(c \div h)$ (actually, its slight variant $B((c \div h)^+)$ as we will explain in a moment). Thus the question "is $U(c \div h) \leq \alpha$" translates into the question "where is located B_{i_α} such that $U(B_{i_\alpha}) = \alpha$ w.r.t. $c \div h$: upward or downward?". We have no logical obstacle to assume that such B_{i_α} can exist in the sequence $B(c \div h)$ we have built. Since we have no other specifications about it, the problem is how to monitor its position w.r.t. $c \div h$. In other words, we need a witness of the inclusion of $c \div h$ in B_{i_α}. Witnessing constitutes the keen functionality involved in learning boolean functions. Hence we will devote a brief excursus to it in view of concluding our reasoning.

Rather, we locate domains within a graduated region.

Sentry points

The witness will be constituted of sample points that are used by the learning algorithm $\mathscr{A}$ not far from how we used the $k+1$-th point in (3.8) to recognize that the threshold line has been raised up to get $\pi = \alpha$. And since witnessing is costly in terms of spread of parameter distribution law, we look for the minimal set of points capable of this task. Moreover, since we do not know which is the c companion of h whose symmetric difference inclusion must be witnessed, we need some additional logic constraint. We characterize these points as exactly what keeps our learning algorithm from computing an h such that $c \div h$ expands outside B_{i_α}. Let us focus for a moment on a single concept, i.e. let us map $c \div h$ onto a new c. In its respect, the above functionality must be tossed in contrast with a whole class of concepts C, and may be dually defined in terms of preventing a given concept c from being fully included (invaded) by another concept within the class. Therefore we call these points either *sentinels* or *sentry points*; they are assigned by a *sentineling function* $\boldsymbol{S}$ to each concept of a class in such a way that:

<small>Stakes to prevent the expansion of the error region</small>

- they are external to the concept c to be sentineled and internal to at least one other including it,
- each concept c' including c has at least one of the sentry points of c either in the gap between c and c', or outside of c' and distinct from the sentry points of c', and
- they constitute a minimal set with these properties.

The technical definition comes from [6].

<small>Formally: sentry function, sentry points, frontiers, detail,</small>

Definition 3.8. *For a concept class* C *on a space* $\mathfrak{X}$, *a sentry function is a total function*[6] $\boldsymbol{S} : \mathsf{C} \cup \{\emptyset, \mathfrak{X}\} \mapsto 2^{\mathfrak{X}}$ *satisfying the conditions*

(i) *Sentinels are outside the sentineled concept* $\big(c \cap \boldsymbol{S}(c) = \emptyset$ *for all* $c \in \mathsf{C}\big)$.

(ii) *Sentinels are inside the invading concept* $\big($Having introduced the sets $c^+ = c \cup \boldsymbol{S}(c)$, *an invading concept* $c' \in \mathsf{C}$ *is such that* $c' \not\subseteq c$ *and* $c^+ \subseteq (c')^+$. *Denoting* up(c) *the set of concepts invading* c, *we must have that if* $c_2 \in$ up(c_1), *then* $c_2 \cap \boldsymbol{S}(c_1) \neq \emptyset\big)$.

(iii) $\boldsymbol{S}(c)$ *is a minimal set with the above properties* $\big($No $\boldsymbol{S}' \neq \boldsymbol{S}$ *exists satisfying* (i) *and* (ii) *and having the property that* $\boldsymbol{S}'(c) \subseteq \boldsymbol{S}(c)$ *for every* $c \in \mathsf{C}\big)$.

(iv) *Sentinels are honest guardians. It may be that* $c \subseteq (c')^+$ *but* $\boldsymbol{S}(c) \cap c' = \emptyset$ *so that* $c' \notin$ up(c). *This however must be a consequence of the fact that all points of* $\boldsymbol{S}(c)$ *are involved in really sentineling* c *against other concepts in* up(c) *and not just in avoiding inclusion of* c^+ *by* $(c')^+$. *Thus if we remove* c', $\boldsymbol{S}(c)$ *remains unchanged* $\big($Whenever c_1 *and* c_2 *are such that* $c_1 \subset c_2 \cup \boldsymbol{S}(c_2)$ *and* $c_2 \cap \boldsymbol{S}(c_1) = \emptyset$, *then the*

---

[6] As usual in theoretical computer science a function $f : A \mapsto B$ does not need to be defined for each element of A, to take into account never-ending computations such as infinite loops. When on the contrary a function is defined on every element of its domain, it is said to be *total*.

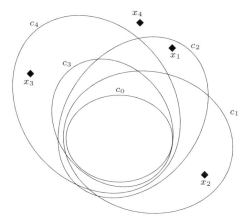

Fig. 3.16. A schematic outlook of outer sentineling functionality: concept c_0 sentineled against c_1, c_2, c_3, c_4 through $\{x_1, x_2, x_3\}$. x_1 and x_3 are used as its own sentry points also by c_1, and x_2 prevents c_1 to invade c_0; x_2 and x_3 are sentry points also of c_2, and x_1 prevents invasion of c_0 by c_2; and x_1 and x_2 are sentry points also of c_4 and x_3 prevents invasion; c_3 does not include $c_0 \cup S(c_0)$, hence it cannot invade c_0. x_4 is a useless point. All marked points are negative examples of c_0.

restriction of S to $\{c_1\} \cup \mathrm{up}(c_1) - \{c_2\}$ is a sentry function on this set).

$S(c)$ *is the* frontier *of c upon* S *and its elements are called either* sentry points *or* sentinels.

With reference to Fig. 3.16, $\{x_1, x_2, x_3\}$ is a candidate frontier of c_0 against c_1, c_2, c_3, c_4. All points are in the gap between a c_i and c_0. They avoid inclusion of $c_0 \cup \{x_1, x_2, x_3\}$ in c_3, provided that these points are not used by the latter for sentineling itself against other concepts. *Vice versa* we expect that c_1 uses x_1 and x_3 as its own sentinels, c_2 uses x_2 and x_3 and c_4 uses x_1 and x_2 analogously. Point x_4 is not allowed as a c_0 sentry point since, like some diplomatic seat, it should be located out of any other concepts just to avoid that it is occupied in case of c_0 invasion.

Definition 3.9. *The frontier size of the most expensive concept to be sentineled with the least efficient sentineling function, i.e. the quantity* $D_C = \sup_{S,c} \#S(c)$, *is called* detail *of* C, *where* S *spans also over sentry functions on subsets of* $\mathfrak{X}$ *sentineling in this case the intersections of the concepts with these subsets. Actually, proper subsets of* $\mathfrak{X}$ *may host sentineling tasks that prove harder than those raising with* $\mathfrak{X}$ *itself.*

Moving to symmetric differences, for another set H of concepts let us consider the class of symmetric differences $c \div \mathsf{H} = \{c \div$

and symmetric differences.

$h, \forall h \in H\}$ *for any c belonging to* C. *The detail of a concept class* H *w.r.t. c is the quantity* $D_{c,H} = D_{c \div H}$. *Vice versa, for a fixed h we consider an augmented sentineling functionality against the entire class* C *where to each h is assigned a minimal set of points necessary to sentinel any $c \div h$, for any $c \in$ C against $c \div$ H. Let us denote by* $D_{(C,H)_h}$ *the cardinality of this sentry set, and define the overall detail of the class* $C \div H = \cup_{c \in C} c \div H$ *as the quantity* $D_{C,H} = \sup_{h \in H}\{D_{(C,H)_h}\}$.

<small>Detail is a complexity index that is often difficult to compute.</small>

A given concept class C might admit more than a single sentry function. As in the case of other class complexity indices such as the Vapnik-Chervonenkis dimension [49], the detail of a class is a parameter extremely meaningful yet difficult to compute.

Master equations for learning boolean functions

Once we have inclusion's witness, we may conclude our reasoning on the scheme presented in Fig. 3.15. Namely we focus on the sequence $B((c \div h)^+)$ having $c \div h \cup S(c \div h)$ within its elements.

<small>If sentry points are included in B_{i_ε},</small>

This sequence exists because we have established a learning algorithm $\mathscr{A}$ which finds in the sample subset $S(c \div h)$ a constraint to the topology of the output h, namely $\mathscr{A}$ cannot compute a h' including h because the sample contains $S(c \div h)$. If, in addition $S(c \div h)$ is contained in B_{i_ε} as well, then this witnesses that whole $c \div h \cup S(c \div h)$ is contained in B_{i_ε}. Assume that $\mathscr{A}$ computes consistent hypotheses; it means that h includes the 1 labeled points of the sample and excludes the 0 labeled points – so that $h(x_i) = c(x_i)$ wherever in the sample. In this case, the implication chain is completed as follows:

i) on the right by the fact that $S(c \div h) \subseteq B_{i_\varepsilon}$ if and only if $K_\varepsilon \geq \#S(c \div h)$, where K_ε is the number of those from among the sampled points which fall in B_{i_ε}, and

ii) on the left by the fact that the event $u_{c \div h} \leq \varepsilon$ is implied by the event $u_{(c \div h)^+} \leq \varepsilon$ and the latter by $(c \div h)^+ \subseteq B_{i_\varepsilon}$.

<small>we are done.</small> Namely:

a. As for the lower bound we have:

$$(u_{c \div h} \leq \varepsilon) \Leftarrow (u_{(c \div h)^+} \leq \varepsilon) \Leftarrow ((c \div h)^+ \subseteq B_{i_\varepsilon})$$
$$\Leftarrow (S(c \div h) \subset B_{i_\varepsilon}) \Leftrightarrow (k_\varepsilon \geq \#S(c \div h)). \quad (3.79)$$

This induces an opposite chain on probabilities, as an extension of inequalities in (3.9):

$$P(U_{c \div h} \leq \varepsilon) \geq P(K_\varepsilon \geq \#S(c \div h)) \geq P^m(K_\varepsilon \geq D_{C,c}). \quad (3.80)$$

b. As for the upper bound, with the same consistency assumption, we obtain:

$$(k_\varepsilon \geq 1) \Leftarrow (u_{c \div h} \leq \varepsilon). \tag{3.81}$$

<small>No growth of $c \div h$ without a sentry point inclusion</small>

The threshold 1 in the left part of (3.81) is due to the fact that h is, in turn, a function $\mathscr{A}$ of a sample specification. Since the symmetric difference of c with the h that $\mathscr{A}$ produces grows with the set of included sample points and *vice versa*, then $(U_{c \div h} < \varepsilon)$ implies that an ε-enlargement region within any $c \div h'$ containing $c \div h$ must include at least one more of the sample points at the basis of h's computation.

The whole probability chain is thus completed by the following formula:

$$\mathrm{P}\left(K_\varepsilon \geq 1\right) \geq \mathrm{P}\left(U_{c \div h} \leq \varepsilon\right) \geq \mathrm{P}\left(K_\varepsilon \geq \mathrm{D}_{\mathsf{C},c}\right). \tag{3.82}$$

Finally, as in (3.9), K_ε follows by definition a Binomial distribution law of parameters m and ε. Thus (3.82) reads in terms of the incomplete Beta function:

<small>A first gauge for error measure c.d.f.</small>

$$I_\varepsilon(1, m) \geq F_{U_{c \div h}}(\varepsilon) \geq I_\varepsilon(\mathrm{D}_{\mathsf{C},c}, m - \mathrm{D}_{\mathsf{C},c} + 1). \tag{3.83}$$

3.2.4 A General Error Distribution Law

How many points fall in $c \div h$? The answer to this question is connected to our learning strategy. The most common strategy is to look for consistent hypothesis. This corresponds to the correct practice of not falsifying what you have really observed. In this way you consider the event of having in B_{i_α} exactly the sentry points, hence at most $\mathrm{D}_{(\mathsf{C},\mathsf{H})_h}$ if you have already computed h, or $\mathrm{D}_{(\mathsf{C},\mathsf{H})}$ if you refer to any future sample. We may have many reasons, however, for deviating from this strategy that we synthesize in the following three: i) the examples labeling is incorrect, ii) we use a hypothesis not belonging to the class of concepts but definitely simpler, and iii) we work with a numerical precision inadequate so that sample points are misplaced w.r.t. the accurate concept. All these cases bring to increase the number of sample points in $c \div h$ with a consequent growth of the gap between the functions binding the $U_{c \div h}$ c.d.f. In particular, we have:

<small>In principle, make hypothesis not contradicting observations, but</small>

1. The first case is referred to as learning from uncertain examples. The framework remains unaltered if the errors are non malicious, i.e. produced by a mechanism that inverts the example labels depending on the toss of a coin having a given probability of beating head. The number of additional points

<small>observations could be unreliable, or</small>

in $c \div h$ may either remain unknown, which allows only hypothetical conclusions, or bounded for instance in percentage, in which case we will augment $D_{(C,H)_h}$ or $D_{(C,H)}$ by a corresponding quantity.

consistency could require too detailed hypotheses, in any case

2. Looking for a simplistic hypothesis may be suggested either by understandability reasons or by true efficiency motivations. On one side, it is clear that a one page long formula is less understandable than a one line long formula (actually we commonly understand formulas made up of up to five symbols). This is the root of approximate theories: they do not explain all cases but give a robust explanation of the bulk of a phenomenon, leaving unexplained cases as exceptions. On the other side, the detail of a concept class grows with the structural complexity of its elements. Thus, having assessed the fact that the points inside or on the border of $c \div h$ degrade the inference accuracy, the play is to find a minimum of the union of the two sets, where approximation generally increases the former and accuracy the latter. In this case, however, the mislabeled points are evident, while the exact number of sentry points is not always computable.

commensurate detail with computation accuracy.

3. Numerical precision is a keen point, though generally disregarded. The fact is that with equations like (3.83) we are defining a distribution law of a continuous random variable as a measure of symmetric differences of functions taking ∞^r specifications for a suitable r. In order to sentinel these functions against the invasion of other ones in the same class we may need the joint sentineling action of more than one sentry point. This generally occurs, however, for invading functions of low-order infinity, whose probability is null or not, depending on the approximation we use for discriminating different functions. For instance, consider the class of bounded rectangles (b_rectangle for short) defined by the inequalities: $\{0 \leq x_i \leq t_i; i = 1, \ldots, n\}$. For a given set $\{t_1, \ldots, t_n\}$, in order to forbid the expansion of the rectangle beyond these thresholds, we need a sentry point on each free (i.e. with a non zero coordinate) edge (see Fig. 3.17(a)). However the set of rectangles having one or more free edges fixed has null measure. Hence it is enough one sentry point exactly looking at one edge to make zero the probability of an expansion. Things become different if we draw rectangles with a poor approximation, say two bits, since in this case the family of rectangles with $n-1$ edges not fixed and the last one bounded into an interval discretized using, say, two bits has probability definitely different from zero. This probability will depend on the actual curves distribution law, hence

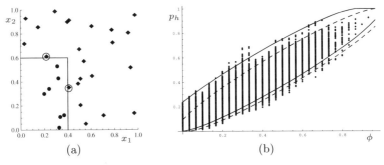

Fig. 3.17. Generating 0.900 confidence intervals for the probability P_h of a bounded rectangle in $\mathfrak{X} = [0,1]^4$ from a sample of 30 elements. (a) The drawn sample and one of its possible labelings in a two-dimensional projection; points inside circles are deputed sentinels. (b) Points: courses of the sample frequency ϕ and probability p_h of falling inside a b_rectangle with a lattice of labeling functions. Plain curves: lines joining the confidence interval extremes for the actual 4-dimensional case. Dashed curves: lines joining the confidence interval extremes for hypothetical 1-dimensional case.

on the actual sample points distribution law, being much higher as much as the distribution in peaked in this interval, like it happens in Fig. 3.17(b).

With the additional notion of *fairly strongly surjective learning function* as an algorithm which computes a hypothesis h on each labeled sample, such that the more sample points lie inside h the more population points there are and *vice versa*, we synthesize the above discussion in the form of the following theorem:

A more meditated gauge for error measure c.d.f.

Theorem 3.1. *For a space $\mathfrak{X}$, assume we are given:*

- *a concept class C on $\mathfrak{X}$;*
- *a set $\mathscr{L}_m$ of labeled samples $\boldsymbol{z}_m \subseteq \mathfrak{X} \times \{0,1\}$;*
- *a fairly strongly surjective function $\mathscr{A} : \mathscr{L}_m \mapsto \mathsf{H}$ computing possibly non consistent hypotheses.*

Consider the families of random sets $\{c \in \mathsf{C} : \boldsymbol{z}_{m+M} = \{(x_i, c(x_i)), i = 1, \ldots, m+M\}$ when $\boldsymbol{z}_{m+M}$ spans the samples in $\mathscr{L}_m$ and the specifications of their random suffixes $\boldsymbol{Z}_M$, with $M \to +\infty$ according to any sampling mechanism. For a given $(\boldsymbol{z}_m, h)$ and $h = \mathscr{A}(\boldsymbol{z}_m)$, denote with:

- *μ_h the detail $\mathrm{D}_{(\mathsf{C},\mathsf{H})_h}$ for the adopted computational approximation,*
- *t_h and ρ_h the number of points misclassified by h and their total probability,*
- *$U_{c \div h}$ the random variable given by the probability measure of $c \div h$ and $F_{U_{c \div h}}$ its c.d.f.*

Then for a given (z_m, h) and each $\beta \in (\rho_h, 1)$:

$$I_\beta(1+t_h, m-t_h) \geq F_{U_{c\div h}}(\beta) \geq I_\beta(\mu_h+t_h, m-(\mu_h+t_h)+1). \tag{3.84}$$

The learning goal

Right in the essence, the quality parameter of learning a boolean function is $U_{c\div h}$, where its degrading is caused by the sum $\mu_h + t_h$ as in (3.84). From the above section we know its distribution law that we may use for the common tasks of inference: interval estimate, point estimate, appreciation of the estimate complexity.

Learning is a matter of stating a good confidence interval on the error measure,

- Interval estimate. The typical goal of learning a boolean function is to design an algorithm insuring a $U_{c\div h}$ confidence interval $(0, \varepsilon)$ with confidence level δ.

  Definition 3.10. *For $\mathscr{Z}_m$, C, and $\mathscr{A} : \mathscr{Z}_m \mapsto \mathsf{H}$, as in Theorem 3.1, C is learnable through $\mathscr{A}$ if for each $\varepsilon, \delta > 0$ an m_0 exists such that for every $m > m_0$ the confidence interval $(0, \varepsilon)$ for the measure $U_{c\div h}$ has at least confidence $1 - \delta$. In formulas:*

  $$P(U_{c\div h} \leq \varepsilon) \geq 1 - \delta. \tag{3.85}$$

It depends on the sample size,

- Learning complexity. Equation (3.84) clearly evidences that learning comes at a cost in terms of sample size. By inverting the inequalities in (3.84), since K_ε follows by definition a Binomial distribution law of parameters m and ε, we obtain the following main result. It fixes an upper bound to the number of examples needed to learn a class of concepts in a distribution-free case.

  Corollary 3.1. *[3] Assume the same hypotheses and notations of Theorem 3.1. If*

  $$m \geq \max\left\{\frac{2}{\varepsilon}\log\frac{1}{\delta}, \frac{5.5(\mu_t + t - 1)}{\varepsilon}\right\}, \tag{3.86}$$

  then $\mathscr{A}$ is a learning algorithm for every z_m and $c \in \mathsf{C}$ ensuring that $P(U_{c\div h} \leq \varepsilon) \geq 1 - \delta$.

A first lesson we draw from the above discussion is that when we want to infer a function we must divide the available examples in two categories, the prominent ones and the mass. As in a professor's lecture, the former fix the ideas, thus binding the difference between concept and hypothesis. The latter are redundant. However, if we produce a lot of examples we are confident that a sufficient number of those belonging to the first category will have been exhibited. The above Corollary allows us to state a relation between the sizes of the two categories. A more operational lesson we can derive from the discussion is that if we adopt the strategy of learning via a consistent hypothesis, as a specific instance of the sufficiency requirement for the involved statistics, the sole benefit coming from the knowledge of the goal concept class lies in the appreciation of $D_{C \div C}$. Its value allows us to compute sample complexity upper bounds as in (3.86). Conversely, a way to contain the complexity lies in limiting this value. This may be obtained in two ways:

and on how many sample elements are burned by the learning algorithm.

1. Looking for approximate hypotheses classes giving rise to a $D_{C \div H}$ less than $D_{C \div C}$. The drawback is that we may have a certain number t_h of mislabeled points in the sample, so that our goal is to minimize the sum $\mu_h + t_h$.
2. Using a great numerical accuracy which renders many of the sentry points useless as they are binding sets of hypotheses of null measure.

We finally remark that computing $D_{C \div H}$ is generally a difficult task.

Until now for the sample size. Another cost comes from the computational load. If we adopt the consistency strategy we easily realize that computing a consistent formula may be a very hard task. For instance, computing a segment consistent with positive and negative points on a line is almost immediate: you take any point in between the sequence of a negative point followed by a positive one as first extreme, and any point in between the sequence of a positive point followed by a negative one as second extreme, and you are done. Moving to the class of circles already requires us a non disregardable effort to write down an algorithm computing a consistent circle. The task may become unbearable when we move to more complex classes of functions. At the end of the day we will be satisfied if we find an algorithm computing sufficiently accurate hypotheses on a class of concepts, doing it in a reasonable running time. Therefore with the usual dichotomy, "polynomial is feasible, non polynomial unfeasible", this is captured by the notion of learnable class.

But also computing the hypothesis may be a heavy job.

Definition 3.11. *A concept class* C *on a set* X *characterized by a complexity index* n *is* learnable *if there exists an algorithm* A *which, on labeled samples* z_m,

Really learnable concepts

1. if for each suffix of z_m there exists a $c \in \mathsf{C}$ computing the labels of its elements, then
2. for each $0 < \varepsilon, \delta < 1$, computes hypotheses h on $\mathfrak{X}$ with measure $U_{c \div h} \leq \varepsilon$ in a time polynomial in $n, 1/\varepsilon, 1/\delta$.

Remark 3.7. Example of complexity indices are the maximum number of propositional variables involved by the elements of a class of boolean formulas, the detail of a class, the dimensionality of the continuous space where the elements of a class of convex polygons are defined, the Vapnik Chervonenkis dimension, etc.

- Point estimators. The best we can do in the shrunk information framework we are working with, is to use the unbiased estimator of $U_{c \div h}$ which from (3.84) reads:

A gauge for the mean error probability as well

$$\frac{1 + t_h}{m + 1} \leq \widehat{U}_{c \div h} = \mathrm{E}\left[U_{c \div h}\right] \leq \frac{\mathrm{D}_{(\mathsf{C},\mathsf{H})_h} + t_h}{m + 1}. \quad (3.87)$$

Two widespread learning algorithms

A few building blocks for learning

In spite of the hard computability of the class detail and of the complexity of the consistency problem, the above theorems say that learning a function is generally a feasible problem, as confirmed by the high learning ability of the human brain. The tools with which humans learn are in general less formal and mostly biased by habits, propensities, etc. The sole exceptions are represented by extremely simple algorithms that solve limited problems but in a clean theoretical way, so that they represent the basic constituents of many more complex learning procedures. From among these building blocks we illustrate here below *support vector machines* and *decision trees*, that will be used in the rest of the book.

Support vector machine

Linear separators

This tool solves the elementary classification task of dividing a sample space with an hyperplane on the basis of a labeled sample. It is the boolean dual of the linear regression problem faced in Section 3.2.1. There we aim to linearly fit sample points, here we search for suitably separating them. Formally the *separating hyperplane* is an hypothesis h in the class H of hyperplanes in $\mathbb{R}^n$ such that all the points $\boldsymbol{x}_i$ with a given label belong to one of the two half-spaces determined by h.

Note that Theorem 3.1 gives a distribution law to $U_{c \div h}$ but not to the hypotheses. Rather, the former distribution holds for whatever hypothesis guaranteeing pledge points in number of $t_h +$

$D_{(\mathsf{C} \div \mathsf{H})_h}$. Thus we have an indeterminacy on the separating hyperplanes that we must remove with other theoretical tools. A choice suggested by symmetry reasons is to look for the optimal margin hyperplane, i.e., the hyperplane maximizing its minimal distance with the sample points. We obtain it by computing the solution $\{\alpha_1^*, \ldots, \alpha_m^*\}$ of a dual constrained optimization problem:

Virtue lies between the two extremes (in medio stat virtus).

$$\max_{\alpha_1,\ldots,\alpha_m} \sum_{i=1}^m \alpha_i - \frac{1}{2} \sum_{i,j=1}^m \alpha_i \alpha_j y_i y_j \boldsymbol{x}_i \cdot \boldsymbol{x}_j \qquad (3.88)$$

$$\sum_{i=1}^m \alpha_i y_i = 0 \qquad (3.89)$$

$$\alpha_i \geq 0 \quad i = 1, \ldots, m, \qquad (3.90)$$

where $\cdot$ denotes the standard dot product in $\mathbb{R}^n$, and then returning a hyperplane (called *separating hyperplane*) whose equation is $\boldsymbol{w} \cdot \boldsymbol{x} + b = 0$, where

$$\boldsymbol{w} = \sum_{i=1}^m \alpha_i^* y_i \boldsymbol{x}_i \qquad (3.91)$$

$$b = y_i - \boldsymbol{w} \cdot \boldsymbol{x}_i \text{ for an } i \text{ such that } \alpha_i^* > 0. \qquad (3.92)$$

Typically only a few components of $\{\alpha_1^*, \ldots, \alpha_m^*\}$ are different from zero, so that the hypothesis depends on a small subset of the available examples (those corresponding to non null αs, that are denoted *support vectors* or SV). To exploit Theorem 3.1, we remark that:

Fact 3.1. *The number of sentry points to bind the expansion of the symmetric difference between a separating hyperplane c and an estimate h of it is at most equal to the dimensionality of the space where the hyperplanes are defined.*

Actually, the key functionality of the support vectors parametrizing the $U_{c \div h}$ distribution is to bind the expansion of the symmetric difference $c \div h$ through forbidding any rotation of h into a h' pivoted along the intersection of c with h. Whatever the dimensionality n of the embedding space, in principle we would need only 1 point on the border of the angle between c and h, provided we know the target concept c, acting as sentry against this expansion [3]. In fact constraining h' to contain the intersection of h with c gives rise up to $n-1$ linear relations on h' coefficients, resulting, in any case, in a single degree of freedom for h' coefficient, i.e. a single sentry point. However, as we do not know c, the chosen sentry point may lie exactly in the intersection between c and h, preventing the former to sentinel the

Sentinelling stands for preventing a hyperplane rotation,

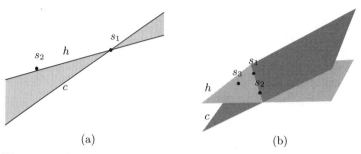

Fig. 3.18. Sentry points in the worst case needed to sentinel the symmetric difference between the hypothesis h and the target concept c in (a) two-dimensional and (b) three-dimensional space.

expansion of the symmetric difference. So we need one more sentry point, and in general, as many points as the dimensionality of the space. Figs. 3.18(a) and (b) illustrate this concept in case $n = 2$ and $n - 3$ respectively, bringing us to the above conclusion in Fact 3.1.

Moreover, concerning the three issues at the beginning of Section 3.2.4

- Numerical accuracy
  The necessity of extra points for sentineling $c \div h$ comes from the fact that sample points fall exactly in the intersection between the two hyperplanes, and there is no way for having the probability of this event different from 0 if both the sample space and its probability distribution are continuous. Hence, we really realize the above linear relations if either the sample space is discrete or the algorithm computing the hyperplane is so badly approximate to work on an actually discretized search space.^{provided you carefully localize it.}

  This is the dual perspective, on the part of sample points accuracy, of the reasoning done therein around the probability measure of the sentry points violation. Moreover, sentineling the expansion of symmetric difference between hypeprplanes is not more expensive than fixing a hyperplane through support vectors. Thus we may conclude:

Fact 3.2. *The number of sentry points of separating hyperplanes computed through SVMs ranges from 1 to the minimal number of involved support vectors minus one, depending on the approximation with which either sample coordinates are stored or hyperplanes are computed.*

- Label accuracy.
  A variant of SVM algorithm, known as *soft-margin classifier* [46], produces hypotheses for which the separability

requirement is relaxed. We do not guarantee the existence of a hyperplane separating the sample and admit a limited number of misclassifications. The optimization problem is essentially unchanged, with the sole exception of the introduction of a parameter $\grave{\alpha}$ ruling the trade-off between maximum margin and classification error. Now (3.90) reads:

$$0 \leq \alpha_i \leq \grave{\alpha} \quad i = 1, \ldots, m. \qquad (3.90')$$

The separating hyperplane equation is still obtained through (3.91-3.92), though the latter equation is computed mediated on indices i such that $0 < \alpha_i^* < \grave{\alpha}$.

- Label uncertainty.
  We may identify the labels with a Bernoulli variable whose parameter – i.e. the probability of tossing 1 – is a function $p(\boldsymbol{x})$ of the point the label refers to. In this case we refer to an augmented space with an additional coordinate represented by the uniform seed u of the label sampling mechanism, provided we know the profile of $p(\boldsymbol{x})$. Under the sole hypothesis of monotonicity when points move from 0- to 1-labeled domains, plus loose regularity conditions (for instance we may assume that p grows linearly or with a hyperbolic tangent profile), a suitable strategy is to divide the augmented space still with a hyperplane. Actually, in this case the sampling mechanism allows for a distribution probability on hyperplanes, so that we may state a somehow hybrid procedure as follows:

  1. generate a bootstrap population of separating surfaces;
  2. identify the candidate hypothesis as the median surface after outliers' removal and reckon its support vectors and mislabeled points (call them *relevant points*);
  3. compute confidence intervals for the misclassification error.

Label uncertainty may introduce an additional dimension to the separation problem

The results of this procedure are shown in Fig. 3.19 contrasted with those produced by the most common computational approach rooted on Vapnik theory [49]. Though they own a slightly different operational meaning, we denote with the same symbol P_{er} the measure of the concept-hypothesis symmetric difference in both approaches for comparison sake. Another approach exploits the geometrical interpretation of the classification problem [9].

Decision trees

Boolean formulas are functions that map from the space $V_n = \{v_1, \ldots, v_n\}$ of propositional variables to $\{0, 1\}$. A formula is

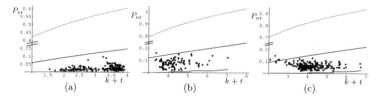

Fig. 3.19. Course of P_{er} with the number of relevant points. Bullets: actual values on random instances; plain curves: upper and lower bounds with Algorithmic Inference; gray curve upper bound with Vapnik theory. The probability profiles are chosen within the family of: (a) vertical planes; (b) skew planes; and (c) hyperbolic tangent surfaces.

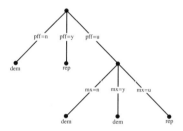

Fig. 3.20. A decision tree classifying the Vote dataset. Leafs: decisions; arcs: vote options on specific matter. pff → physician fee freeze, mx → mx missile, dem → democrat, rep → republican and x, y, u as defined in Section 1.5.9.

Discriminating on the basis of logic conditions, satisfied by an assignment from $X_n = \{0,1\}^n$ to the set of the propositional variables if the result of the formula is 1. A *clause* is the disjunction $(\ell_i \vee \ell_j \vee \cdots \vee \ell_k)$ of literals (a literal ℓ being an affirmed v or a negated v̄ propositional variable, see Section 1.3.) and a *Conjunctive Normal Form* – CNF for short – is the conjunction of a set of clauses. A *monomial* is the intersection $(\ell_i \wedge \ell_j \wedge ... \wedge \ell_k)$ of literals, and **possibly expressed in a standard form,** *Disjunctive Normal Form* – DNF for short – is the union of a set of monomials. Every boolean formula may be represented either through a CNF or a DNF.

Though these standard forms may prove much long, hence very difficult to synthesize in our brain, and in spite of the fact that the latter proves more complex to learn according to Definition 3.10, in the everyday life we are used to discuss boolean formulas in terms of decision trees that represent an extension of DNF to many valued (in place of only $\{0,1\}$) inputs. A typical description of a decision tree is in Fig. 3.20. Your formula is satisfied if there is a path in the tree that represents your assignment to propositional variables. You may sequentially check your pattern variable by variable. Actually, the highest learning complexity of the procedure for learning the tree is made harmless by the small size we look for it. There is a certain number of criteria for downsizing the tree, just in light of the balance accuracy-complexity we discussed earlier. One of the most famous and employed goes under the name of C4.5 [40] and its evolution [39].

Though in some cases the symmetric difference between concept and hypothesis proves exceedingly large, generally C4.5 works well and efficiently, as a sharp witness of the fact that the regularity of the distribution functions ruling the examples and a good approximation in the computation in contrast with the required learning accuracy reduce the number of necessary sentry points and the complexity of the learning task.

<small>with the ususal trade-off between accuracy, complexity and length.</small>

3.3 Conclusions

Taking out operational benefits from a sample of data results in moulding the uncertainty about data into uncertainty on other quantities that we assume to be more useful to us. As stated in Chapter 1, the gap between two consecutive among m sample data $(x_{(i)}, x_{(i+1)})$ is filled up by an indiscernible set of values cumulating a probability of order $1/(m+1)$. You may transform these blobs and map them into new random variable spaces, for instance parameter spaces where the indeterminacy concerns properties about the observed data. The indeterminacy grain remains $1/(m + 1)$; our ability stands to render it in a form that is more suitable to our operational framework, i.e. to describe by means of *tight* distribution laws properties of the observed phenomenon we are concerned with. This passes through a selection of a good statistic as indeterminacy transfer tool, and skillful computations to optimize the transfer. Actually, in the simple cases we may rely on well behaving statistics for filling up univocally the former task, say $\sum_i x_i$ for dealing with the parameter p of a Bernoulli variable. On the contrary, in many other cases we are called to identify a suboptimal solution taking into account more or less explicit ancillary informations; for instance symmetry assumptions on the example distribution push us to find support vector machines as linear separators. Most solutions of this chapter boils around the use of the parameters' (like P) distribution law. *Per se* it requires no other techniques than the inversion of quantiles or averaging of some parameter functions. Rather, the most sophisticated concepts concern the identification of meaningful statistics, hence of spendable properties of available data, having sufficient statistic as the template property, and the way of ordering the random variable specifications at the basis of any probabilistic reasoning. While ordering looks an innate attitude embedded in observing data, things become difficult when functions are directly involved. This happens in this chapter when we want to learn about functions, such as regression curves or concepts, and will happen in further chapters when we want to use functions to order points, for instance for clustering them. While we like to stress the statistic value of the learning functionality, having computational aspects as both premises and consequences, in any case, the synthesis of the parameter distribution law, hence the uncertainty nullifying, is definitely a matter of optimization task. In line with the book general thread, also with this task we are mainly concerned about the *what* rather than the *how* we optimize. In turn *what* concerns the use we will do of the data and the knowledge base we have available. We underline, in any case, the clear separate role we may devote to the two (statistic and optimization) tasks in this book. With more conventional approaches on the contrary, we hit directly to point estimators (hence parameter synthesis) having statistic identification as a result of a cost function optimization, such as unbiasedness or variance minimization, and interval estimates as possibly puzzling corollaries.

3.4 Exercises

1. Implement the alternate sampling mechanism for a Binomial distribution described in Algorithm 3.2.
2. Show that the random variable associated with the parameter p of a Binomial distribution is an observable property according to Definition 3.1.
3. Show that the statistic associated to the property described in Exercise 2 is well behaving according to Definition 3.2.
4. Derive an analytical proof of the equations (3.4) and (3.5).
5. Write the version of Algorithm 3.3 dealing with the double exponential distribution (3.3) and implement it in order to obtain the graph of the c.d.f. of the random variable Λ corresponding to the parameter λ.
6. Deduce the equivalent of equations (3.7-3.9) for a Binomial distribution, considering its integer parameter as fixed. Then, obtain the corresponding of Figs. 3.5 and 3.6.
7. Obtain a twisting argument for the parameter μ of a Poisson distribution.
8. Redraw the equivalent of Fig. 3.8 for different confidence levels.
9. Analytically derive equations (3.28-3.29).

10. Redraw the equivalent of Fig. 3.8 for the parameter μ of a Gaussian distribution with $\sigma = 3$ and $m = 10$ and $\sum_{i=1}^{m} x_i$ ranging between -10 and $+10$.
11. Compare the unbiased and weakly unbiased estimators for the real-valued parameter A of a uniform distribution in $[0, a]$ with those obtained through the method of moments and the maximum likelihood method.
12. Give an example of a family of nested confidence regions.
13. Analytically derive equations (3.48-3.50).
14. Rewrite the CD block of the procedure described in Section 3.2.1 to compute regression line confidence regions when ε is a specification of a Gaussian variable having null mean.
15. Elaborate on the reason why the score described in the NR block of the procedure in Section 3.2.1 assigns low values to lines close to the envelope's frontier.
16. Compute equations (3.64-3.65) for the data in Tables 1.1 and 1.2.
17. Apply the bootstrap procedure in order to jointly infer the mean and the variance of a Gaussian distribution.
18. Define a sentry function for the class of concepts described in the following table:

	x_1	x_2	x_3
c_1	$-$	$-$	$-$
c_2	$+$	$-$	$-$
c_3	$+$	$+$	$-$
c_4	$+$	$+$	$+$

where $\mathfrak{X} = \{x_1, x_2, x_3\}$, $C = \{c_1, c_2, c_3, c_4\}$ and $x_i \in c_j$ if the corresponding matrix entry contains the symbol $+$.

19. Find a sentry function for the class of rectangles in $\mathbb{R}^2$.
20. Repeat Exercise 19 for the class of rectangles in $\mathbb{R}^n$, for fixed $n \in \mathbb{N}$.
21. Apply equations (3.82-3.83) to the class of b_rectangles in $\mathbb{R}^2$.
22. Try to deduce the optimal margin hyperplane (3.91) and (3.92) from the dual constraint optimization problem defined by (3.88-3.90) (you may find the solution in Section 12.1.1).
23. Run C4.5 algorithm to the SMSA dataset chosing as target value the age adjusted mortality index quantized in three values. Compare the optimal tree found by the algorithm with the pruned one. Then generate the set of rules on the basis of the obtained decision tree. (The lastest version of C4.5 can be downloaded on the WEB [40].)

Further Reading

Reading inside the randomness of data is an old attitude which dates millennia. Knowing the fate on the basis of random events (thunder direction, pebbles arrangement, and so on) has been a common aim of human kind and a profitable business of diviners [13]. This is a *positive* attitude with respect to uncertainty that in a more scientific fashion is adopted for instance by Fisher in terms of statistical inference – describe the distribution law of a parameter of the future population, for instance its mean, *in light* of the observed sample [17] – and formalized as *predictive inference* by De Finetti [15]. In the same thread we find the Bayesian approach, which requires, however, an *a priori* distribution of the questioned parameter that may be improved (into an *a posterior* distribution) after the observed sample [34, 20]. The *negative* attitude toward randomness is paradigmatically represented by the Laplace approach. He took probability as an instrument for repairing defects in knowledge [30]. Accordingly, the conventional statistical framework starts with an ideal world where variables are spread with fixed distribution laws and challenges the user to discover them, hence their shapes and their parameters, *in spite* of the randomness of the few specifications of these variables he is allowed to observe [14, 28]. Progress in computer science in the last decades brought an even more positive perspective on randomness, assuming it as added value of many algorithms aimed at solving highly computationally complex problems [41, 27]. In the absence of actually more profitable techniques that visit a huge set of alternative solutions, it guarantees do not getting biased by a subset of them. It is the strategy you may adopt facing a trivium without any hint on the road to take. You toss a die with the solace that, maintaining this strategy in all trivia of your life you will succeed one over three times, in spite of any unfavorable disposition of the optimal way in each of them any opponent may have thought against you [36].

Algorithmic Inference approach looks at randomness not as a drawback but as an inherent aspect of data. A continuous version of a log of the mentioned die tossing is exactly the seed we use of any random variable. It exactly avoids getting stuck in some bias that may not be dealt with in terms of structural property of the data, hence their explaining function. The mature reader may remember long disputes held around halves the past century concerning the interpretation of the parameter distribution in terms of fiducial distribution [18], structural probabilities [22], priors/posteriors [42], and so on. With our approach this brings to the notion of compatible explanations of the data that has some analogies with the mentioned framework of parallel universes (multiverses) [26]. Only one explanation will run among the compatible ones in the future population. All parallel worlds run with the other approach, though only one is visited by us. These disputes and distinctions may prove immaterial when we pursue the main commitment by elaborating only what may be run by feasible algorithms, hence by our computer in the limited space and the small time we are allowed with [6] . With these constraints what is conventionally called a *pseudo-random* number [24] becomes an actual random number. So that we may really generate the parameters distribution laws starting from uniform random number generators. This means that we may handle in a very comfortable way many inference problems that are commonly considered solvable only at a cost of heavy approximations [2].

With our *population bootstrap* method and the twisting argument analytical counterpart, we may process every element of the parameter population, for instance *peeling* out those that are not of our interest [45, 1]. In this way we are able to extract from data their structure, i.e. the algorithms describing their main ensemble features, an attitude that covers statistical learning [49, 4, 5].

Note that the first issue of the learning functionality in recent years was demanded to subsymbolic methods that will be discussed in Chapters 8 and 9:

> I recognize that a function computes what expected, but I don't know why.

This the typical feature of a trained neural network. Nowadays this approach has been mostly abandoned in favor of a more readable one:

> I don't pretend to learn highly complex functions, but I want to understand what I have learnt.

This calls for simple, mainly symbolic algorithms whose statistical bases may be elaborated. This is essentially the goal of this chapter, and this book in general, where simplicity extends also to the generation of the random variables. We omitted a treatment of the vast topic of test of hypotheses. We did it simply because this important matter is perfectly covered in a number of other books and in practice as well within conventional statistical frameworks. It matches exactly their structure: you have a prefixed probabilistic model running the world, you don't know it but have a hypothesis about it. Decide if the hypothesis is congruent with (could have generated) the sample you have observed [19, 31].

References

1. Apolloni, B., Bassis, S., Gaito, S., Malchiodi, D.: Appreciation of medical treatments by learning underlying functions with good confidence. Current Pharmaceutical Design 13(15), 1545–1570 (2007)
2. Apolloni, B., Bassis, S., Gaito, S., Malchiodi, D.: Bootstrapping complex functions. Nonlinear Analysis (in press, 2007)
3. Apolloni, B., Chiaravalli, S.: Pac learning of concept classes through the boundaries of their items. Theoretical Computer Science 172, 91–120 (1997)
4. Apolloni, B., Kurfess, F. (eds.): From Synapses to Rules – Discovering Symbolic Rules from Neural Processed Data. Kluwer Academic/Plenum Publishers, New York (2002)
5. Apolloni, B., Malchiodi, D.: Gaining degrees of freedom in subsymbolic learning. Theoretical Computer Science 255, 295–321 (2001)
6. Apolloni, B., Malchiodi, D., Gaito, S.: Algorithmic Inference in Machine Learning. International Series on Advanced Intelligence, 2nd edn., vol. 5, Magill, Adelaide (2006)
7. Apolloni, B., Bassis, S., Gaito, S., Iannizzi, D., Malchiodi, D.: Learning continuous functions through a new linear regression method. In: Apolloni, B., Marinaro, M., Tagliaferri, R. (eds.) Biological and Artificial Intelligence Environments, pp. 235–243. Springer, Heidelberg (2005)
8. Apolloni, B., Iannizzi, D., Malchiodi, D., Pedrycz, W.: Granular regression. In: Apolloni, B., Marinaro, M., Nicosia, G., Tagliaferri, R. (eds.) WIRN 2005 and NAIS 2005. LNCS, vol. 3931. Springer, Heidelberg (2006)
9. Apolloni, B., Malchiodi, D., Natali, L.: A modified svm classification algorithm for data of variable quality. In: Apolloni, B., Howlett, R.J., Jain, L. (eds.) KES 2007, Part III. LNCS (LNAI), vol. 4694, pp. 131–139. Springer, Heidelberg (2007)
10. Aratyn, H., Rasinariu, C.: A Short Course in Mathematical Methods with Maple. World Scientific Publishing Company, Singapore (2005)

11. Bogdanoff, D.A., Pierce, D.A.: Bayes-fiducial inference for the weibull distribution. Journal of the American Statistical Association 68(343) (1973)
12. Carpenter, J., Bithell, J.: Bootstrap confidence intervals: when, which, what? A practical guide for medical statisticians. Statistics in Medicine 19, 1141–1164 (2000)
13. Cicero, M.T.: De divinatione. Iliades (1938) 2000 b.c.
14. Cramér, H.: Mathematical Methods of Statistics. Princeton University Press, Princeton (1958)
15. De Finetti, B.: Theory of Probability. In: A Critical Introductory Treatment, vol. 2. John Wiley & Sons, New York (1975)
16. Dubois, D., Prade, H.: Possibility Theory. Plenum Press, London (1976)
17. Fisher, M.A.: The fiducial argument in statistical inference. Annals of Eugenics 6, 391–398 (1935)
18. Fisher, M.A.: Statistical Methods and Scientific Inference. Oliver and Boyd, Edinburgh and London (1956)
19. Fisher, R.A.: Mathematics of a lady tasting tea. In: Newman, J.R. (ed.) The world of mathematics, pp. 1512–1521. Simone & Schuster, New York (1956) Original work published in 1935
20. Florens, J.P., Mouchart, M.M., Rolin, J.M.: Elements of bayesian statistics. Marcel Dekker, Inc., New York (1990)
21. Fraser, D.A.S.: Statistics. An Introduction. John Wiley & Sons, London (1958)
22. Fraser, D.A.S.: Structural probability and a generalization. Biometrika 53(1/2), 1–9 (1966)
23. Glivenko, V.: Sulla determinazione empirica delle leggi di probabilità. Giornale dell'Istituto Italiano degli Attuari 3, 92–99 (1933)
24. Goldreich, O.: Modern cryptography, probabilistic proofs, and pseudorandomness. Algorithms and combinatorics. Springer, Berlin (1999)
25. Gomez, C. (ed.): Engineering and Scientific Computing with Scilab. Birkhauser, Basel (1999)
26. William, J.: Pragmatism's conception of truth. Lecture 6 in Pragmatism: A New Name for Some Old Ways of Thinking, p. 83 (1907)
27. Karp, R.M., Rabin, M.O.: Efficient randomized pattern-matching algorithms. IBM Journal of Research and Development 31(2), 249–260 (1987)
28. Kendall, M.G., Stuart, A.: The advanced theory of statistics. In: Inference and relationship, vol. 2. Charles Griffin, London (1961)
29. Kolmogorov, A.N.: Three approaches to the quantitative definition of information problems. Problems of Information Transmission 1(1), 1–7 (1965)
30. de Laplace, P.S.M.: Essai philosophique sur les probabilités. Veuve Courcier, Paris (1814)
31. Lehman, E.L.: Testing Statistical Hypotheses, vol. 4. John Wiley & Sons Inc., Chapman & Hall, London (1984)
32. Levin, L.A.: Laws of information (nongrowth) and aspects of the foundation of probability theory. Problems of Information Transmission 10(3), 206–210 (1974)

33. Levin, L.A.: One-way functions and pseudorandom generators. Combinatorica 4(7), 357–363 (1987)
34. Lindley, D.V.: Fiducial distributions and bayes' theorem. Journal of the Royal Statistical Society; Series B 20, 102–107 (1958)
35. Mood, A.M., Graybill, F.A., Boes, D.C.: Introduction to the Theory of Statistics. McGraw-Hill, New York (1974)
36. Motwani, R., Raghavan, P.: Randomized Algorithms. Cambridge University Press, Cambridge (1995)
37. Neyman, J.: Su un teorema concernente le cosiddette statistiche sufficienti. Giornale dell'Istituto Italiano degli Attuari 6, 320–324 (1935)
38. Pestman, W.R.: Mathematical Statistics An Introduction. Paperback De Gruyter (1998)
39. Quinlan, R.: Data Mining Tools: See5 and C5.0, RuleQuest Research (2007),
http://www.rulequest.com/see5-info.html
40. Qunilan, J.R.: C4.5: programs for machine learning. Morgan Kaufmann Publishers, San Mateo, California (1993)
41. Rabin, M.O.: Probabilistic algorithm for testing primality. Journal of Number Theory 12, 128–138 (1980)
42. Ramsey, F.P.: The foundations of mathematics. Proc. of London Mathematical Society (1925)
43. Roger, H.: Theory of recoursive functions and effective computability. McGraw-Hill, New York (1967)
44. Rohatgi, V.K.: An Introduction to Probablity Theory and Mathematical Statistics. Wiley Series in Probability and Mathematical Statistics. John Wiley & Sons, New York (1976)
45. Rousseeuw, P., Hubert, M.: Regression depth (with discussion). Journal of American Statistical Association 94, 388–433 (1999)
46. Schölkopf, B., Burges, C.J.C., Smola, A.J. (eds.): Advances in kernel methods: support vector learning. MIT Press, Cambridge (1999)
47. Student. The probable error of a mean. Biometrika 6(1), 1–25 (1908)
48. Valiant, L.G.: A theory of the learnable. Communications of the ACM 11(27), 1134–1142 (1984)
49. Vapnik, V.: Statistical Learning Theory. John Wiley & Sons, New York (1998)
50. Wilks, S.S.: Mathematical Statistics. Wiley Publications in Statistics. John Wiley, New York (1962)
51. Wolfram, S.: The Mathematica book, 5th edn. Wolfram Media (2003)

Part II
The Development of Fuzzy Sets

Introduction

If Hansel cannot find information from outside, say because it is clouded and he cannot see how the pebble sequence continues, he must find it inside his brain. We speak in general of *experience* or *intuit* or *creativity* in these cases, to mean that this information comes from a sound synthesis of previous data, eventually processed with the millennia by the human kind and stored in the genome, possibly compressed in relatively recent time in the synapses of a single mind, or what else. This means that Hansel has some rules available to make decision exactly from the few pebbles he sees in its surrounding, where generally these rules lead to a spectrum of decisions and some criterion to differentiate between them. Probability too could represent a criterion, but the peculiarity of the criteria we will discuss in this part is that they associate instances within a set to a given decision, say the appropriateness of possible slopes of the pebble ground to a given direction of the next move, rather than a single instance to different decisions, thus the preference of the next move directions in respect to a specific ground slope. The fact is that at the end of the day we need the latter criterion to make our final decision, but dispose only of the former. Hence drawing in our mind in order to asses the allowed criterion – that we will denote as the *membership function* of the set of instances vaguely associated with a given decision A, hence to the *fuzzy set* A – we must endow it with features that prove to be mostly suitable to arrive to a concrete decision-making tool.

    Next chapter will be devoted to the task of identifying and discussing these features and characterizing families of fuzzy sets accordingly. In Chapter 5, we will discuss basic methods of adapting some of these families to the data we have available, a task that we are used to denote as *membership function estimation*. The process in which we start from data and arrive at decisions will be treated in Part V.

Degrees to many elements of a single attribute

in order to assign degrees to many attributes for a single element.

4 Construction of Information Granules: From Data and Perceptions to Information Granules

We focus on the development of fuzzy sets by presenting various ways of designing fuzzy sets and determining their membership functions. From a logical perspective, moving from Definition 1.4 we have no many other conceptual tools than the following elementary relationships:

Fuzziness is a non univocal category with a slim axiomatic basis,

Definition 4.1. *For the fuzzy sets A and B in the universe of discourse $\mathscr{X}$, the following relationships are defined:*

$A \subseteq B$ *if* $\mu_A(x) \leq \mu_B(x)$, *for each x*   *(inclusion)*   (4.1)

$A = B$ *if* $\mu_A(x) = \mu_B(x)$, *for each x*   *(identity)*   (4.2)

$A = \overline{B}$ *if* $\mu_A(x) = 1 - \mu_B(x)$, *for each x*   *(complementation)*   (4.3)

and a few obvious corollaries such as the fact that if $A = \overline{B}$ then $\overline{A} = B$.

Therefore, the subject of elicitation and interpretation of fuzzy sets is of paramount relevance from the conceptual, algorithmic, and application-driven standpoints. We put the development of fuzzy sets in the framework of granular computing and show linkages of fuzzy sets with rough sets. It essentially occurs along the line of their generalizations into more abstract constructs usually referred to as higher order fuzzy sets, in particular fuzzy sets of 2-nd order. Various implementation and conceptual issues arising around numeric values of membership functions give rise to a collection of concepts of granular membership grades. In the simplest scenario they give rise to an idea of interval-valued fuzzy sets. More refined versions of the construct produce type-2 fuzzy sets and fuzzy sets of higher type.

admitting alternative descritpions/elicitations.

4.1 Semantics of Fuzzy Sets: Some Insights

Fuzzy sets are reflective of information granules we introduce and deal with them as generic processing entities. There has been a great deal of investigation on the underlying concept and a certain level of controversy around fuzzy sets most of which must have emerged because of attempts to bring the discussion within the realm of probability and statistics. While statistics could be viewed as a helpful vehicle to use for estimation purposes, it is essential to stress that fuzzy sets require an own specific framework. Therein, we make use of a finite number of fuzzy sets leading to some essential vocabulary reflective of the underlying domain knowledge. In particular, we are concerned with the related semantics, calibration capabilities of membership functions and the locality of fuzzy sets.

Fuzzy sets are knowledge in pills.

The limited capacity of a short term memory [17] strongly suggests that we could easily and comfortably handle and process 7 ± 2 variations of a conceptual entity. The observation sounds reasonable – quite commonly in practice we witness situations in which this holds. For instance, when describing linguistically quantified variables, say error or change of error, we may use seven generic concepts (descriptors) labeling them as positive *large*, positive *medium*, positive *small*, *around* zero, negative *small*, negative *medium*, negative *large*. When characterizing speed of a vehicle on a highway, we may talk about its quite intuitive descriptors such as *low*, *medium* and *high* speed. In the description of an approximation error, we may typically use the concept of a *small* error around a point of linearization (in all these examples, the terms are indicated in italics to emphasize the granular character of the constructs and the role being played there by fuzzy sets). While embracing very different tasks, the descriptors used there exhibit a striking similarity. All of them are information granules, not numbers (whose descriptive power becomes very limited). For instance, in modular software development when dealing with a collection of modules (procedures, functions and alike), the list of their parameters is always limited to a few items which is again a reflection of the limited capacity of the short term memory. The excessively long parameter list is strongly discouraged due to the possible programming errors and rapidly increasing difficulties of an effective comprehension of the software structure and ensuing of control flow.

The basic goal is to quantize variables into levels,

not too many in any case,

In general, the use of an excessive number of terms does not offer any advantage. To the contrary: it remarkably clutters our description of the phenomenon and hampers further effective usage of such concepts we intend to establish to capture the essence of the domain knowledge. With the increase in the number of fuzzy sets, their semantics become also negatively impacted. Fuzzy sets may be built into a hierarchy of terms (descriptors) but at each level of this hierarchy (when moving down towards higher specificity that is an increasing level of detail), the number of fuzzy sets is kept at a certain limited number.

While fuzzy sets capture the semantics of the concepts, they may require some calibration depending upon the specification of the problem at hand. This flexibility of fuzzy sets should not be treated as any shortcoming but rather viewed as a certain and fully exploited advantage. For instance, a term *low* temperature comes with a clear meaning yet it requires a certain calibration depending upon the environment and the context it was placed into. The concept of *low* temperature is used in different climate zones and is of relevance in any communication between people yet for each of the community the meaning of the term is different thereby requiring some calibration. This could be realized e.g. by shifting the membership function along the universe of discourse of temperature, affecting the universe of discourse by some translation, dilation and alike. As a communication vehicle, linguistic terms are fully legitimate and as such they appear in different settings. They require some refinement so that their meaning is fully understood and shared by the community of the users.

but adapted to their meaning.

4.2 Transferring Meaning into Fuzzy Sets

As mentioned in the introduction of this part, fuzzy sets are reflective of domain knowledge and opinions of experts. The ways of formalizing them may be different, being compliant with different perspectives.

4.2.1 Fuzzy Set as a Descriptor of Feasible Solutions

With this perspective we relate membership function to the level of feasibility of individual elements of a family of solutions associated with the problem at hand. Let us consider a certain function $f(x)$ defined in $\mathfrak{X}$, that is $f : \mathfrak{X} \to \mathbb{R}$, where $\mathfrak{X} \subseteq \mathbb{R}$. Our intent is to determine its maximum, namely $x_{\text{opt}} = \arg\max_x f(x)$. On a basis of the values of $f(x)$, we can form a fuzzy set A describing a collection of feasible solutions that could be labeled as optimal. Being more specific, we use the fuzzy set to represent an extent (degree) to which some specific values of x could be sought as potential (optimal) solutions to the problem. Taking this into consideration, we relate the membership function of A with the corresponding value of $f(x)$ cast in the context of the boundary values assumed by f. For instance, the membership function μ_A of A could be expressed in the following form:

The more you are accepted in a chess club the more your gaming score moves closer to the maximum,

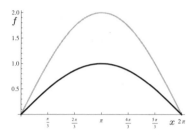

Fig. 4.1. Function $f = 2\sin(x/2)$ (gray curve) and the induced membership function μ_A (black curve).

$$\mu_A(x) = \frac{f(x) - f_{\min}}{f_{\max} - f_{\min}}. \qquad (4.4)$$

The boundary conditions are straightforward: $f_{\min} = \min_x f(x)$ and $f_{\max} = \max_x f(x)$ where the minimum and the maximum are computed over $\mathfrak{X}$. For "other values of x, where f attains its maximal value $\mu_A(x)$ is equal 1, and around" equal 1 and around this point the membership values are reduced when x is likely to be a solution to the problem with $f(x) < f_{\max}$. The form of the membership function depends upon the character of the function under consideration. The following examples illustrate the essence of the construction of membership functions.

but how to evaluate the gap is questionable.

Example 4.1. Let us consider the case study of determining a maximum of the function $2\sin(x/2)$ defined in $[0, 2\pi]$. The minima of f in the range of the arguments between 0 and 2π coincide exactly with these points, while the maximum of f is reached at $x^* = \pi$. According to (4.4) the membership function of x to the solution of the optimization problem is $\mu_A(x) = \sin(x/2)I_{[0,2\pi]}(x)$, as shown in Fig. 4.1.

Linearization, its nature and description of quality fall under the same banner as the optimization problem. When linearizing a function around some given point, a quality of such linearization can be articulated in a form of some fuzzy set. Its membership function attains 1 for all those points where the linearization error is equal to 0 (in particular, this holds at the point around which the linearization is carried out). The following example illustrates this point.

Example 4.2. We are interested in the linearization of the function $g(x) = 0.5\sin(x)$ around $x_0 = \pi/4$ and in assessing the quality of this linearization in the range $[0, 7\pi/2]$. The linearization formula reads as $y - y_0 = g'(x_0)(x - x_0)$ where $y_0 = g(x_0)$ and $g'(x_0)$ is the derivative of $g(x)$ at x_0. Given the form of the

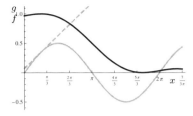

Fig. 4.2. Function f representing the quality of linearization (gray dashed curve) of the nonlinear function $g(x) = 0.5\sin(x)$ (gray curve) around the point $x_0 = \pi/4$ and the induced fuzzy set A (black curve). The membership function μ_A is defined over $\left[0, \frac{7}{2}\pi\right]$.

function, $g(x) = 0.5\sin(x)$, the linearized version of the function reads as $y = \frac{1}{2\sqrt{2}} + \frac{1}{2\sqrt{2}}(x - \frac{\pi}{4})$. Next let us define the quality of this linearization by taking the absolute value of the difference between the original function and its linearization (linearization error), $f(x) = \left|g(x) - \frac{1}{2\sqrt{2}} - \frac{1}{2\sqrt{2}}(x - \frac{\pi}{4})\right|$. As the fuzzy set A describes the quality of linearization, its membership function takes into account the following expression:

$$\mu_A(x) = 1 - \frac{f(x) - f_{\min}}{f_{\max} - f_{\min}}, \quad (4.5)$$

where $f_{\max} \approx f(5.50) \approx 2.37$ and evidently $f_{\min} = 0.0$. When at some z, $f(z) = f_{\min}$, this means that $\mu_A(z) = 1$ which in the sequel indicates that the linearization at this point becomes perfect as no linearization error has been generated. The plot of the membership function μ_A is presented in Fig. 4.2. We note that the higher quality of approximation is achieved for the arguments higher than the point at which the linearization has been completed. From the application standpoint, this implies that the shape of the membership function is illustrative when it comes to the use of the resulting linearized function.

> How much the linearization of a function is a good approximation?

4.2.2 Fuzzy Set as a Descriptor of the Notion of Typicality

Fuzzy sets address an issue of gradual typicality of elements to a given concept. They stress the fact that there are elements that fully satisfy the concept (are typical for it) and there are various elements that are allowed only with partial membership degrees. The form of the membership function is reflective of the semantics of the concept. Its details could be captured by adjusting the parameters of the membership function or choosing its form depending upon experimental data. For instance, consider a fuzzy

> The more you are accepted in a society the more you comply with its template.

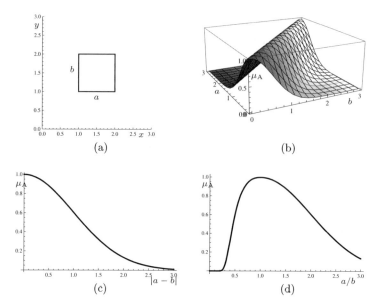

Fig. 4.3. Perception of geometry of squares (a) and its quantification in the form of membership functions of the concept of fuzzy square (b-d) for different universes of discourse as explained in the text.

set of squares. Formally, a rectangle includes a square shape as its special example when the sides are equal, $a = b$, see Fig. 4.3(a). What if $a = b+\varepsilon$, where ε is a very small positive number? Could this figure be sought as a square? It is very likely so. Perhaps the membership value of the corresponding membership function could be equal to 0.99. Our perception, which comes with some level of tolerance to imprecision, does not allow us to tell apart this figure from the ideal square.

How much close is a rectangle to a square?

Higher differences between the values of the sides of the rectangle could result in lower values of the membership function. The definition of the fuzzy set square could be formed in a number of ways. Prior to the introduction of the definition or even visualization of the membership function, it is important to formulate a space over which this fuzzy set will be defined. There are several intuitive alternatives worth considering:

(a) for each pair of values of the sides (a and b), collect an experimental assessment of membership of the rectangle to the category of squares. Here the membership function is defined over a Cartesian space of the lengths of sides of the rectangle. While selecting a form of the membership we require that it assumes the highest value at $a = b$ and is gradually reduced when the arguments start getting more different (see Fig. 4.3(b));

(b) we can define an absolute distance $|a - b|$ between a and b and form a fuzzy set over the space $\mathfrak{X} \subset \mathbb{R}_+$ defined as $\mathfrak{X} = \{x, x = |a - b|\}$. The semantic constraints translate into the condition of $\mu_A(0) = 1$. For higher values of x we may consider monotonically decreasing values of μ_A (see Fig. 4.3(c));

(c) we can envision ratios $x = a/b$ of a and b and construct a fuzzy set over the space of $\mathfrak{X} \subset \mathbb{R}_+$ such that $\mathfrak{X} = \{x, x = a/b\}$. Here we require that $\mu_A(1) = 1$. We also anticipate lower values of membership grades when moving to the left and to the right from $x = 1$. Note that the membership function could be asymmetric so we allow for different membership values for the same length of the sides, say $a = 6, b = 5$ and $a = 5$ and $b = 6$ (the effect could be quite apparent due to the visual effects when perceiving some geometric phenomena). The previous model of $\mathfrak{X}$ as outlined in (a) cannot capture this effect (see Fig. 4.3(d)).

Once the form of the membership function has been defined, it could be further adjusted by modifying the values of its parameters on a basis of some experimental findings. They come in the form of ordered triples or pairs, say (a, b, μ_A), $(|a - b|, \mu_A)$ or $(a/b, \mu_A)$ depending on the previously accepted definition of the universe of discourse. The membership values μ_A are those available from the expert offering an assessment of the likeness of the corresponding geometric figure.

4.2.3 Membership Functions for Visualizing Solution Preferences

A simple electric circuit shown in Fig. 4.4(a) is helpful in illustrating the underlying idea in which fuzzy sets are used as an effective visualization mechanism. Consider the problem of optimization (maximization) of power dissipated on the external resistance of the circuit. The voltage source E is characterized by some internal resistance equal to r. The external resistance R is the one on which we want to maximize power dissipation. By straightforward calculations we determine the power dissipated on R to be given in the form:

$$P = i^2 R = \left(\frac{E}{R+r}\right)^2 R, \qquad (4.6)$$

Your acceptability can be directly determined by technical reasons,

where i is the current in the circuit. The maximization of P with respect to R is determined by zeroing the derivative of P, i.e., imposing that $dP/dR = 0$, which leads to the optimal value of the resistance R_{opt}. Through simple derivations we obtain $R_{\text{opt}} = r$. It

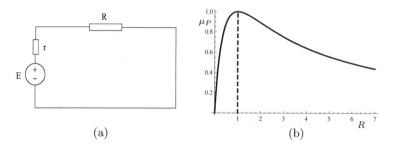

Fig. 4.4. (a) A simple electric circuit and the problem of maximization of power dissipation on the external resistance R; (b) membership function of the optimal power dissipation μ_P on external resistance R; the maximal value is achieved for $R = 1$ (the internal resistance r is assumed here to be equal to 1).

becomes evident that while the condition $R = r$ produces the maximum of P, this solution is not technically feasible as there is a substantial level of power dissipation on the internal resistance. If we plot the relationship of P versus R, see Fig. 4.4(b), and treat it as a membership function of R (which requires a simple normalization of P by dividing it by the maximal value obtained for $R = r$), we note that the shape of this relationship is highly asymmetric: when increasing the value of resistance over the optimal value (R_{opt}), the membership function changes quite smoothly and the reduction of the membership grades is quite limited.

On the other hand, when moving towards lower values of R such that $R < R_{\text{opt}}$, the reduction in the membership grades is quite substantial. We can say that the membership function of the optimal resistance offers a highly visible and very much intuitive quantification of the notion of optimality. The asymmetric shape of the resulting fuzzy set delivers some guidance in the selection of possible suboptimal solution, for instance connected to the resistances available in the market catalog, while the membership degree serves as an indicator of the suitability (degree of optimality) of the individual value of R.

4.2.4 Nonlinear Transformation of Templates

In many problems, we encounter a family of fuzzy sets defined in the same space. The family of fuzzy sets $\{A_1, A_2, \ldots, A_c\}$ is referred to as *referential* fuzzy sets. To form a family of semantically meaningful descriptors of the variable at hand, we usually require that these fuzzy sets satisfy the requirements of unimodality, limited overlap, and coverage. Technically, all of these features are reflective of our intention to provide this family of fuzzy sets with some semantics. These fuzzy sets could be sought as generic descriptors (say *small, medium, high*, etc.) being described by some typical membership functions. For instance, those could be uniformly distributed, triangular or Gaussian fuzzy sets (see Appendix C for a list of commonly used membership functions) with some standard level of overlap between the successive terms (descriptors).

or may be adapted from templates,

As mentioned, fuzzy sets are usually subject to some calibration depending upon the character of the problem at hand. We may use the same terms of *small, medium*, and *large* in various contexts yet their detailed meaning (viz. membership degrees) has to be adjusted. For the given family of referential fuzzy sets, their calibration could be accomplished by taking the space $\mathfrak{X} = [a,b]$ over which they are originally defined and transforming it into itself, that is $[a,b]$, through some nondecreasing monotonic and continuous function $\phi(x;\boldsymbol{\theta})$, where $\boldsymbol{\theta}$ is a vector of some adjustable parameters bringing the required flexibility of the mapping. The nonlinearity of the mapping is such that some regions of $\mathfrak{X}$ are contracted and some of them are stretched (expanded) and in this manner capture the required local context of the problem. This affects the membership functions of the referential fuzzy sets $\{A_1, A_2, \ldots, A_c\}$ whose membership functions are expressed now as $\mu_{A_i}(\phi(x;\boldsymbol{\theta}))$. The construction of the mapping ϕ is optimized taking into account some fixed points concerning membership grades given at some points of $\mathfrak{X}$. More specifically, the fixed points come in the form of the input-output pairs:

through any kind of monotone transform.

$$\begin{aligned} x_1 &- (\mu_1(1), \mu_2(1), \ldots, \mu_c(1)) \\ x_2 &- (\mu_1(2), \mu_2(2), \ldots, \mu_c(2)) \\ &\vdots \qquad\qquad \vdots \\ x_m &- (\mu_1(m), \mu_2(m), \ldots, \mu_c(m)) \end{aligned} \qquad (4.7)$$

where the k-th input-output pair consists of x_k which denotes some point in $\mathfrak{X}$ while $(\mu_1(k), \mu_2(k), \ldots, \mu_c(k))$ are the numeric values of the corresponding membership degrees. The objective is to construct a nonlinear mapping optimizing it with respect to the available parameters $\boldsymbol{\theta}$. More formally, we could translate the problem into the minimization of the following sum of squared errors:

$$\sum_{k=1}^{m}\sum_{i=1}^{c} (\mu_i(k) - \mu_{A_i}(\phi(x_k;\boldsymbol{\theta})))^2. \qquad (4.8)$$

One of the feasible mapping comes in the form of a piecewise linear function shown in Fig. 4.5(a). Here the vector of the

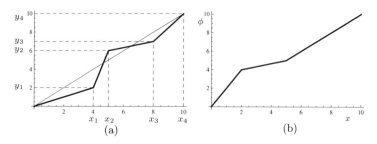

Fig. 4.5. Piecewise linear transformations function ϕ (bold curve); shown is also a linear mapping not affecting the universe of discourse and not exhibiting any impact on the referential fuzzy sets (plain curve). The proposed piecewise linear mapping is fully invertible.

adjustable parameters $\boldsymbol{\theta}$ involves a collection of the split points $\boldsymbol{x} = x_1, x_2, \ldots, x_m$ and the associated displacements $\boldsymbol{y} = y_1, y_2, \ldots, y_m$; hence $\boldsymbol{\theta} = (\boldsymbol{x}, \boldsymbol{y})$. The regions of expansion or compression are used to affect the referential membership functions and adjust their values given the fixed points.

Example 4.3. We consider some examples of nonlinear transformations of Gaussian fuzzy sets through the piecewise linear transformations (here $c = 3$) shown in Fig. 4.5(b) described as follows:

$$\phi(x) = 2x I_{[0,2)}(x) + \left(4 + \frac{1}{3}(x-2)\right) I_{[2,5)} + x I_{[5,10]}. \quad (4.9)$$

Note the fact that some fuzzy sets become more specific while others are made more general and expanded over some regions of the universe of discourse. This transformation leads to the membership functions illustrated in Fig. 4.6. For instance, the membership function of the leftmost fuzzy set reads as:

$$\mu_{A_3}(x) = e^{-\frac{(2x-1)^2}{2\sigma^2}} I_{[0,2)}(x) + e^{-\frac{\left(4+\frac{1}{3}(x-2)-1\right)^2}{2\sigma^2}} I_{[2,5)} + e^{-\frac{(x-1)^2}{2\sigma^2}} I_{[5,10]}, \quad (4.10)$$

with $\sigma = 1.5$.

Example 4.4. Considering the same nonlinear mapping as used before, two triangular fuzzy sets are converted into fuzzy sets described by piecewise membership functions as shown in Fig. 4.7.

4.2.5 Shadowed Sets as a Three-Valued Logic Characterization of Fuzzy Sets

Fuzzy sets offer a wealth of detailed numeric information conveyed by their detailed numeric membership grades (membership

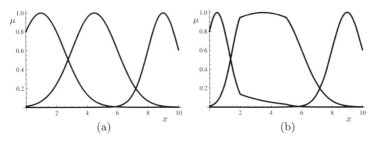

Fig. 4.6. Examples of original membership functions (a) and the resulting fuzzy sets (b) after the piecewise linear transformation shown in Fig. 4.5(b).

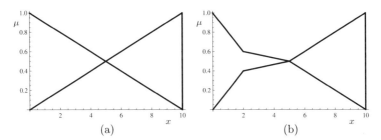

Fig. 4.7. Two triangular fuzzy sets (a) along with their transformed counterpart (b) according to the piecewise linear transformation shown in Fig. 4.5(b).

functions). This very detailed conceptualization of information granules can clearly act as a two-edge sword. On the one hand we may enjoy a very detailed quantification of elements to a given concept (fuzzy set). On the other hand, those membership grades could be somewhat overwhelming and introduce some burden when it comes to a general interpretation. It is also worth noting that numeric processing of membership grades comes sometimes with quite substantial computing overhead. To alleviate these problems while capturing the essence of fuzzy sets, we introduce a category of information granules characterized by only three quantification levels of the membership function called *shadowed sets*. Formally speaking, a shadowed set A defined in some space $\mathfrak{X}$ is a set-valued mapping expressed in the following form:

$$A : \mathfrak{X} \to \{0, [0,1], 1\}. \tag{4.11}$$

The co-domain of A consists of three components, that is 0, 1, and the entire unit interval [0, 1]. They can be treated as quantifications of membership of elements to A. These three quantification levels come with an apparent interpretation. All elements for which A(x) assume value 1 are called a *core* of the shadowed

In some cases it is enough to know what does really belong to a set, what not and a gap where we cannot state it with certainty.

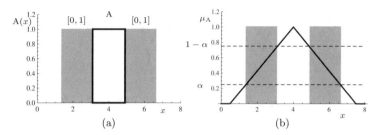

Fig. 4.8. (a) An example of a shadowed set A; note shadows formed around the core of the construct; (b) the concept of a shadowed set induced by some fuzzy set; note the range of membership grades (located between α and $1 - \alpha$) generating a shadow.

set – they embrace all elements that are fully compatible with the concept conveyed by A. The elements of $\mathfrak{X}$ for which A(x) attains zero are fully excluded from A. The elements of $\mathfrak{X}$ for which we have assigned the unit interval are completely uncertain, i.e., we are not in a position to allocate any numeric membership grade. Therefore we allow the usage of the unit interval, which reflects uncertainty meaning that any numeric value could be permitted here. In essence, such element could be excluded (we pick up the lowest possible value from the unit interval), exhibit partial membership (any number within the range from 0 and 1) or could be fully allocated to A. Given this extreme level of uncertainty (viz. nothing is known and all values are allowed), we call these elements *shadows* and hence the name of the shadowed set. An illustration of the underlying concept of a shadowed set is included in Fig. 4.8(a). One can view this mapping as an example of a three-valued logic as encountered in the classic model introduced by Lukasiewicz [13]. Having this in mind, we can think of shadowed sets as a *symbolic* representation of *numeric* fuzzy sets. Obviously, the elements of co-domain of A could be labeled using symbols (say, certain, shadow, excluded; or a, b, c and alike) endowed with some well-defined semantics.

> It looks like an indistinguishable shadow around the border of the set.

Accepting the point of view that shadowed sets are algorithmically implied by some fuzzy sets, we are interested in the transformation mechanisms translating fuzzy sets into the corresponding shadowed sets. The underlying concept is the one of uncertainty condensation or *localization*. While in fuzzy sets we encounter intermediate membership grades located in-between 0 and 1 and distributed practically across the entire space, in shadowed sets we *localize* the uncertainty effect by building constrained and fairly compact shadows. By doing so we could remove (or better to say, re-distribute) uncertainty from the rest of the universe of discourse by bringing the corresponding low and high membership grades to

> Shadowing may be a way of compressing a membership function information,

0 and 1 and then compensating these changes by allowing for the emergence of uncertainty regions. This transformation could lead to a certain optimization process in which we complete a total balance of uncertainty. To illustrate this optimization procedure, let us start with a continuous, symmetric, unimodal and usual membership function μ_A. In this case we can split the problem into two tasks by considering separately the increasing and decreasing portion of the membership function, see Fig. 4.8(b). For the increasing portion of the membership function, we reduce low membership grades to 0, elevate high membership grades to 1 and compensate these changes (which in essence lead to an elimination of partial membership grades) by allowing for a region of the shadow where there are no specific membership values assigned but where we admit the entire unit interval as feasible membership grades. Computationally, we form the following balance of uncertainty preservation that could be symbolically expressed as:

possibly maintaining some kind of consitency.

$$\text{Reduction of membership} + \text{Elevation of membership} = \text{shadow}. \tag{4.12}$$

Again referring to Fig. 4.8(b), once given the membership grades below α and above $1 - \alpha$ with $\alpha \in (0, 1/2)$, we express the components of the above relationship in the following form (we assume that all integrals do exist):

- reduction of membership (here low membership grades are reduced to zero):

$$v_r = \int_{x:\mu_A(x) \leq \alpha} \mu_A(x) dx; \tag{4.13}$$

- elevation of membership (where high membership grades elevated to 1):

$$v_e = \int_{x:\mu_A(x) \geq 1-\alpha} (1 - \mu_A(x)) dx; \tag{4.14}$$

- shadow:

$$v_s = \int_{x:\alpha < \mu_A(x) < 1-\alpha} dx. \tag{4.15}$$

The minimization of the absolute difference:

$$v(\alpha) = |v_r + v_e - v_s| \tag{4.16}$$

completed with respect to α is given in the form of the following optimization problem:

$$\alpha_{\text{opt}} = \arg \min_{\alpha} v(\alpha), \tag{4.17}$$

where $\alpha \in (0, 1/2)$.

142 Construction of Information Granules

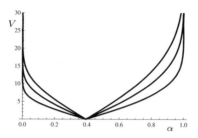

Fig. 4.9. Performance index v treated as a function of α, when $\sigma = 2, 3$ and 4, from lower to higher curves.

Example 4.5. When dealing with triangular membership functions $\mu_{A\{a,b,c\}}$ (see Appendix C) we derive the optimal value of α by minimizing (4.17). Whatever the value of the parameters (edge, slope and vertex) are, the optimal value of α is equal to $\sqrt{2} - 1 \approx 0.414$. For the symmetric parabolic membership functions $\mu_{A\{a,(a+c)/2,c\}}$ the optimization leads to the value of α approximately equal to 0.391. For the Gaussian membership function $\mu_A = \mu_{A\{0,\sigma,1\}}$ centered at 0, we get the optimal value of α resulting from the relationship (for symmetry reasons the calculations here concern the decreasing part of the membership function defined over $[0, \infty)$):

$$v(\alpha) = 2 \left| \int_{\sigma\kappa_\alpha}^{\infty} \mu_A(x)dx + \int_0^{\sigma\kappa_{1-\alpha}} (1 - \mu_A(x))dx - \int_{\sigma\kappa_{1-\alpha}}^{\sigma\kappa_\alpha} dx \right|, \quad (4.18)$$

with $\kappa_z = \sqrt{-\log z}$. Then the optimal value of α is approximately equal to 0.395 and it does not depend upon the spread σ. The plot of V at Fig. 4.9, reveals that V exhibits a global minimum.

Let us move on to the most general case in which we do not impose any assumptions as to the form of the membership function. Rather, we consider discrete membership values $\{u_1, u_2, \ldots, u_m\}$. Denote the minimal and maximal value in this set by $u_{\min}$ and $u_{\max}$, respectively. The overall reduction of lower membership grades is expressed in the form of the sum $\sum_{k \in \mathfrak{U}_r} u_k$ where $\mathfrak{U}_r = \{k | u_k \leq \alpha\}$. The elevation of higher membership grades to 1 leads to the expression $\sum_{k \in \mathfrak{U}_e} (1 - u_k)$ with $\mathfrak{U}_e = \{k | u_k \geq u_{\max} - \alpha\}$. For the shadows we consider the cardinality of the set $\mathfrak{U}_s = \{k | u_k \in (\alpha, u_{\max} - \alpha)\}$. Then the above conditions translate into the following optimization problem:

$$\min_\alpha v(\alpha) = \min_\alpha \left| \sum_{k \in \mathfrak{U}_r} u_k + \sum_{k \in \mathfrak{U}_e} (1 - u_k) - \#\mathfrak{U}_s \right|, \quad (4.19)$$

where the range of feasible values of α is given as $\left[u_{\min}, \frac{u_{\min}+u_{\max}}{2}\right]$ and $\#\mathfrak{U}$ denotes the cardinality of set $\mathfrak{U}$.

4.2.6 Interval-Valued Fuzzy Sets

When defining or estimating membership functions or membership degrees, one may argue that characterizing membership degrees as single numeric values could be counterproductive and even somewhat counterintuitive given the nature of fuzzy sets themselves. Some remedy could be sought along the line of capturing the semantics of fuzzy sets through intervals of possible membership grades rather than single numeric entities. This gives rise to the concept of so-called *interval-valued* fuzzy sets regarded as a counterpart of the confidence regions introduced for regression functions in Chapter 3. Formally, an interval-valued fuzzy set $A = (A_-, A_+)$ is defined by two mappings from $\mathfrak{X}$ to the unit interval, where A_- and A_+ are the lower and upper bound of membership grades, with $A_-(x) \leq A_+(x)$ for all $x \in \mathfrak{X}$. The bounds are used to capture an effect of a lack of uniqueness of numeric membership: not knowing the detailed numeric values we admit bounds of possible membership grades. Hence the name of the interval-valued fuzzy sets is very much descriptive of the essence of the construct. The broader the range of the membership values, the less specific we are about membership degree of the element to the information granule. An illustration of the interval-valued fuzzy set is included in Fig. 4.10. In particular, when $A_-(x) = A_+(x)$, we end up with a *standard* fuzzy set. *Vice-versa* in the Atanassov [2, 3] construct with $A_+(x)$ we denote a degree of membership of x to A while with $A_-(x)$ a degree of non-membership of x to A, and require:

$$A_+(x) + A_-(x) \leq 1. \qquad (4.20)$$

> Maybe we need more descriptional power of a single membership function,

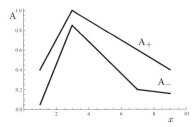

Fig. 4.10. An illustration of an interval-valued fuzzy set; note that the lower and upper bound of possible membership grades could differ quite substantially across the universe of discourse.

With this relationship we want to underline that there exists a *gap* between a strength of membership (which is a sort of some *positive* assessment) and a strength of non-membership (which could be viewed as a type of *negative* assessment).

Finally, it is worth to note that with shadow set the width of the membership values is either 0 or 1. The additional information in respect to sharp set is represented by the extension of the region in the universe of discourse where the interval width is 1.

> maybe we need a more articulated description.

4.3 Fuzzy Sets of Higher Order

Let us recall that a fuzzy set is defined in a certain universe of discourse $\mathfrak{X}$ so that for each element of it we come up with the corresponding membership degree which is interpreted as a degree of compatibility of this specific element with the concept conveyed by the fuzzy set under discussion. We may extend the compatibility measure to *metaelements* constituted by (sharp or fuzzy) sets on own turn. In this way we obtain different variants of higher order fuzzy sets, like in the following.

4.3.1 Second Order Fuzzy Sets

Let us consider a concept of a comfortable temperature which we define over a finite collection of some generic fuzzy sets A_1, $A_2, \ldots, A_c$, say *around* 10°C, *warm, hot, cold, around* 20°C, etc. We could easily come to a quick conclusion that the term comfortable sounds more *descriptive* and hence semantically more advanced in comparison to the generic terms using which we describe it. A 2-nd order fuzzy set is provided in Fig. 4.11. We can write down the membership of comfortable temperature in the vector form as $(0.7, 0.1, 0.9, 0.8, 0.3)$. It is understood that the corresponding entries of this vector pertain to the generic fuzzy sets we started with when forming the fuzzy set. To make

> We may have a weighted mixture of fuzzy sets.

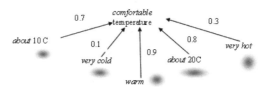

Fig. 4.11. An example of 2-nd order fuzzy set of comfortable temperature defined over a collection of generic fuzzy sets (graphically displayed as small clouds); shown are also corresponding membership degrees.

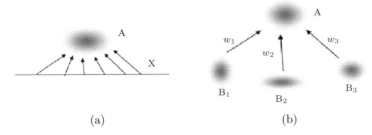

Fig. 4.12. Contrasting fuzzy sets of order 1 (a) and order 2 (b). Note the role of reference fuzzy sets played in the development of order 2 fuzzy sets.

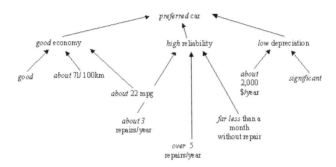

Fig. 4.13. Fuzzy set of order 2 for the concept of *preferred* car; note a number of descriptors quantified in terms of fuzzy sets and contributing directly to its formation.

a clear distinction, fuzzy sets studied so far could be referred to as fuzzy sets of the first order. Fig. 4.12 graphically emphasizes the difference between fuzzy sets of first and second orders. For the latter we can use the notation $B = (w_1, w_2, w_3)$ given that the reference fuzzy sets are A_1, A_2, and A_3. Fuzzy sets of order 2 could be also developed on a Cartesian product of some families of generic fuzzy sets. Consider, for instance, a concept of a preferred car. To everybody this term could mean something else yet all of us agree that the concept itself is quite complex and definitely multifaceted. We easily include several aspects such as economy, reliability, depreciation, acceleration, and others. For each of these aspects we might have a finite family of fuzzy sets, say when talking about economy, we may use descriptors such as *about* $10l/100km$ (or expressed in mpg), *high* fuel consumption, *about* 30 mpg, etc. For the given families of generic fuzzy sets in the vocabulary of generic descriptors we combine them in a hierarchical manner as illustrated in Fig. 4.13. In a similar

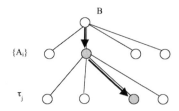

Fig. 4.14. The process of linguistic approximation of B in terms of the elements of the vocabulary and the collection of the linguistic modifiers.

way, we can propose fuzzy sets of higher order, say 3-rd order or higher. They are formed in a developed manner. While conceptually appealing and straightforward, their applicability could become an open issue. One may not venture in allocating more effort into their design unless there is a legitimate reason behind the further usage of fuzzy sets of higher order. Nothing prevents us from building fuzzy sets of 2-nd order on a family of generic terms that are not only fuzzy sets. One might consider a family of information granules such as sets over which a certain fuzzy set is being formed.

4.3.2 Linguistic Approximation

We may modify fuzzy sets through another fuzzy set.

An additional feature of the mentioned vocabulary $A_1, A_2, \ldots, A_c$ may be represented by a finite collection of so-called linguistic modifiers (hedges) $\tau_1, \tau_2, \ldots, \tau_p$. Consider the general category made up of two modifiers realizing operations of concentration and dilution. Their semantics relates to the linguistic adjectives of the form *very* (concentration) and *more or less* (dilution). Given the semantics of A_is and the available linguistic modifiers, the objective of the representation scheme is to capture the essence of a new fuzzy set B. Given the nature of the objects and the ensuing processing being used here we refer to this process as a linguistic approximation. There are several scenarios of the realization of the linguistic approximation. The scheme shown in Fig. 4.14 comprises of two phases: first we find the best match between B and A_is (where the quality of matching is expressed in terms of some distance or similarity measure). At the next step we refine the construct by applying one of the linguistic modifiers. The result of the linguistic approximation comes in the form B $\sim \tau_i(A_j)$ with the indexes i and j determined through the optimization of the matching mechanism.

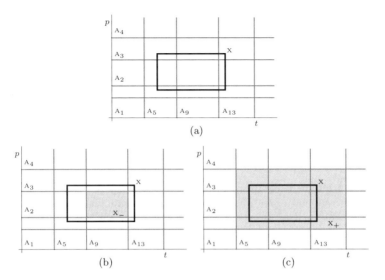

Fig. 4.15. A collection of generic information granules forming the vocabulary and their use in the problem description. Environmental conditions X result in some interval of possible values (a). In the sequel, this gives rise to the concept of a rough set with the roughness of the description being captured by the lower and upper bounds (approximations) as illustrated in (b-c).

4.3.3 Rough Fuzzy Sets and Fuzzy Rough Sets

Let us map the mentioned Cartesian product of features in a Cartesian grid so that the description approximation of a set may lead to lower and upper bounds in the grid. Hence we speak of *rough sets*.

Consider, for instance, a description of environmental conditions expressed in terms of temperature and pressure. For each of these factors, we fix several ranges of possible values where each of such ranges comes with some interpretation such as *values below, values in-between, values above*, etc. By admitting such selected ranges in both variables, we construct a grid of concepts formed in the Cartesian product of the spaces of temperature and pressure. Being more descriptive, this grid forms a vocabulary of generic terms (information granules) using which we would like to describe all new information granules. As illustrated in Fig. 4.15, there is a finite family of those, say $A_1, A_2, \ldots A_{12}$.

Now let us consider that the environmental conditions monitored over some time have resulted in some values of temperature and pressure ranging in-between some lower and upper bound as illustrated in Fig. 4.15. Denote this result as X. It becomes obvious that when describing it in terms of the information granules

> Shadow may be at level of linguistic terms that we call granules.

> The border of the set falls in the gap between

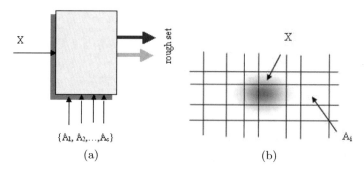

Fig. 4.16. (a) Rough set as a result of describing X in terms of some fixed vocabulary $\{A_1, A_2, \ldots, A_c\}$; the lower and upper bounds are results of the description; (b) the development of the fuzzy rough set.

of the vocabulary, we end up with a collection of elements that are fully included in X. They form a lower bound of description of X when being completed in presence of the given vocabulary. Likewise, we may identify elements of the vocabulary that have a nonempty overlap with X and in this sense constitute an upper bound of the description of the given environmental conditions. Along with the vocabulary, the description forms a certain rough set. More formally, we describe an upper bound by enumerating elements of A_is that have a nonzero overlap with X, that is:

<small>what possibly belongs to the set, and</small>

$$X_+ = \{A_i | A_i \cap X \neq \emptyset\}. \quad (4.21)$$

More specifically, in Fig. 4.15 we have $X_+ = \{A_6, A_7, A_8, A_{10}, A_{11}, A_{12}, A_{14}, A_{15}, A_{16}\}$. The lower bound of X involves all A_i such that they are fully included within X, namely:

<small>what surely belongs to the set.</small>

$$X_- = \{A_i | A_i \subset X\}. \quad (4.22)$$

Here $X_- = \{A_{11}\}$. The lower and upper boundaries (approximation) are reflective of the resulting imprecision caused by the conceptual incompatibilities between the concept itself and the existing vocabulary, see Fig. 4.16(a).

It is interesting to note that the vocabulary used in the above construct could comprise information granules being expressed in terms of any other formalism, say fuzzy sets. Quite often we can encounter constructs like rough fuzzy sets and fuzzy rough sets in which both fuzzy sets and rough sets are put together. These constructs rely on the interaction between their constituents. Let us consider a finite collection of sets A_i and use them to describe some fuzzy set X. In this scheme, we arrive at the concept of a certain fuzzy rough set (refer to Fig. 4.16(b)). The upper bound of this fuzzy rough set is computed as in the previous case (4.21)

<small>In some cases, the granule shadow may be detailed through a membership function.</small>

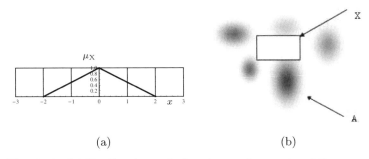

Fig. 4.17. (a) Family of generic descriptors, fuzzy set and its representation in the form of some rough set; (b) the concept of a rough fuzzy set.

yet given the membership function of X the detailed calculations return membership degrees rather than $\{0,1\}$ values. Given the binary character of A_is the above expression for the upper bound reads in the form:

$$X_+(A_i) = \sup_x [\min(A_i(x), \mu_X(x))] = \sup_{x \in \mathrm{supp}(A_i)} \mu_X(x). \quad (4.23)$$

The lower bound of the resulting fuzzy rough set is taken in the form

$$X_-(A_i) = \inf_x [\max(1 - \mu_X(x), A_i(x))]. \quad (4.24)$$

Example 4.6. Let us consider a universe of discourse $\mathfrak{X} = [-3, 3]$ and a collection of intervals regarded as basic descriptors, see Fig. 4.17(a). The fuzzy set A with a triangular membership function distributed between -2 and 2 gives rise to some rough set with the lower and upper approximation of the form $X_+ = [0, 0.5, 1, 1, 0.5, 0]$, $X_- = [0, 0, 0, 0, 0, 0]$. We can also consider another combination of information granules in which $\{A_i\}$ is a family of fuzzy sets and X is a set, see Fig. 4.17(b). This leads us to the concept of rough fuzzy sets.

Alternatively, we can envision a situation in which both $\{A_i\}$ and X are fuzzy sets. The result comes with the lower and upper bound whose computing follows the formulas presented above.

4.4 Type-2 Fuzzy Sets

A different way of merging the fuzzy sets in the vocabulary is through a *type-2 fuzzy set* representing an appealing generalization of interval-valued fuzzy sets. Instead of intervals of numeric values of membership degrees, we allow for the characterization of membership by fuzzy sets themselves. Consider a certain element of the universe of discourse, say x. The membership of x to A is captured by a certain fuzzy set formed over the unit interval. This construct generalizes the fundamental idea of a fuzzy set and helps us relieve from the restriction of having single numeric values describing a given fuzzy set [16, 11].

<small>**Fuzzy sets having fuzzy sets as elements of the universe of discourse.**</small>

With regard to these forms of generalizations of fuzzy sets, there are two important facets that should be taken into consideration. First, there should be a clear motivation and a straightforward need to develop and use them. Second, it is imperative that there is sound membership determination procedure in place using which we can construct the pertinent fuzzy set. To elaborate on these two issues, let us discuss a situation in which we deal with several databases populated by data coming from different regions of the same country. Using them we build a fuzzy set describing a concept of high income where the descriptor *high* is modeled as a certain fuzzy set. Being induced by some locally available data, the concept could exhibit some level of variability with the regions, yet we may anticipate that all membership functions might be quite similar as being reflective of some general commonalities, see Fig. 4.18(a). In any case, like with the total probability theorem (1.12), we need to merge the different concepts, yet lacking tools as conditioning event probabilities that we use in the probabilistic framework. While a very structured solution to this problem will be given in Part V, we come up here with some aggregation of the individual fuzzy sets, see Fig. 4.18(b). In the case that the individual estimated membership functions are triangular or similar, with reference to Appendix C, to simplify notation let the corresponding estimated trapezoidal membership functions $\mu_{A_{\{a_i,b_{1i},b_{2i},c_i\}}}(x)$ be denoted by $\mu_{A_i}(x)$, where i ranges over a number c of databases. A first aggregation alternative leads to the emergence of an interval-valued fuzzy set A. Its membership function assumes interval values where for each x the interval of possible values of the membership grades is given in the form $[\min_i\{\mu_{A_i}(x)\}, \max_i\{\mu_{A_i}(x)\}]$. It is worth noting that both the upper and lower bounds associated with the intervals formed in this way do not form any longer triangular or trapezoidal membership functions.

<small>**We may come to interval fuzzy sets,**</small>

Example 4.7. Let us consider four triangular fuzzy sets $A_{\{2,5,9\}}$, $A_{\{1,7,11\}}$, $A_{\{1.5,8,10\}}$, and $A_{\{0,7,10\}}$. The resulting membership function of the interval-valued fuzzy set A is illustrated in Fig. 4.19. One should note that A is not necessarily either a triangular or a trapezoidal fuzzy set as we may encounter a substantial level of diversity among fuzzy sets.

Type-2 Fuzzy Sets   151

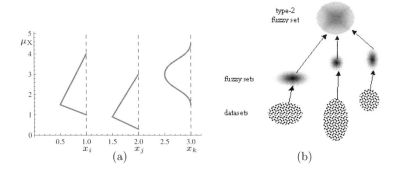

Fig. 4.18. (a) An illustration of type-2 fuzzy set; for each element of $\mathfrak{X}$ there is a corresponding fuzzy set of membership grades; (b) a scheme of aggregation of fuzzy sets induced by 3 datasets.

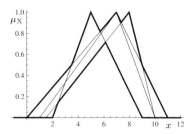

Fig. 4.19. Interval-valued fuzzy set resulting from the aggregation of triangular fuzzy sets.

The form of the interval-valued fuzzy set may be advantageous in further computing yet the estimation process could be very conservative leading to very broad ranges of membership grades (which is particularly visible when dealing with different data and fuzzy sets induced on their basis). Being aware of the drawbacks of the conservative way in which the membership function of the interval-valued fuzzy set has been estimated, we may refine the estimation process and instead rely on the minimal and maximal membership grades; then take advantage of the statistical characteristics of the collection of the membership grades. This implies that for each x we collect the membership grades and apply to them the estimation procedure as described in Section 3.1.3. Subsequently, this leads to the triangular or trapezoidal membership functions defined in $[0, 1]$. In essence, in this way we have constructed a certain type-2 fuzzy set.

or to more complex solutions.

4.5 Conclusions

We have discussed various approaches and elaborated on algorithmic aspects of the design of fuzzy sets. Following this discussion, we outline the following general guidelines supporting the development of fuzzy sets.

(a) Highly visible and well-defined semantics of information granules. No matter what the detailed determination technique is, one has to become cognizant of the semantics of the resulting fuzzy sets. Fuzzy sets are interpretable information granules of a well-defined meaning and this aspect needs to be fully captured. Given this, the number of information granules has to be maintained quite small with their number being restricted to 7 ± 2 fuzzy sets.

(b) Fuzzy sets are context-sensitive constructs and as such require careful calibration. This feature of fuzzy sets should be treated as their genuine advantage. The semantics of fuzzy sets can be adjusted through shifting fuzzy sets or/and adjusting their membership functions. The nonlinear transformation we introduced here helps complete an effective adjustment of the membership functions making use of some "standard" membership functions. The calibration mechanisms being used in the design of the membership function are reflective of human-centricity of fuzzy sets.

(c) The development of fuzzy sets can be carried out in a stepwise manner. For instance, a certain fuzzy set can be further refined, if required in the problem at hand. This could lead to several more specific fuzzy sets that are associated with the fuzzy set formed at the higher level. Being aware of the complexity of the granular descriptors, we should resist temptation of forming an excessive number of fuzzy sets at a single level as such fuzzy sets could be easily lacking any sound interpretation.

4.6 Exercises

1. Describe through fuzzy sets the fuel consumption of a car. Decide the number of sets and their membership function. Then locate inside them your car.
2. We maximize a function $f(x) = (x-6)^4$ in the range $[3, 10]$. Suggest a membership function describing a degree of membership of the optimal solution which maximizes $f(x)$. What conclusion could you derive based on the obtained form of the membership function?

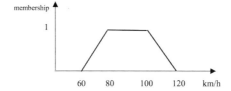

Fig. 4.20. A fuzzy set of a safe speed on an average highway.

3. Looking at 20 elements representing the digit 4 in then MNIST database, formulate a membership function to the set *number 4* of a generic digit as a function of specific features (such as number and slope of horizontal traits, number of non-white pixels, etc).
4. Starting from the triangular fuzzy sets on $\mathfrak{X}$: $A_{low} = \{5, 7, 9\}$, $A_{medium} = \{8, 12, 15\}$, $A_{high} = \{10, 20, 22\}$, decide a piecewise linear transform $\phi(x)$ such that: $\mu_1(10) = 0.2, \mu_2(10) = 0.6, \mu_3(10) = 0.3$.
5. Transform the fuzzy set A having Gaussian membership function with $\mu = 7$ and $\sigma = 2$ into a shadow set such that $v(\alpha) = 0.2$ in (4.16).
6. Repeat Exercise 5 when the fuzzy set A has triangular membership function with $a = 2, b = 6, c = 9$.
7. Design an interval-valued fuzzy set describing a comfortable temperature in a meeting room such that the membership interval never exceeds in percentage 10% of its minimum value.
8. Draw in a square paper the contour of Cyprus island. Then provide an upper-bound and a lower-bound of the island by listing squares in the paper.
9. Consider a fuzzy set of a *safe* speed maintained on an average highway (see Fig. 4.20). How could this membership be affected when re-defining this concept in the following settings of: (a) autobahn (note that on these German highways there is no speed limit), and (b) a snowy country road. Elaborate on the impact of various weather conditions on the corresponding membership function. From the standpoint of the elicitation of the membership function, how could you transform the original membership function to address the needs of the specific context in which it is planned to be used?
10. Use a type-2 fuzzy set to describe a wonderful sea location having weather, seascape and accommodation as evaluation criteria.

5 Estimating Fuzzy Sets

Willing to use fuzzy set models described in the previous chapter, we face with the similar problem typically encountered with the probabilistic framework of adapting models to the operational problem at hand. This passes through the identification of the model free parameters from a set of experimental data, a task that we call estimation by analogy with statistical inference. *Vice versa*, the difference between fuzzy estimation and statistical inference stands in the weight we give the data: uniform with statistics, *ad hoc* with fuzzy methods. This reflects either in the shape of the membership function, or in the function mapping data into parameters, or both. As for the former, the methods for shaping the functions are not requested to mimic through these functions histograms of huge sample as for probability models. Thus no any constraint is given to the area subtended by a membership function, rather its identification is mainly demanded to the criteria discussed in the previous chapter. As for the data weight, we are in any case drawn to affect them with equal weights in absence of counter-indications. The fact is that these weights are shared by local subgroups in place of the whole dataset. The main scheme is the following: on the one hand you (either implicitly or explicitly) decide a set of points around which to pivot the membership functions of a given fuzzy set. We refer to them as pivotal points as being a counterpart of the sample points in statistical frameworks. On the other hand, on each point you collect one or more membership degrees to the set. Finally you extract parameters of the membership function by mixing with equal weights specific subsets of pairs (point, degree). We pass from subgroups of a single element whose membership to a given fuzzy set is expressly asked to an expert, to same subgroups on which opinions are gathered by many experts, to local subsets of a dataset on which to compute statistics.

A specific weight to each data item

Pivotal point in place of samples

Statistics in background

In this chapter, we will expound some keen methods having the goal of fixing the membership function of either a single fuzzy set or jointly a series of fuzzy sets concerning in any case a single observation variable. Estimation of joint variables' membership and their functional relationships will be discussed in Part V under the banner of fuzzy set *learning*.

5.1 Vertical and Horizontal Schemes of Membership Estimation

The vertical and horizontal modes of membership estimation are two standard approaches used in the determination of fuzzy sets. They reflect distinct ways of looking at fuzzy sets whose membership functions at some finite number of points are quantified by experts. In the former the points are explicitly decided by the user, in the latter they represent the solution of the inverse problem of finding points with a given membership degree, for a set of degrees decided by the user.

- In the horizontal approach we identify a collection of elements in the universe of discourse $\mathfrak{X}$ and request that an expert answers the question:

    does x belong to set A?

    The answers are expected to come in a binary ("yes"-"no") format. The set A defined in $\mathfrak{X}$ could be any linguistic notion, say *high* speed, *low* temperature, etc. Given m experts whose answers $\{x_i\}$ for a given point of $\mathfrak{X}$ form a mix of "yes"-"no" replies, we count the number of "yes" answers and compute the ratio of the positive answers ($x_i = 1$) versus the total number of replies m, that is $\sum_{i=1}^{m} x_i/m$. This ratio is treated as a membership degree of the set at the given point of the universe of discourse. When all experts accept that the element belongs to the set, then its membership degree is equal to 1. Higher disagreement between the experts (quite divided opinions) results in lower membership degrees. The set A defined in $\mathfrak{X}$ requires collecting results for some other elements of $\mathfrak{X}$ and determining the corresponding ratios as outlined in Fig. 5.1(a). Note that the elements of $\mathfrak{X}$ need neither to be evenly distributed w.r.t. a given set A, nor to sum to 1 their membership to a series of fuzzy sets A_i. As the replies on each questioned x follow a binomial distribution of parameters p_x and m, then we could come up with a confidence interval of the individual membership grade $\mu_A(x)$, using the estimation methods discussed in Section 3.1.3. In particular, in the approximation of the variable $\overline{X} = \sum_{i=1}^{m} X_i$ with a Gaussian variable of mean p_x and standard deviation $\sigma_x = \sqrt{\frac{p_x(1-p_x)}{m}}$, and p_x with $\overline{x}$, we may associate to $\mu_A(x)$ the δ confidence interval:

$$\left(p_x - z_{1-\delta/2}\sigma_x, p_x + z_{1-\delta/2}\sigma_x\right), \qquad (5.1)$$

where $z_{1-\delta/2}$ is the $1 - \delta/2$ quantile of the Normal distribution for a suitable δ. In essence, when the confidence intervals are taken into consideration, the membership estimates become intervals of possible membership values and this leads to the set of so-called interval-valued fuzzy sets seen in Section 4.2.6. We may conventionally set $z_{1-\delta/2}$ to 1. By assessing the width of the estimates, we could control the execution of the experiment: when the ranges are too long, one could re-design the experiment and monitor closely the consistency of the responses collected in the experiment.

> You locally estimate the membership function, possibly with high accuracy.

Example 5.1. Let us consider responses of 10 experts who came up with the following assessment of the concept $A =$ (*high* interest rate percentage) with the number of "yes" responses collected as follows:

$x(\%)$	2	3	5	8	10
no. of "yes" replies	0	2	4	7	10

Following these responses, the membership function and its confidence values σ producing confidence intervals are given below:

$x(\%)$	2	3	5	8	10
$\mu_A(x)$	0.0	0.2	0.4	0.7	1.0
σ_x	0.0	0.126	0.155	0.144	0.0

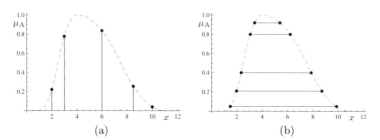

Fig. 5.1. (a) A horizontal method for the estimation of the membership function; observe a series of estimates determined for selected elements of $\mathfrak{X}$; (b) a vertical approach of membership estimation through the reconstruction of a fuzzy set through its estimated α-cuts. Dotted lines show a possible smooth fit of the membership function values.

The advantage of the method comes with its simplicity as the technique relies explicitly upon a direct counting of responses. The method is convincing. The concept is also intuitively appealing. The probabilistic nature of the replies helps build confidence intervals that are essential to the assessment of the specificity of the membership quantification. A certain drawback is related with the local character of the construct: as the estimates of the membership function are completed separately for each element of the universe of discourse, they could exhibit a lack of continuity when moving from a certain point to its neighbor. This concern is particularly valid in the case when $\mathfrak{X}$ is a subset of real numbers.

Just a confidence interval for a Bernoulli variable parameter

- The vertical mode of membership estimation is concerned with the estimation of the membership function by focusing on the determination of the successive α-cuts. The experiment focuses on the unit interval of membership grades. The experts involved in the experiment are asked the questions of the form:

The membership levels are fixed by you,

what are the elements of $\mathfrak{X}$ which belong to fuzzy set A at degree not lower than α?

the extremes of corresponding α-cuts by the experts,

where α is a certain level (threshold) of membership grades in $[0, 1]$. Formally:

Definition 5.1. *Given a fuzzy set* A *in* $\mathfrak{X}$, *the* α-*cut of* A, *denoted as* A_α, *is defined as:*

$$A_\alpha = \{x \in \mathfrak{X}, \mu_A(x) \geq \alpha\}. \qquad (5.2)$$

Moreover, A_0 *and* A_1 *are called the* support *and* core *of* A *respectively.*

The essence of the method is illustrated in Fig. 5.1(b). Note that the satisfaction of the inclusion constraint is obvious: we envision that for higher values of α, it is very likely that the expert is going to identify more confined subsets of $\mathfrak{X}$; the vertical approach leads to the fuzzy set by combining the estimates of the corresponding α-cuts.

Given the nature of this method, we are referring to the collection of random sets as these estimates appear in the successive stages of the estimation process. The elements are identified by the expert as they form the corresponding α-cuts of A. By repeating the process for several selected values of α we end up with the α-cuts and using them we reconstruct the fuzzy set. The simplicity of the method is its genuine advantage. The request to the expert, however, is more complex and demanding than in the previous method. Indeed, formulating his hypothesis on the i-th α-cut, the expert must necessarily take into account, at least in terms of inclusion relationship, previous $(i-1)$-th cuts. Like in the horizontal method of membership estimation, a possible lack of continuity is a certain disadvantage one has to be aware of. Here the selection of suitable levels of α needs to be carefully investigated. Similarly, an order at which different levels of α are used in the experiment could impact the estimate of the membership function.

with a monotonicity constraint.

Both the horizontal and vertical methods of membership estimation are straightforward as the sound membership estimation technique. They are simple and easy to arrange, however they are not free from limitations. The most essential comes with a lack of continuity of the method. Each element x_i of the universe of discourse $\mathfrak{X}$ is investigated in isolation and when assigning to it some membership grade this allocation is independent from the assignment being completed for other elements of $\mathfrak{X}$. The estimation process suffers from the "tunneling" effect as no algorithmic provisions are being made to alleviate this negative effect. The expert could position himself in a way allowing him to record inconsistent estimates. It is also worth stressing that the lack of consistency is not easy to spot and eliminate, especially when $\mathfrak{X}$ consists of a large number of elements. Some consistency checking could be instrumented when the universe of discourse is regarded as a subset of real numbers, $\mathfrak{X} \subseteq \mathbb{R}$, as in this case we can establish a linear order of x_is in which the membership values are determined. A simple mechanism of consistency checking can be introduced in the following manner:

The experts must feel free but consistent;

(a) for the horizontal approach, we can monitor the monotonicity in the successive membership values: we anticipate that there should not be any substantial jumps between two neighboring elements of $\mathfrak{X}$, otherwise any jumps of this nature could be reflective of possible inconsistency in the estimation procedure;

hence membership functions must be smooth

(b) for the vertical approach the consistency criterion is more explicit. We require that for increasing sequence of α levels, the corresponding α-cuts meet the monotonicity requirement $\alpha_1 < \alpha_2$ implies $A_{\alpha_1} \supseteq A_{\alpha_2}$. If this condition is violated then the resulting estimates do not form any fuzzy set and the estimation process has to be refined.

and reflecting degree monotonicity versus α-cuts inclusion.

The above mechanisms are generally overridden by the overall shape, say linear, parabolic, Gaussian and alike, of membership function whose parameters we are estimating (see Appendix C). In their respect the above estimates enter as reference points of some optimization criterion, such as the MSE between the

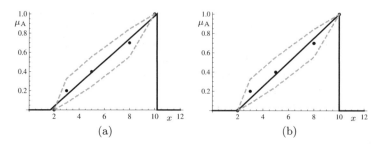

Fig. 5.2. Triangular membership function with vertical right edge minimizing MSE with the observed points (a) and with the two extreme points (b). Gray dashed lines provide interval-valued fuzzy sets.

estimated degrees and those computed by the overall membership function in the pivot points.

Example 5.2. Continuing Example 5.1, if we assume a triangular membership function for A with a vertical right edge, then we may interpolate the points $(x, \mu_A(x))$ as in Fig. 5.2(a). A simpler strategy could consist in locating only two points of the slope, provided we are extremely confident on the verdict of the experts in these points (see Fig. 5.2(b)).

Example 5.3. The interval-valued fuzzy set represented by gray dashed lines in Fig. 5.2 has been computed from the membership function and its confidence values σ, as described in Example 5.1.

5.2 Maximum Likelihood Estimate

Pivot points are representative of a surrounding. As discussed in Section 4.3 we may affect the pivot points directly with a local membership function denoting how a point we decided as a pivot shares features with its surrounding as a representative of the fuzzy set A under question. In this way we are implicitly asking the expert to declare both the representativeness of the point w.r.t. A and its exclusiveness in terms of how surrounding points would play the same role. Starting from these information granules, one around each pivot point, the estimation problem results on how to integrate them within a metagranule represented by A. We solve it requiring that pivot points reckon membership degrees to both specific granule and A, denoting a transfer of the degree decreed by the expert from local to global membership functions. As they are granule representatives, we move memberships from points to granules. The joint membership of the latter plays the analogous role of sample points' likelihood in a statistical framework, introducing similar optimization tasks to identify parameters. To fully extend the solutions found in Section 3.1.3 we refer to the fuzzy metric definition in Appendix A.3 focusing (for the sake of exemplification) to some T-norm and co-norms reported therein.

We introduce fuzzy sets since we have no sharp ordering relationships in a given space $\mathfrak{X}$. Hence we consider it a universe of discourse and equip subsets of it with weight systems that we call membership functions. This allows to state comparison between fuzzy sets through the simple relationships in Definition 4.1.

But we cannot state any sharper relations because we cannot define on its elements true norm operations endowed with the identity of indiscernible items (see N4 in Definition A.2). Rather we have surrogates in items T4-T5 or S4-S5 in Definition A.4, depending on whether for merging A and B weights in (1.12) privilege the biggest or the smallest of the two membership functions. Weight systems that are based on the sole values of μ_A and μ_B and privilege the smallest are called T-norms. *Vice versa*, if they privilege the biggest are called T-co-norms or S-norms. The two families of norms are associated to the logical operations of intersection and union we define on sharp set, as a shift from crisp verdicts. Namely, for degenerate *fuzzy sets* A and B having membership function equal to 1 along their supports and 0 elsewhere, any T-norm gives rise to set intersection and any S-norm to set union. Thus, norms are a way of dealing with analogous operations with fuzzy sets.

Even though in a vague way, you must specify if you are in favor of the weakest or strongest membership degree.

In Fig. 1.10(b) we have synthesized the granules observed as in Fig. 1.10(a) through the Gödel S-norm $\perp_{\max}$.

The underlying idea was to describe the membership of any point x to the union of the granules, hence to an S-norm. But which S-norm, for instance within the set of those in Appendix A.3 is most suitable? As usual with fuzzy sets we have no a univocal answer to this question. However we may take some general directions from limit reasoning. Thus, consider what happens with the number of observed granules growing to infinity. The Gödel S-norm $\perp_{\max}$ has the property of being the minimal S-norm in the sense that for every (a, b) and S-norm selection (denote it $\perp_{S-\text{norm}}$), $\perp_{\max}(a, b) \leq \perp_{S-\text{norm}}(a, b)$. In particular, it is the sole idempotent S-norm, in the sense that $\perp_{\max}(a, a) = a$. With these properties it is impossible to have a membership function different from the constant 1 when grouping the non numerable set of granules centered in each element of the fuzzy set A. Namely, in case of a continuous universe of discourse and continuous granule membership functions, the A membership function on each point x

A fuzzy set synthesizes the granularity of its elements

with a membership function that does not explode.

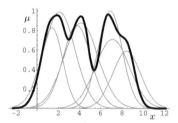

Fig. 5.3. Same graph as in Fig. 1.10 but with an extended product norm.

comes from the extension of the S-norm to non numerable infinite many points. If $\perp_{\text{S-norm}}(a,b) > \perp_{\max}(a,b)$ by a quantity of the same order of a and b, then the sole solution, apart from points where both a and b are disregardable, is $\mu_A(x) = 1$. We may remedy this drawback with a displacement utility allowing to stop the ever increasing process raised by the above S-norm extension. The object is to maintain $\perp_{\text{S-norm}}(a', b'') \leq a$ where a, a', b'' refer to membership functions of A and B evaluated on point x, x', x'', respectively. In this way, even an infinite composition of this operation does not imply the membership function saturation. For instance, the membership function in Fig. 5.3 is obtained through an extended product norm and displacing points twice their distance from the granule centers.

The shift trick

5.2.1 In Search of the Maximum Compliance between Pivotal Granules and a Model

With a few granules $G_1, \ldots, G_m$ you may try to infer the fuzzy set membership function in a very specular way we do with distribution laws, but being centered on local properties. Here we extend to the use of any norm the procedure discussed in Section 1.2 with the Gödel norm.

If you have a bias on the membership function you may infer its parameters.

According to the above definition of a fuzzy set A as a function of its granules, we assumed the A membership function to be an S-norm of granule membership functions. Moreover, as mentioned before, with expanding a sample point into a granule we are forming an equivalence relation between the points within a granule. Thus we are looking for the representative point in each granule, i.e. that point that did really contribute with its membership function, to fix the values of A membership function. In case of S-norm $\perp_{\max}$ coinciding with the maximum of its arguments, this process is straightforward: you find such a point in the cross x of granule and fuzzy set membership function, given the idempotence of this operator: resume the S-norm of all

Which is the granule's virtual point that determines the membership function to the set?

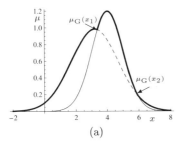 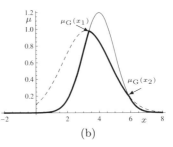

Fig. 5.4. Membership functions of a granule G (plain curve) and a fuzzy set A (dashed curve). (a) Membership function of the union between A and B computed through the Gödel norm; (b) membership function of the intersection with the same norm. The fixed points x_1 and x_2 are candidates to contributing to A membership function.

granules exactly in $\mu_A(x)$, then this value is a fixed point w.r.t. the S-norm of $\mu_A(x)$ and $\mu_G(x)$ (see Fig. 5.4(a)). If you have more crosses, hence more candidate points and related membership degrees you select their S-norm again, by definition, in order to compute the membership of the entire granule to the fuzzy set, hence the point with maximum membership function. Note that in this case also the complementary T-norm is idempotent and coincides with the S-norm when the arguments are equal (see Fig. 5.4(b)). It says that $\mu_A(x)$ is also the joint membership function to both G and A. In case of other norms we have the problem of shifting the points contributing to generate $\mu_A(x)$. Namely, we focus on the mentioned joint membership function, and assume now the relaxed fixed point relationship:

The one constituting a fixed point w.r.t. the membership function

or a quasi fixed point.

$$\mu_A(x) = \perp_{\text{S-norm}}(\mu_A(x'), \mu_G(x'')), \quad (5.3)$$

where $\mu_A(x')$ represents in turn the S-norm contribution of all granules, G included, that determines $\mu_A(x)$. We maintain the compliance focus between granule and fuzzy set in the intersection(s) $(x, \mu_A(x) = \mu_G(x))$ of the two membership functions, while its quantitative appreciation is based on displaced x' and x''. Of course we don't know x', since we don't know the infinitely many other granules. So we are free to select it in such a way that (5.3) holds. We may state on our convenience the reciprocal positions of x' and x'', with a general requirement of inducing the shortest drift from x. Like before the determined pivot points x are affected by a membership function given by the T-norm of $\mu_A(x')$ and $\mu_G(x'')$ to denote the double membership, and we apply an S-norm to the found T-norms to identify the membership of the granule G to A.

A sequence of S-norm, T-norm, S-norm, and T-norm to determine the fuzzy likelihood of a candidate membership function.

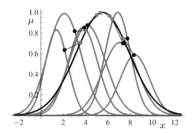

Fig. 5.5. Fuzzy set compatible with the observed granules according to product norm. Black and gray disks denote the shifted coordinates and their degrees concurring to determine the membership function. Thin lines join corresponding couples of points.

Membership as a compatibility measure

With the dual perspective at the overall property describing the data, you read the membership of the granule G to the set A in own turn as a compatibility measure of A with G that quantifies the compliance. To produce compatibility of A with the entire set of granules, or *vice versa* the likelihood of the latter, we take the T-norm of the membership functions of the information granules. If μ_A has free parameters we use them to maximize this compatibility, in a way that is the companion of maximum likelihood in the statistical framework.

Example 5.4. Let us consider the set family $\mathscr{F}_A$ comprising Gaussian shape functions defined in Example 1.5, where each bell is characterized by the triple (h, ν, σ) denoting respectively the height, central position and width (given in terms of standard deviation) of the bell. To this family they belong both the membership functions of granules, each characterized by a triple (h_i, ν_i, σ_i) and the one of the fuzzy set synthesizing the former, whose (h_A, ν_A, σ_A) is the object of our inference. In respect to the problem solved in Example 1.5 here we adopt the *product* norm defined in Appendix A.3. Hence, to solve the compatibility optimization problem rooted in (5.3), with the i-th granule we obtain x' and x'' as shifts from x making these points farther than x respectively to ν_A and ν_i – as done with Fig. 5.3 in the previous section. This entails $x' = \nu_A + \rho(x - \nu_A)$ and $x'' = \nu_i + \rho(x - \nu_i)$ so that $\mu_A(x') + \mu_{G_i}(x'') - \mu_A(x')\mu_{G_i}(x'') = \mu_A(x) = \mu_{G_i}(x)$. In this way we obtain the picture in Fig. 5.5, where displaced points of both granule and fuzzy set concur to determine the most compatible membership function as explained before. Unlike Fig. 5.4 we explicitly consider both intersections of μ_A with μ_{G_i}, hence four displaced points, producing a single granule compatibility value. The joint granules' compatibility optimization has been carried out by a locally incremental method in the $\{h_A, \nu_A, \sigma_A\}$ search space.

Note that the procedure is exactly the same if the membership function shape is a function of an external variable as well, as it occurs in regression problems.

5.3 Saaty's Priority Method of Pairwise Membership Function Estimation

The priority method introduced by Saaty forms another interesting and technically sound alternative used to estimate the membership function. To explain the essence of the method, let us consider a collection of elements $x_1, x_2, \ldots, x_m$ (those could be, for instance, some alternatives whose allocation to a certain fuzzy set is sought) for which given are membership grades $\mu_A(x_1), \mu_A(x_2), \ldots, \mu_A(x_m)$. Let us organize them into a so-called reciprocal matrix of the following form:

Better comparing two sole elements at time

$$R = [r_{ij}] = \begin{bmatrix} \frac{\mu_A(x_1)}{\mu_A(x_1)} & \frac{\mu_A(x_1)}{\mu_A(x_2)} & \cdots & \frac{\mu_A(x_1)}{\mu_A(x_m)} \\ \frac{\mu_A(x_2)}{\mu_A(x_1)} & \frac{\mu_A(x_2)}{\mu_A(x_2)} & \cdots & \frac{\mu_A(x_2)}{\mu_A(x_m)} \\ \vdots & \vdots & & \vdots \\ \frac{\mu_A(x_m)}{\mu_A(x_1)} & \frac{\mu_A(x_m)}{\mu_A(x_2)} & \cdots & \frac{\mu_A(x_m)}{\mu_A(x_m)} \end{bmatrix} = \begin{bmatrix} 1 & \frac{\mu_A(x_1)}{\mu_A(x_2)} & \cdots & \frac{\mu_A(x_1)}{\mu_A(x_m)} \\ \frac{\mu_A(x_2)}{\mu_A(x_1)} & 1 & \cdots & \frac{\mu_A(x_2)}{\mu_A(x_m)} \\ \vdots & \vdots & & \vdots \\ \frac{\mu_A(x_m)}{\mu_A(x_1)} & \frac{\mu_A(x_m)}{\mu_A(x_2)} & \cdots & 1 \end{bmatrix}$$

(5.4)

Noticeably, the diagonal values of R are equal to 1. The entries that are symmetrically positioned w.r.t. the diagonal satisfy the condition of reciprocality, that is $r_{ij} = 1/r_{ji}$. Furthermore an important transitivity property holds, that is $r_{ik}r_{kj} = r_{ij}$ for all indices i, j, and k. This property holds because of the way in which the matrix has been constructed. To design the estimation method please refer to the definition of eigensystem in Appendix D and the related properties of reciprocal matrices. Namely, let us multiply the matrix by the vector of the membership grades $\boldsymbol{\mu}_A = (\mu_A(x_1), \mu_A(x_2), \ldots, \mu_A(x_m))^T$. For the i-th row of R (that is the i-th entry of the resulting vector of results) we obtain:

$$[R \cdot \boldsymbol{\mu}_A]_i = \begin{bmatrix} \frac{\mu_A(x_i)}{\mu_A(x_1)} & \frac{\mu_A(x_i)}{\mu_A(x_2)} & \cdots & \frac{\mu_A(x_i)}{\mu_A(x_m)} \end{bmatrix} \begin{bmatrix} \mu_A(x_1) \\ \mu_A(x_2) \\ \cdots \\ \mu_A(x_m) \end{bmatrix} = m\mu_A(x_i),$$

(5.5)

From the preference matrix you obtain automatically the membership degrees of all elements,

for $i = 1, 2, \ldots, m$. Overall once completing the calculations for all i, this leads us to the expression $R \cdot \mu_A = m\mu_A$. This means that μ_A is the eigenvector of R associated with an R eigenvalue which is equal to m (see Appendix D.1 for algebraic definitions). As a further puzzle of this cadre we have that the maximum eigenvalue $\lambda_{\max}$ of an $m \times m$ matrix A such that $a_{ij} = 1/a_{ji}$ is m if and only if the above transitivity property $a_{ik}a_{kj} = a_{ij}$ holds, being $\lambda_{\max} \geq m$ in any case. Hence in order to derive the membership values $\mu_A(x_i)$ from R, on one side we look for the eigenvector corresponding to its highest eigenvalue. On the other, we check $\lambda_{\max}$ to be not exceedingly higher than m.

The starting points of the estimation process are entries of the reciprocal matrix which are obtained through collecting results of pairwise evaluations offered by an expert, designer or user. Prior to making any assessment, the expert is provided with a finite scale with values spread in-between 1 to 7. Some other alternatives of the scales such as those involving 5 or 9 levels could be sought as well. If x_i is strongly preferred over x_j when being considered in the context of the fuzzy set whose membership function we would like to estimate, then this judgement is expressed by assigning high values of the available scale, say 6 or 7. If we still sense that x_i is preferred over x_j yet the strength of this preference is lower in comparison with the previous case, then this is quantified using some intermediate values of the scale, say 3 or 4. If no difference is sensed, the values close to 1 are the preferred choice, say 2 or 1. The value of 1 indicates that x_i and x_j are equally preferred. On the other hand, if x_j is preferred over x_i, the corresponding entry assumes values below one. Given the reciprocal character of the assessment, once the preference of x_i over x_j has been quantified, the inverse of this number is plugged into the entry of the matrix that is located at the (j, i)-th coordinate. As indicated earlier, the elements on the main diagonal are equal to 1. Next the maximal eigenvalue is computed along with its corresponding eigenvector. The normalized version of the eigenvector is then the membership function of the fuzzy set we considered when doing all pairwise assessments of the elements of its universe of discourse. The pairwise evaluations are far more convenient and manageable in comparison to any effort we make when assigning membership grades to all elements of the universe in a single step. Practically, the pairwise comparison helps the expert focus only on two elements once at a time thus reducing uncertainty and hesitation while leading to the higher level of consistency. In addition, once decided which one is preferred, the graduation concerns only positive scores. The assessments, however, are not free of bias. In particular, with the mentioned

better if you use a few preference levels,

mathematical tools we may control the satisfaction of the transitivity requirement in terms of the difference $\lambda_{\max} - m$. We prefer to normalize this difference, thus regarding the ratio:

$$v = \frac{\lambda_{\max} - m}{m - 1} \qquad (5.6)$$

as an index of inconsistency of the data; the higher its value, the less consistent are the collected experimental results. If the value of v is too high, exceeding a certain superimposed threshold, the experiment may need to be repeated. Typically if v is less than 0.1 the assessment is sought to be consistent, while higher values of v call for the re-examination of the experimental data and a re-run of the experiment. To quantify how much the experimental data deviate from the transitivity requirement, we may also calculate the absolute differences between the corresponding experimentally obtained entries of the reciprocal matrix, namely r_{ik} and $r_{ij}r_{jk}$. The sum expressed in the form:

provided some consistency between preferences does exist.

$$v'_{ik} = \sum_{j=1}^{m} |r_{ij}r_{jk} - r_{ik}| \qquad (5.7)$$

serves as a useful indicator of the lack of transitivity of the experimental data for the given pair of elements x_i and x_k. If required, we may repeat the experiment if the above sum takes high values. The overall sum $v' = \sum_{i,k}^{m} v'_{ik}$ becomes then another global evaluation of the lack of transitivity of the experimental assessment.

Example 5.5. Let us estimate the membership function of the concept *hot* temperature for the space of temperatures consisting of $10, 20, 30, 45$ degrees Celsius. The scale in which the pairs of these elements are evaluated consists of 5 levels (say, $1, 2, \ldots, 5$). The experimental results of the pairwise comparison are collected in the reciprocal matrix R,

$$R = \begin{bmatrix} 1 & 1/2 & 1/4 & 1/5 \\ 2 & 1 & 1/3 & 1/4 \\ 4 & 3 & 1 & 1/3 \\ 5 & 4 & 3 & 1 \end{bmatrix}.$$

Calculating the maximal eigenvalue, we obtain $\lambda_{\max} = 4.114$ which is slightly higher than the dimension ($m = 4$) of the reciprocal matrix. The corresponding eigenvector is equal to $(0.122, 0.195, 0.438, 0.869)^T$ which after normalization gives rise to the membership function of *hot* temperature to be equal to $(0.14, 0.22, 0.50, 1.00)^T$. The value of the inconsistency index v is equal to $(4.114 - 4)/3 = 0.038$ and is far lower than the threshold of 0.1.

Example 5.6. Now let us consider some modified version of the previously discussed reciprocal matrix with the following entries:

$$R = \begin{bmatrix} 1 & 2 & 1/4 & 1/5 \\ 1/2 & 1 & 1/3 & 4 \\ 4 & 3 & 1 & 1/3 \\ 5 & 1/4 & 3 & 1 \end{bmatrix}.$$

Now the maximal eigenvalue is far higher than the dimensionality of the problem, $\lambda_{\max} = 6.119$. In this case, the lack of consistency becomes reflected in the high value of the inconsistency index $v = (6.119 - 4)/3 = 0.706$. Given this high level of inconsistency, it is not advisable to compute the corresponding eigenvector. If one would intend to fix the problem of the quite essential lack of transitivity, then a thorough analysis of the lack of transitivity is necessary. We determine the results for the pairs of indexes (i and k) as articulated by (5.7). In this way we highlight those assessments that tend to be highly inconsistent. These are the candidates whose evaluation has to be revisited.

5.4 Fuzzy Sets as Granular Representatives of Numeric Data

_{The fuzzy set version of the bias-variance tradeoff}

In general, a fuzzy set A is reflective of numeric data that are put together in some context. Using its membership function we attempt to embrace them in a concise manner. The development of the fuzzy set is supported by the following experiment-driven and intuitively appealing rationale:

(a) first, we expect that A reflects (or matches) the available experimental data to the highest extent, and
(b) second, the fuzzy set is kept specific enough so that it comes with a well-defined semantic.

_{So general as to embrace a phenomenon,}

_{so specific as to provide information.}

These two requirements point at the multiobjective nature of the construct: we want to maximize the coverage of experimental data (as articulated in (a)) and minimize the spread of the fuzzy set (as captured by (b)). Given the objective we wish to accomplish, we will be referring to it as a *principle of justifiable information granularity*. These two requirements give rise to a certain optimization problem. Furthermore, which is quite legitimate, we assume that the fuzzy set to be constructed either has a unimodal membership function or its maximal membership grades occupy a contiguous region in the universe of discourse in

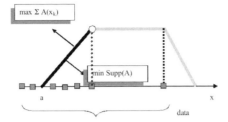

Fig. 5.6. Optimization of the linear increasing section of the membership function of A; highlighted are the positions of the membership function originating from the realization of the two conflicting criteria of coverage and representation.

which this fuzzy set has been defined. This helps us build a membership function separately for its rising and declining sections. The core of the fuzzy set is determined first. Next, assuming the simplest scenario when using the linear type of membership functions, the essence of the optimization problem boils down to the rotation of the linear section of the membership function around the upper point of the core of A (for the illustration refer to Fig. 5.6). The point of rotation of the linear segment of this membership function is marked by an empty circle. By rotating this segment, we intend to maximize (a) and minimize (b).

Before moving on with the determination of the membership function, we concentrate on the location of its numeric representative. We prefer to work with the median of the experimental data $x_1, x_2, \ldots, x_m$ with fuzzy sets framework in place of their average, since: i) on one side we cannot rely on an equally weighting of the data suggested by probabilistic models; and ii) on the other, this statistic is less affected by outliers. Let us recall that the median is an order statistic and is formed on the basis of an ordered set of numeric values. In the case of an odd number of data in the dataset, the point located in the middle of this ordered sequence is the median. When we encounter an even number of data in the granulation window, instead of picking up an average of the two points located in the middle, we consider these two points to form a core of the fuzzy set. Thus depending upon the number of data points, we either end up with a triangular or trapezoidal membership function. These are very simple forms of membership functions whose fuzzy sets are often called *fuzzy numbers*. They denote numbers with some approximation and no special additional knowledge available to moulding uncertainty in a more complex way. In detail, fuzzy numbers are fuzzy subsets of the real line having a peak or plateau with membership grade 1; moreover their membership function increases towards the peak and decreses away from it.

First fix the core of the fuzzy set in the median of its declared elements;

Having fixed the modal value of A (that could be a single numeric value b or a certain interval $[b_1, b_2]$), the optimization of the spreads of the linear portions of the membership functions are carried out separately for their increasing and decreasing portions. We consider the increasing part of the membership function (the decreasing part is handled in an analogous manner). Referring to Fig. 5.6, the two requirements guiding the design of the fuzzy set are transformed into the corresponding multiobjective optimization problem as outlined as follows:

then negotiate its province,

(a) maximize the experimental evidence of the fuzzy set; this implies that we tend to cover as many numeric data as possible, viz. the coverage has to be made as high as possible. Graphically, in the optimization of this requirement, we rotate the linear segment up (counterclockwise) as illustrated in Fig. 5.6. Formally, we identify *experimental evidence* with the sum of the membership grades $\sum_k \mu_A(x_k)$ – where μ_A is a linear membership function and x_k is located to the left of the modal value – and look for its maximization;

(b) simultaneously, we would like to make the fuzzy set as specific as possible so that is comes with some well defined semantics. This requirement is met by making the support of A as small as possible, that is $\min_a |b - a|$, with a denoting the lower bound of fuzzy number.

by maximizing a membership density.

To accommodate the two conflicting requirements, we combine (a) and (b) in the form of the ratio that is maximized w.r.t. the unknown parameter of the linear section of the membership function:

$$a_{\text{opt}} = \arg\max_{a \neq b} v(a) = \arg\max_{a \neq b} \frac{\sum_k \mu_A(x_k)}{|b - a|}. \quad (5.8)$$

The linearly decreasing portion of the membership function is optimized in the same manner. The overall optimization process returns the parameters of the fuzzy number in the form of lower and upper bound (denoted here by a and c, respectively) and its core (a single modal value b or the interval $[b_1, b_2]$). We can write down such fuzzy numbers as $A_{\{a,b_1,b_2,c\}}$. We exclude a trivial solution of $a = b_1$ in which case the fuzzy border of the set collapses to a single numeric entity.

A probabilistic metaphor may help.

Since the method is based on the sole coordinates of the questioned points, its probabilistic flavor is evident. Focusing on the right slope of the triangle/trapezium, on any x, $\mu_A(x)$ looks like the probability of finding a point with coordinate greater than x yet member of A. Actually, we do no assumptions on how pivotal points x_k are produced. But if they are sampled from the random variable X with a sampling mechanism linked to A the

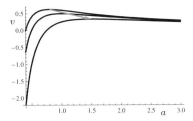

Fig. 5.7. Plots of v versus a for selected values of $\tau = 0.8, 1.0, 1.5$, respectively from higher to lower curve. Arrow joins the points $(a, v(a))$ for $a = \tau$.

above interpretation is statistically correct. Consider the simple case where X is uniform over the range $[0, \tau], \tau > 0$, that is $p(x) = 1/\tau$ over $[0, \tau]$ and 0 otherwise, and the linear membership function of A is the one of the form $\mu_A(x) = \max\{0, 1 - x/a\}$, so that the modal coordinate of μ_A is equal to zero. With $a > \tau$, we may compute the optimization target over the entire X population as follows:

$$v(a) = \frac{1}{a} \int_0^\tau \mu_A(x) p(x) \mathrm{d}x = \frac{1}{a\tau} \int_0^\tau \left(1 - \frac{x}{a}\right) \mathrm{d}x =$$
$$\frac{1}{a\tau}\left(\tau - \frac{\tau^2}{2a}\right) = \frac{2a - \tau}{2a^2}. \quad (5.9)$$

The plot of v with a is shown in Fig. 5.7, where τ is the curve parameter. As a matter of fact, a derivative w.r.t. a of (5.9) equal to 0 when a is equal to τ denotes $a_{\mathrm{opt}} = \tau$. Moreover, the form of the relationship $v = v(a)$ is highly asymmetric; while the values of a higher than the optimal value (a_{opt}) leads to a very slow degradation of the performance (v changes slowly), the rapid changes in v are noted for values of a which are lower than the optimal value.

Example 5.7. We show the details on how the data driven triangular fuzzy set is being formed. The dataset under discussion consists of the following numeric data:

$$\{-2.00 \ \ 0.80 \ \ 0.90 \ \ 1.00 \ \ 1.30 \ \ 1.70 \ \ 2.10 \ \ 2.60 \ \ 3.30\}.$$

The values of the performance index obtained during the optimization of the left and right-part slope of the triangular membership function and being viewed as a function of the intercept are shown in Fig. 5.8. The performance index shows a clear maximum for both linear parts of the membership function. The final result comes in the form of a triangular fuzzy set that we interpret as a fuzzy number. It is uniquely described by its bounds and the modal value, altogether described as the triangular fuzzy

Simple geometrical forms

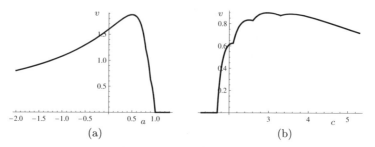

Fig. 5.8. The values of the performance index v optimized for the linear sections of the membership function; in both cases we note a clearly visible maximum occurring at both sides of the modal value of the fuzzy set that determine the location of the bounds of the membership function ($a = 0.50$ and $c = 2.50$).

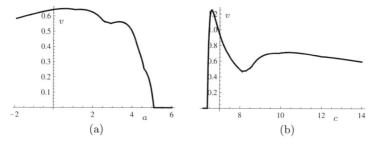

Fig. 5.9. Performance index v computed separately for the linearly increasing and decreasing portions of the optimized fuzzy set.

set $A_{\{0.50, 1.30, 2.50\}}$. Fig. 5.10(a) shows that the compromise between the above goals (a) and (b) nicely captures the core part of the numeric data.

Example 5.8. Consider now another dataset. It comes with a far higher dispersion (some points are sitting at the tails of the entire distribution):

{1.1 2.5 2.6 2.9 4.3 4.6 5.1 6.0 6.2 6.4 8.1 8.3 8.5 8.6 9.9 12.0}.

The plots of the optimized performance index v are shown in Fig. 5.9. The optimized fuzzy set comes in the form of the trapezoidal membership function $A_{\{3.33, 6.0, 6.2, 10.4\}}$ (see Fig. 5.10(b)). The location of several data points that are quite remote from the modal values makes substantial changes to the form of the membership function in which the left-hand side slope is pushed towards higher values of the arguments.

Example 5.9. The analogon of the sampling mechanism for injecting knowledge into the experimental scenario comes with many

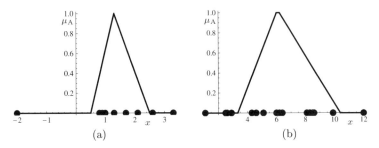

Fig. 5.10. Optimized fuzzy sets for the datasets in Example 5.7 (a) and 5.8 (b).

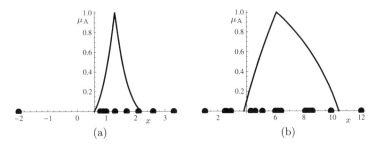

Fig. 5.11. Optimized fuzzy sets for the datasets in Example 5.7 (a) and 5.8 (b) with monomial membership functions $\mu_A(x) = x^\alpha$, with $\alpha = 3$ and 0.01 respectively.

options. A former one is constituted by the form of the membership function, where, in place of the linear type (viz. their linearly increasing or decreasing sections), any monotonically increasing or decreasing functions could be sought. In particular, a polynomial (monomial, to be more precise) type of relationships, say x^α with α being a positive real, could be of interest. For instance, we produce in Fig. 5.11 the companions of the membership functions in Fig. 5.10, for values of α different from 1.

Another option concerns functional relationships between pivotal points that increase the complexity of their distribution, like in the following two examples.

Example 5.10. Let us consider the geometrical shape in Fig. 5.12. Fixing the coordinates of the central point to the axes origin we may assume the shape to represent a fuzzy circle, viz. a circle whose radius is a fuzzy set R (fuzzy number). Hence, picking radiuses at equally distributed angles we obtain a set of variously concentrated radius values from which to compute their membership function. The optimization scheme is still governed by (5.8).

Fuzzy parameters of fuzzy geometrical forms

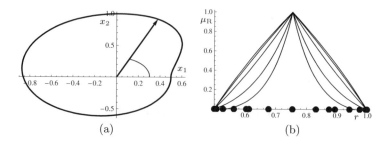

Fig. 5.12. Example of figures to be represented as fuzzy circles (a) and the monomial membership of the fuzzy radius determined by (5.8) for different values of the degree $\alpha = 0.01, 0.5, 1, 2, 6$ and 10 from concave to convex shapes (b).

Fuzzy distance from a geometrical form

Example 5.11. Consider a situation displayed in Fig. 5.13(a). There are a number of definitions which aim to capture thet distance of a point from a geometrical form. For instance the distance between a point x and a domain B can be defined by:

$$d_{\min}(x, B) = \min_{y \in B} d(x, y),$$

which aims at capturing the minimum distance d between x and the elements of B for a suitable metric. Analogously we could focus on the maximum or any intermediate value. With a fuzzy counterpart we aim at considering all these values with a suitable weight. A feasible strategy consists in fixing pivotal points $\{y_1, \ldots, y_m\}$ in the domain, computing their distances $\{d_1, \ldots, d_m\}$ from x and identifying a predefined membership shape, say piecewise linear, like in the previous examples. In this way we obtain the fuzzy set fsD in Fig. 5.13(b). Note that even though the object (geometrical figure) has clearly delineated boundaries (there is no uncertainty as to their position), the fuzzy set of distance is reflective of the complexity and non-uniqueness of the definition itself. We may obviously generalize this procedure to compute a distance between two domains A and B (see Fig. 5.14(a)). On one side, we can use the Hausdorff distance which is defined by:

$$d_H(A, B) = \max\{\max_{x \in A}(\min_{y \in B} d(x, y)), \max_{y \in B}(\min_{x \in A} d(x, y))\}. \quad (5.10)$$

On the other, with our procedure we: i) sample pivotal points in the two domains producing a collection of points inside, ii) compute distances between pairs of them $d(x_i, y_j)$, where $x_i \in A$ and $y_j \in B$, and iii) use those to form a fuzzy set of distance D (see Fig. 5.14(b)).

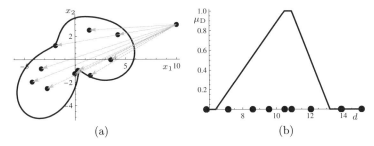

Fig. 5.13. Computing a fuzzy set of distance between a point and some geometric figure A; note a sample of points located within the bounds of the figure (a) and induced fuzzy set (b).

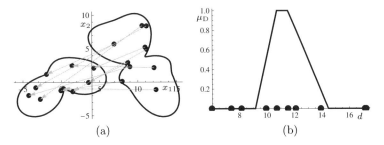

Fig. 5.14. Determining a fuzzy set of distance between two planar figures A and B. Same notation as in Fig. 5.13.

5.5 Fuzzy Sets That Reduce the Descriptional Length of a Formula

As mentioned before we appreciate a formula as suitable if it is understandable, and we understand a formula if it is based on 3 to 5 variables. Hence, once we are arrived at a certain analytical conclusion we pay a great effort to *simplify* the theory, i.e. to reduce the length of its formulas (descriptional length), even at a cost of missing the explanation of special cases that we assume as outliers. This is the reason why we usually try to describe phenomena through a linear or quadratic relation on a few variables.

Short formulas to be understandable formulas

Coming to Boolean formulas a most suitable expression of them is through canonical forms consisting of either CNF or DNF, as seen in Section 1.3. Consider the goal of synthesizing a set of *positive* and *negative* points (see Definition 1.8) through disjunction of monomials (DNF). We use formulas for describing them in a succinct way, with the obvious goal of having formulas at least shorter than the list of points. We achieve this goal simply exploiting the following fact.

Fact 5.1. *Given a set $\mathfrak{X} = \{0,1\}^n$ and two Boolean formulas f_1 and f_2 on it, we say that f_1 is* slimmer *than f_2 if the support σ_1 of the former is properly included in the support σ_2 of the latter, hence $\sigma_1 \subset \sigma_2$. The slimmest DNF consistent with a set E^+ of positive points in $\mathfrak{X}$ and a set E^- of negative points in the same set is constituted by the disjunction of $\#E^+$ monomials, where $\#E$ is the cardinality of E, such that for each $\boldsymbol{x} \in E^+$ there exists a monomial $\mathsf{m} = \ell_1 \ldots \ell_n$ such that if $x_i = 1$, then $\ell_i = v_i$, else $\ell_i = \overline{v}_i$* [1].

<small>Short formulas versus slim formulas: another bias variance tradeoff,</small>

<small>that may be improved by working with fuzzy supports of the formulas.</small>

Since it is the slimmest consistent formula, it will contain all positive points in its support, yet excluding all negative points. The drawback of this solution is its descriptional length. It consists of $\#E^+$ monomials, each described by n variables. Besides the obvious shortening of this length, reducing the number of these variables in the monomials has the additional benefit of generating formulas with larger support possibly containing more than one positive point, which allows to eliminate some monomials completely. The cost may be represented by the inclusion of negative points in these supports. To relieve this drawback we may consider the formula we obtain after simplification as a fuzzy set. Namely, with reference to a monomial m, we consider a set of nested domains (call it *focal set*) representing the support of the formulas we obtain after a progressive removal of literals. We give to each domain a mass proportional to the number of negative points it includes in its gap with the previous element in the focal set, and mass 0 to the original monomial assumed as the core set. To have the sum of these measures equal 1, we normalize the measure by dividing it by the total number of points falling in the gap between the last and first elements of the nested sequence. Then we identify the support of the fuzzy set with the union of the nested elements, and compute the membership function of a point to it as the sum of masses of the focal elements including it. This gives rise to a stepwise function decreasing from 1 in the core monomial to 0 outside its relaxation (see Fig. 5.15). Coming to the formulas, we distinguish a crisp monomial constituted by the core of the fuzzy set and a fuzzy frontier constituted by the union of the other focal sets. We also transfer each value of the membership function steps to the literal whose removal generated the focal set and identify the fuzzy frontier directly with the ordered sequence of the literals.

Definition 5.2. *Given a monomial m, denote with $\mathrm{set}(\mathsf{m})$ the set of literals joined by it. For an ordered sequence $\boldsymbol{\ell} = (\ell_1, \ldots, \ell_\nu)$*

---

[1] Fact 5.1 somehow extends the framework in Definition 1.8 to non monotone formulas.

Fuzzy Sets That Reduce the Descriptional Length   177

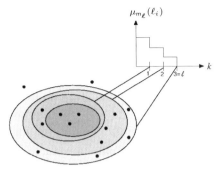

Fig. 5.15. The fuzzy border of a monomial m after three subsequent literal removals. Dark region $\rightarrow$ m $= \ell_1\ell_2\ell_3\ldots\ell_r$; diminishing gray regions $\rightarrow$ progressive enlargements after removal of ℓ_1, ℓ_2, and ℓ_3 from set(m); $\mu_{m_\ell}(\ell_k) \rightarrow$ membership function as in Definition 5.2.

of length ν of literals from set(m), *let us denote by ℓ^k its prefix of length k. Let* $m_{\ell^0} =$ m, *and* m_{ℓ^k} *denote the monomial obtained by removing the literal ℓ_k from the representation of* $m_{\ell^{k-1}}$. *Let us denote $\sigma(\ell^k)$ the cardinality of the subset of E (the example set) belonging to* $m_{\ell^k}\setminus$m. *We define the* (fuzzy) *membership function* $\mu_{m_\ell}(\ell_k)$ *of a literal ℓ_k w.r.t.* m_ℓ *as follows:*

$$\mu_{m_\ell}(\ell_k) = 1 - \frac{\sigma(\ell^k)}{\sigma(\ell)}. \tag{5.11}$$

The monotonicity of the membership function induces a dummy metric where points annexed by (i.e. after the removal of) one literal in ℓ are farther from the crisp monomial than the ones annexed by previous literals. According to this metric we introduce the notion of radius of the fuzzy frontier as a mean distance of its support from the core in a pseudoprobabilistic interpretation of the membership function. Actually, although we exploit the local features of this function in the statement of the cost function, we may interpret $\mu_{m_\ell}(\ell_k)$ as a probability estimate of finding points that belong to the fuzzy frontier outside the enlargement induced by ℓ_k, in a dummy framework where points may or may not belong to the sole monomial we are dealing with. That could be considered as a complement to 1 of a sort of conditional cumulative distribution function. We lack, however, the *prior* distribution of the conditioning variable that could allow us to assess the global probabilities (summing to 1) of a point to belong to the different DNF monomials. With this interpretation however we may define an expected distance from the core m as the sum of the complement to 1 of the corresponding cumulative distribution function (the membership function indeed). Namely:

Definition 5.3. *Given a monomial* m_i *and an ordered sequence* $\ell_i = (\ell_1, \ldots, \ell_\nu)$ *of length ν of literals from* $\mathrm{set}(\mathsf{m}_i)$, *denoting* m_{ℓ_i} *the monomial* $\mathsf{m}_{i\ell_i}$ *for short, we call* $\mathsf{m}_{\ell_i} \setminus \mathsf{m}_i$ *the fuzzy frontier of* m_i, *and*

$$\rho_i = \sum_{k=1}^{\nu} \mu_{\mathsf{m}_{\ell_i}}(\ell_k) \qquad (5.12)$$

its radius.

Remark 5.1. With (5.12) we have a dual mode of defining a shadow set, where the shadowed region around the core of the set plays the usual topological role coupled with a quantity ρ meaning the thickness of the shadow. Note that the membership functions we defined are not univocal for a given expansion of a monomial from m to m'. They strictly depend on the history of literals removals we followed. We may further enrich this history by considering also some reinsertion of literals whenever this is suggested by the cost optimization algorithm. A reinsertion just deletes an item from ℓ preserving the relative order of the remaining ones.

We exploit the local properties of our fuzzy sets by simply cumulating the radiuses of the formulas belonging to the DNF in a cost function C aimed at balancing them with the total length of the formula. To sum up, the cost function we want to minimize w.r.t. the border formula f is:

$$\mathsf{C}_f = \lambda \sum_{i=1}^{m} l_i + (1-\lambda) \sum_{i=1}^{m} \rho_i, \qquad (5.13)$$

where:

- $\sum_i l_i$ is the length of the formula, being l_i the number of literals in the i-th monomial; and
- λ is the free parameter balancing the costs.

Example 5.12. Consider the Vote dataset. To adapt it to a Boolean problem we associated favorable votes with 1, unfavorable ones to 0, and split the records each time we met an unexpressed vote, substituting it once with 1 and once with 0, producing a total of 439 binary records. The data contain some pairs of contradictory examples made up of same vote patterns but different labels. This phenomenon is rooted in the meaning of the data and amplified by the mentioned splitting. It accentuates the fuzzy values of the discovered rules, but is not bearable by consistent formulas. Therefore we simply resolve the contradictions in the sole training set in favor of the majority label (randomly in case of parity).

We toss the fuzzy extended formulas with a *cross-validation* strategy. Namely, we formed 50 different random partitions of the file into training and test sets, where the former is the 70% of the whole file. We obtained a mean value of 4.45% and 2.433% of positive and negative points erroneously classified, respectively, with standard deviations of 1.864 and 1.018.

5.6 Fuzzy Equalization

While in Section 5.4 we have considered singularly each cluster as direct consequence of absence of measures (for instance conditioning probabilities) connecting each other, here we consider a way of smearing the uncertainty between fuzzy sets $\{A_1, \ldots, A_c\}$ under a vague fairness assumption requiring that a fuzzy set has to some extent be reflective of the existing numeric evidence $\sum_k \mu_{A_i}(x_k)$. The problem of fuzzy equalization, that we anticipate to be substantially supported by the existing numeric data, can be outlined as follows:

Given is a finite collection of numeric data $\{x_1, x_2, \ldots, x_m\}$, $x_i \in \mathbb{R}$, that we consider to be arranged in a non-decreasing order, that is $x_1 \leq x_2 \leq \ldots \leq x_m$, construct a family (partition) of triangular fuzzy sets $A_1, A_2, \ldots, A_c$: i) with a *semantic overlap* of $1/2$ between neighboring fuzzy sets, i.e. so that the upper vertex of a triangle coincides with the lower vertices of the neighboring two (see Fig. 5.16), and ii) so that each of them comes with the same experimental evidence. The latter property translates into relations:

<small>A fair distribution of evidence in place of probability</small>

$$\begin{aligned}
\sum_{k=1}^{m} \mu_{A_1}(x_k) &= \tfrac{m}{2(c-1)} \\
\sum_{k=1}^{m} \mu_{A_2}(x_k) &= \tfrac{m}{(c-1)} \\
&\vdots \\
\sum_{k=1}^{m} \mu_{A_{c-1}}(x_k) &= \tfrac{m}{(c-1)} \\
\sum_{k=1}^{m} \mu_{A_c}(x_k) &= \tfrac{m}{2(c-1)},
\end{aligned} \qquad (5.14)$$

where for the first and the last fuzzy set (A_1 and A_c) we require that this evidence is $1/2$ of the one required for all remaining fuzzy sets. The essence of this construct is illustrated in Fig. 5.16.

We can propose the following procedure to build the fuzzy sets $A_1, A_2, \ldots, A_{c-1}$ satisfying (5.14); the simplicity of the algorithm does not assure that the same numeric requirement holds for

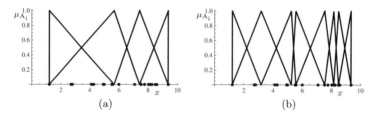

Fig. 5.16. A collection of triangular fuzzy sets with equal experimental support provided by the numeric dataset as in Example 5.13 when the number of fuzzy sets equals 4 (a) and 8 (b).

A_c. We elaborate on this process in more detail later on. The equalization is concerned with the determination of the modal values of the fuzzy sets. We start with A_1 and move to the right by choosing a suitable value of a_2 so that the sum of membership grades $\sum_{k=1}^{m} \mu_{A_1}(x_k)$ is equal to $\frac{m}{2(c-1)}$. The determination of the value of a_2 could be completed through a stepwise increment of its value. The modal value of A_1 is equal to the minimal value encountered in the dataset, that is $a_1 = x_1$. We assume here that the boundaries x_1 and x_m are not outliers; otherwise they have to be dropped and the construct should be based upon some other extreme points in the dataset. The experimental evidence of A_2 is made equal to $\frac{m}{(c-1)}$ by a proper choice of the upper bound of its membership function, namely a_3. Note that, as the value of a_2 has been already selected, this implies the following level of experimental evidence accumulated so far:

obtained through an approximate algorithm,

$$\sum_{x_k \in [a_1,a_2]} \mu_{A_2}(x_k) = \sum_{x_k \in [a_1,a_2]} (1 - \mu_{A_1}(x_k)) =$$

$$\sum_{x_k \in [a_1,a_2]} 1 - \sum_{x_k \in [a_1,a_2]} \mu_{A_1}(x_k) = m_2 - \frac{m}{2(c-1)} = m'_2. \quad (5.15)$$

Given this, we require that the value of a_3 is chosen so that the following equality holds:

$$\sum_{x_k \in [a_2,a_3]} \mu_{A_2}(x_k) + m'_2 = \frac{m}{c-1}. \quad (5.16)$$

Note that depending upon the distribution of numeric data the resulting fuzzy set A_2 could be highly asymmetric. To determine the parameters of the successive fuzzy sets, we repeat the same procedure moving towards higher values of x_k and determining the values of $a_3, a_4, \ldots, a_{c-1}$.

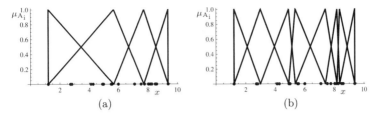

Fig. 5.17. Fuzzy set equalization obtained by optimizing $v(a)$ through (5.17). Same notation as in Fig. 5.16.

Example 5.13. Consider the following dataset, where $m = 20$ points are randomly distributed in $[0.5, 9.5]$:

$$\{1.21\ \ 2.72\ \ 2.81\ \ 4.12\ \ 4.27\ \ 4.92\ \ 4.99\ \ 5.51\ 5.56\ \ 5.99$$
$$7.09\ \ 7.53\ \ 7.76\ \ 8.03\ \ 8.15\ \ 8.26\ \ 8.49\ \ 8.50\ \ 8.59\ \ 9.37\}.$$

Fig. 5.16 provides the output of the above procedure when the number of fuzzy sets c equals 4 and 8 respectively.

One notes that the last fuzzy set A_c does not come with the required level of experimental evidence as we do not have any control over the sum of the corresponding membership grades. To alleviate this shortcoming, one may consider a replacement of the algorithm (whose advantage resides with its evident simplicity) by the minimization of the performance index v over the vector of the modal values $\boldsymbol{a} = (a_2, a_3, \ldots, a_{c-1})$:

or even with an exact optimization algorithm.

$$v(\boldsymbol{a}) = \left(\sum_{k=1}^{m}\mu_{A_1}(x_k) - \frac{m}{2(c-1)}\right)^2 + \left(\sum_{k=1}^{m}\mu_{A_2}(x_k) - \frac{m}{(c-1)}\right)^2 + \ldots + \left(\sum_{k=1}^{m}\mu_{A_{c-1}}(x_k) - \frac{m}{(c-1)}\right)^2 + \left(\sum_{k=1}^{m}\mu_{A_c}(x_k) - \frac{m}{2(c-1)}\right)^2,$$

(5.17)

that is $\boldsymbol{a}_{\text{opt}} = \arg\min_{\boldsymbol{a}} v(\boldsymbol{a})$. Fig. 5.17 shows how this procedure provides a slightly different fuzzy set equalization than the former as for the dataset described in Example 5.13, in any case preventing from giving too much/low weight to the last fuzzy set.

5.7 Conclusions

Willing to affect observed points with a further coordinate denoting the value of its semantic, with Statistics we deduce it from the data original coordinates, while with Fuzzy Sets approach we work directly on the new coordinate. This makes the latter missing the facility of attributing same value to all data as a counterpart of an equipartition through them of the sample space for a proper metric. The immediate benefit is that the user is relatively free of determining a value for the additional coordinate on his own in terms of membership grade to a fuzzy set. It may result however in an excess of degrees of freedom in the methods for inferring the various membership functions. In turn, this drawback is reduced by imposing the satisfaction of specific criteria – someone dictated by logic constraints, others by the operational framework (say, by aesthetics). Thus a propensity for an inference method is linked mainly to the option on these criteria. In the chapter we expounded some of them, ranging from: i) simple consistency checks with vertical and horizontal methods, to ii) maximum compliance between granules and gathering fuzzy set, iii) priority transitivity checking, iv) bias variance balances, and v) various fuzziness equalization criteria.

The multiplicity of methods denotes the delicacy of the problem we are called to solve. It leaves room to many variants and nuances induced by the specific operational framework, with a general and strict request of soundness and rigor, however.

5.8 Exercises

1. In the horizontal mode of constructing a fuzzy set of safe speed on a highway, the "yes"-"no" evaluations provided by the panel of 9 experts are the following:

x (km/h)	20	50	70	80	90	100	110	120	130	140	150	160
No. of "yes" responses	0	1	1	2	6	8	8	5	5	4	3	2

   Determine the membership function and assess its quality by computing the corresponding confidence intervals with confidence level $\delta = 0.1$. Interpret the results and identify the points of the universe of discourse that may require more attention.

2. In the vertical mode of membership function estimation, we are provided with the following experimental data:

α	0.3	0.4	0.5	0.6	0.7	0.8	0.9	1.0
range of $\mathfrak{X}$	$[-2, 13]$	$[-1, 12]$	$[0, 11]$	$[1, 10]$	$[2, 9]$	$[3, 8]$	$[4, 7]$	$[5, 6]$

   Plot the estimated membership function and suggest its analytical expression.

3. Solve the problem in Example 5.4 shifting x' and x'' w.r.t. x as follows: $x' = x''$ such that $\mu_A(x') + \mu_{G_i}(x'') - \mu_A(x')\mu_{G_i}(x'') = \mu_A(x) = \mu_{G_i}(x)$. The seven granules have parameters $\{\nu, \sigma, h\}$ as follows:

Granule	$\{\nu, \sigma, h\}$
G_1	$\{3.76, 1.74, 0.86\}$
G_2	$\{4.15, 2.59, 0.88\}$
G_3	$\{8.43, 1.72, 0.59\}$
G_4	$\{7.10, 2.51, 0.71\}$
G_5	$\{2.10, 2.21, 0.99\}$
G_6	$\{1.40, 1.00, 0.83\}$
G_7	$\{6.89, 1.29, 1.00\}$

Check that the result is as in Fig. 5.18

4. The results of pairwise comparisons of 4 objects being realized in the scale of $\{1, 2, \ldots, 5\}$ are given in the following matrix form:
$$\begin{bmatrix} 1 & 5 & 2 & 4 \\ 1/5 & 1 & 3 & 1/3 \\ 1/2 & 1/3 & 1 & 1/5 \\ 1/4 & 3 & 5 & 1 \end{bmatrix}$$

What is the consistency of the scoring? Evaluate the effect of the lack of transitivity. Determine the membership function of the corresponding fuzzy set.

5. In the method of pairwise comparisons, we use different scales involving various levels of evaluation, typically ranging from 5 to 9. What impact could the number of these levels have on

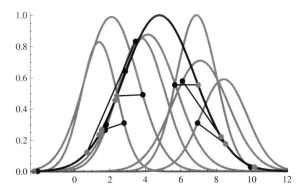

Fig. 5.18. Fuzzy set compatible with the observed granules. Same strategy as in Fig. 5.5, but with different shift function.

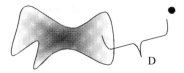

Fig. 5.19. Forming a fuzzy set of distance between a geometric figure with fuzzy boundaries and a point.

the produced consistency of the results? Could you offer any guidelines as how to achieve high consistency? What would be an associated tradeoff one should take into consideration here?

6. Construct a fuzzy set of *large* numbers for the universe of discourse of integer numbers ranging from 1 to 10. You may exploit the experimental pairwise comparison described in the form:

$$f(x,y) = \begin{cases} x - y & \text{if } x > y \\ 1 & \text{if } x = y. \end{cases}$$

(for $x < y$ we consider the reciprocal of the above expression).

7. Apply the principle of justifiable information granularity to draw a triangular membership function from the set $\{4.5, 1.4, 0.8, 1.0, 1.0, 4.4, 8.4, 2.2\}$. Repeat the exercise substituting quadratic (in place of linear) edges in the triangle.

8. In the calculations of the distance between a point and a certain geometric figure (as discussed in Section 5.4), we assumed that the boundary of the figure is well defined. How could you proceed with a more general case when the boundaries are not clearly defined, viz. the figure itself is defined by some membership function (see Fig. 5.19)? In other words, the figure is fully characterized by some membership function $\mu_A(\boldsymbol{x})$ where $\boldsymbol{x}$ is a vector of coordinates of $\mathbb{R}^n$, for arbitrary n. If $\mu_A(\boldsymbol{x}) = 1$, the point fully belongs to the figure while lower values of $\mu_A(\boldsymbol{x})$ indicate that $\boldsymbol{x}$ is closer to the boundary of A.

9. Construct a fuzzy set describing a distance between the point $(5,5)$ from the circle $x^2 + y^2 = 4$.

10. The method of membership estimation shown in Section 5.4 is concerned with one-dimensional data. How could you construct a fuzzy set over multidimensional data? Consider using one-dimensional constructs first.

11. Draw 7 equalized fuzzy sets from the dataset in Example 5.13 using the procedures introduced in Section (5.6).

Further Reading

The design of fuzzy sets and the generalization of fuzzy sets have been an open issue for some time. The issue of origin of membership functions, both in terms of their shapes and values of their parameters was under careful investigation and the introduced techniques were carefully scrutinized [26]. As of now, fuzzy sets come with a wealth of estimation techniques and the main categories were discussed in this chapter. While some of the algorithms rely on statistics of collected data, which might lead to some confusion (as it was quite apparent at early years of fuzzy sets), it is important to stress that the probability and fuzzy sets are somehow orthogonal in terms of the underlying concepts and methodological development [18, 29]. This does not preclude that they could result in some hybrid concepts which benefit from the joint usage of fuzzy sets and probability. Semantically, the resulting constructs show advantages in case we encounter entities whose probability is given, yet their characterization in terms of membership functions brings some new insights. The papers [14, 8, 12, 15, 25] along with the constructs of random fuzzy sets and probabilistic sets are reflective of this interesting and practically relevant tendency of hybrid constructs. A nice structured overview of set measures, having: i) *possibility* and *necessity* measures at their extremes [30], ii) probability in between, iii) membership function as density variant, and iv) focal elements of rough sets [19] as possible atomic support, is provided in [6].

The generalizations of fuzzy sets have been pursued at different directions. The one, intuitively appealing, concerns the generalization of membership functions where we depart from numeric values of membership grades and generalize them to intervals leading to interval-valued fuzzy sets [9, 24], fuzzy sets (type-2 fuzzy sets) [16, 11] and L-fuzzy sets [10] with membership elements defined over lattices, just to stress the most visible approaches.

The second direction deals with a description of fuzzy sets over more abstract universes of discourse. In particular, one can point here to fuzzy sets of second order which are defined over a family of some fuzzy sets. The concept can be generalized to fuzzy sets of n-th order; however one has to realize that the increased representation flexibility usually comes with significant computing overhead.

With regard to the estimation of membership functions, there is a significant diversity of the methods that support the construction of membership functions. In general, one can clearly distinguish between user-driven and data-driven approaches with a number of methods that share some features specific to both data- and user-driven techniques, and hence are located somewhere in-between. The first one is reflective of the domain knowledge and opinions of experts. In the second one, we consider experimental data whose global characteristics become reflected in the form and parameters of the membership functions. In the first group we can refer to the pairwise comparison (Saaty's approach [23, 22], see Section 5.3) as one of the representative examples. To deepen the method the reader may refer to the comprehensive treatise by Saaty [22] and a number of generalizations. The method has delivered an important feature of evaluating consistency of the results in this way offering the designer some mechanism of quantifying the membership degrees. There were a significant number of applied studies along this line as well [7, 21, 28]. Methods within data-driven approach have been only partially treated in this chapter, since the great role played by fuzzy clustering which can be regarded as a methodology and a suite of algorithmic tools to support the construction of information granules from data [1]. The books by Bezdek [4] can serve a good reference material to study with this regard. One may point at the use of techniques of neurocomputing in the estimation of membership functions [5]. We will deal with these methods and derivatives in Chapters 9 and 10.

The concept of representation of numeric data in terms of information granules which comes under the principle of justifiable granularity has been introduced by Pedrycz and Vukovich [27]. Fuzzy equalization links fuzzy sets with the existing experimental evidence provided by numeric data by assuring that information granules come with the same level of support of data [20].

References

1. Apolloni, B., Bassis, S.: Algorithmic inference: from information granules to subtending functions. Nonlinear Analysis (in press, 2007)
2. Atanassov, K.: Intuitionistic fuzzy sets. Fuzzy Sets & Systems 20, 87–96 (1986)
3. Atanassov, K.: Intuitionistic Fuzzy Sets. Physica-Verlag, Heidelberg (1999)
4. Bezdek: Pattern Recognition with Fuzzy Objective Function algorithms. Plenum Press, New York (1981)
5. Bose, N., Yang, C.: Generating fuzzy membership function with self-organizing feature map. Pattern Recognition Letters 27(5), 356–365 (2006)
6. Dubois, D., Prade, H.: Possibility Theory. Plenum Press, London (1976)
7. Duran, O., Aguilo, J.: Computer-aided machine-tool selection based on a Fuzzy-AHP approach. Expert Systems with Applications 34(3), 1787–1794 (2008)
8. Florea, M.C., Jousselme, A.L., Grenier, D., Bosse, E.: Approximation techniques for the transformation of fuzzy sets into random sets. Fuzzy Sets and Systems 159(3), 270–288 (2008)
9. Gehrke, M., Walker, C., Walker, E.: Some comments on interval-valued fuzzy sets. Int. Journal of Intelligent Systems 11, 751–759 (1996)
10. Goguen, J.: L-fuzzy sets. J. Mathematical Analysis and Applications 18, 145–174 (1967)
11. Mendel, J., John, R.: Type-2 fuzzy sets made simple. IEEE Trans. Fuzzy Systems 10, 117–127 (2002)
12. Karimi, I., Hullermeier, E.: Risk assessment system of natural hazards: A new approach based on fuzzy probability. Fuzzy Sets and Systems 158(9), 987–999 (2007)
13. Lukasiewicz, J.: Elements of Mathematical Logic. Pergamon Press, Oxford (1966)
14. Trussell, H., Civanlar, M.: Constructing membership functions using statistical data. Fuzzy Sets & Systems 18(1), 1–13 (1986)

15. Denoeux, T., Masson, M.: Inferring a possibility distribution from empirical data. Fuzzy Sets & Systems 157(3), 319–340 (2006)
16. Mendel, J.: Uncertain Rule-based Fuzzy Logic Systems. Prentice-Hall, Upper Saddle River (2001)
17. Miller, G.A.: The magical number seven plus or minus two: some limits of our capacity for processing information. Psychological Review 63, 81–97 (1956)
18. Nguyen, H.T., Wu, B.: Random and fuzzy sets in coarse data analysis. Computational Statistics and Data Analysis 51(1), 70–85 (2006)
19. Pawlak, Z.: Rough Sets – Theoretical Aspects of Reasoning about Data. Kluwer Academic Publishers, Boston (1991)
20. Pedrycz, W.: Fuzzy equalization in the construction of fuzzy sets. Fuzzy Sets & Systems 119, 329–335 (2001)
21. Pendharkar, P.: Characterization of aggregate fuzzy membership functions using saaty's eigenvalue approach. Computers & Operations Research 30(2), 199–212 (2003)
22. Saaty, T.: The Analytic Hierarchy Process. McGraw-Hill, New York (1980)
23. Saaty, T.: Scaling the membership functions. European Journal of Operational Research 25(3), 320–329 (1986)
24. Sambuc, R.: Fonctions Phi-floues, Application l'aide an diagnostic en pathologie thyroidienne. PhD thesis, University of Marseille, France (1975)
25. Trutschnig, W.: A strong consistency result for fuzzy relative frequencies interpreted as estimator for the fuzzy-valued probability. Fuzzy Sets and Systems 159(3), 259–269 (2008)
26. Turksen, I.: Measurement of membership functions and their acquisition. Fuzzy Sets & Systems 40(1), 5–138 (1991)
27. Vukovich, G., Pedrycz, W.: On elicitation of membership functions. IEEE Trans. on Systems, Man, and Cybernetics, Part A 32(6), 761–767 (2002)
28. Wu, M.C., Lo, Y.F., Hsu, S.H.: A fuzzy CBR technique for generating product ideas. Expert Systems with Applications 34(1), 530–540 (2008)
29. Zadeh, L.A.: Toward a perception-based theory of probabilistic reasoning with imprecise probabilities. Journal of Statistical Planning and Inference 102, 233–264 (2002)
30. Zadeh, L.A.: Fuzzy sets as a basis for a thaory of possibility. Fuzzy Sets System 1(1), 3–28 (1978)

Part III
Expanding Granula into Boolean Functions

Introduction

Imagine you are committed by your old parent to tide up the attic of your home. You have there plenty of old things accumulated over the time. Of many of them you never suspected the existence, of others you neither remember the utility nor imagine a possible use. In any case, tidying up implies that you dispose in some suitable order the items, eventually grouping some of them into shelves. In some sense, you must reinvent the meaning of these objects – like the first man Adam put by God in the world – hence to give a suitable code to each of them. The final aim is to substitute the broad bullets you have around the points representing them in a feature space with the support of a Boolean function including the bullets and figuring as an expansion of them.

<small>Making order in a mess of data</small>

A main distinction we will do on whether you have already an idea of the different label to attribute to the objects or not. In the former case your question is around why the objects have the label they have, so that you can attribute the same labels to new objects in a suitable way, a question that is afforded by the *discriminant analysis*. In the latter it is up to you to group also the objects at hand, a task in the province of the *cluster analysis*. In a broad sense you speak of *classification* when you use the results of one of the two analyses to give next questioned objects a label.

<small>with or without external help</small>

For sure we will differently accomplish our tedious tidying order depending whether there is enough light in the attic or not, maybe because it is a cloudy day or not, there is only a small window, and so on. Different light intensity and direction indeed allow to establish different perspectives on the objects, hence different features to be discovered. This is a key point, since a good representation of items is a strong premise for a good understanding of what they really are. Thus, while the first of the two next chapters affords the bulk of the clustering, we will devote the second one to devising representations suitable for clustering. We also discuss some methods of discriminant analysis at the end of the first chapter by contrasting them with analogous methods of clustering.

6 The Clustering Problem

An obvious strategy you may follow in grouping objects is that objects located on the same shelf are much similar to each other, whilst objects belonging to different shelves are very dissimilar. These groups of objects will be named *clusters*, understanding that they become *classes* once we acknowledge their utility and reward them for this by giving them a name. For instance, everybody knows what rice is, since it is a suitable and desirable food all over the world. On the contrary, we have as yet no commonly accepted classification of unsuitable e-mails.

A grouping rule emerging directly from data

In its essential fashion we may formally enunciate a cluster analysis problem as follows.

Definition 6.1 (Clustering Problem). *Starting from a set of m objects that we identify through a set of vectorial patterns $\{y_1, \ldots, y_m\}$, we must suitably group them into k clusters $\{d_1, \ldots, d_k\}$.*

The adverb *suitably* means that this problem may admit a vast variety of solutions. It occurs when the attic tidier is either you or your uncle simply because both of you want to make it in a great way (hence you solve an optimization problem), but according to a different optimality criterion and possibly with a different optimization method (but scrambling the items never looks like an optimization method!).

6.1 The Roots of Cluster Analysis

We propose to resume the most of the above methods into the following framework. In the absence of an *a priori* meaning of the classes we want to discover, we locate the clusters in the feature space with the sole commitment of proving very different each other according to a metric we state among them. Let us denote by d_i the i-th cluster, by D_i the decision of attributing a generic pattern y to it, and by $\ell[D_i, d_j]$ a cost function which appraises a penalty we incur in moving an item from cluster D_i to a different cluster d_j, assuming null the penalty $\ell[D_i, d_i]$ of not moving the item. The above commitment may be obtained by *maximizing* with respect to the cluster assignment function $D(y)$, applying to item y, a global penalty C associated to all assignments. Namely, C is computed as a double sum: on each item it sums a cumulative cost that in turn adds up the losses we pay with decision $D(y)$ versus shifts to all different clusters. Namely, we want to partition the pattern space $\mathfrak{Y}$ into k subsets through a decision rule $D(y) : \mathfrak{Y} \to \{d_j\}$ such that:

Fix a penalty,

$$D = \arg\max_{\widetilde{D}} \mathsf{C}[\widetilde{D}] = \arg\max_{\widetilde{D}} \left\{ \sum_{y \in \mathfrak{Y}} \sum_{j=1}^{k} \ell\left[\widetilde{D}(y), d_j\right] \right\}. \quad (6.1)$$

and identify a law maximally penalizing the transgressors.

Note how with this strategy the loss function is not an ethical punishment tool for a *mistake* you did; it is only a way to select the class to which to efficiently attribute the pattern at hand. We identify efficiency with sensitivity to a cluster change, whatever the future meaning of the cluster, and we want a decision rule maximizing it. We could call this an *agnostic approach*.

In alternative,

With other approaches, that we could denote *aprioristic*, one prefers to start with a well defined model of the data and their class membership, apart from some free parameters $\boldsymbol{\theta}$. In this case we have a dual optimization problem, in that we want to minimize an overall error cost representing the average loss you pay connected to the probability that the decision you take about the label of an item is wrong. Namely, $\lambda[D_i|\widetilde{d}_j]$ is the cost of the decision D_i when the *true* class label – assigned by the model but hidden to you – is $\widetilde{d}_j$. Like in the mixture of Gaussian model in Fig. 6.1, a pattern y may belong to more than one class, with certain probabilities, so that we may conversely deal with a conditional probability distribution $p_{\Lambda|\boldsymbol{\theta}}$ over the class labels given the free parameters. We obtain the risk $R[D_i|y]$ by averaging λ with respect to the conditional probabilities $p_{\Lambda|y,\boldsymbol{\theta}}(d_j)$ and the overall risk R of the clustering rule by averaging $R[D(y)|y]$ over Y. Thus, the optimization goal reads:

assume a truth

and derive a rule maximally rewarding the most compliancy.

$$D = \arg\min_{\widetilde{D}_\theta} \sum_{y \in \mathfrak{Y}} \sum_{j=1}^{k} \lambda\left[\widetilde{D}(y)|\widetilde{d}_j\right] p(\widetilde{d}_j|y, \boldsymbol{\theta}). \quad (6.2)$$

Both optimization goals – maximum spread between clusters with the former, minimum dispersion within cluster with the latter – resolve into the rules of a decision theory, possibly referred to an $\mathfrak{Y}$ spanning the space of all possible patterns.

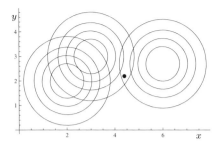

Fig. 6.1. Contour plot of Gaussian densities subtending a clustering analysis problem.

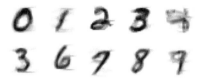

Fig. 6.2. Average feature profiles produced by the clustering procedure on MNIST dataset.

The dependency of the rules on the data to be clustered derives from their actual implementation. With the aprioristic approach (6.2) generates complete rules when parameter $\boldsymbol{\theta}$ is known. Otherwise we are called to jointly minimizing R with respect to both D and an estimate $\boldsymbol{\theta}$ as a function of a set of patterns $\{\boldsymbol{y}_1, \ldots, \boldsymbol{y}_m\}$ like in Definition 6.1. In this case $\mathfrak{Y}$ coincides with this set and the related sum in (6.2) reckons exactly m addends. We may view the patterns as a training set over which the rules are assessed in order to be applied to any further item of an extended $\mathfrak{Y}$. With the agnostic approach we cannot avoid to appeal statistics in order to fix decision rules since we may not know their free parameters by definition. Hence $\mathfrak{Y}$ necessarily coincides with a specific set of patterns in (6.1). The generalization of the rules to an extended set goes as above.

Agnostic and aprioristic approaches

In this book we will adopt the agnostic approach. In most common clustering problems the two approaches coincide. However, this is not always true and ours allows for a more operational reading of both problems and their solutions. A typical figure of clustering output of a subset of 100 items of MNIST dataset is shown in Fig. 6.2. In the figure are reported the average feature profiles, i.e. for each square the gray scale coming from averaging the colors of the corresponding squares of the handwritten digits grouped with a given label between 0 and 9. You see that feature profiles of 3 and 5 are extremely similar, which, on the

196   The Clustering Problem

No truth if you cannot discover it.

part of Martians, tells us that they have no way (at least with a same technology enabling this clustering) of distinguishing between the two digits. Said in other words, if we want to dialogue with Martians we must adopt another digit representation, for instance the Roman one where 3 is III and 5 is V.

Actually, with (6.1) we enunciated just an approach. To obtain results as above we must fill up significant operational aspects that we may raster into *rules*, *metrics* and *representation*.

Metrics. Let us start with the second category, which complies better with our intuition. The general feeling is to associate the loss function with a dissimilarity notion. Thus, having located the items to be clustered into a feature space, we mainly perceive points that are close each other as similar, hence to be grouped into a same cluster, and points that are far as dissimilar, hence to be associated to different clusters. For instance,

At the basis of penalty there is a metric,

- for ℓ related to the Euclidean distance:

$$\ell[D(\boldsymbol{y}), d_j] = \begin{cases} 0 & \text{if } D(\boldsymbol{y}) = d_j \\ (\boldsymbol{y} - \boldsymbol{v}_{d_j})^T(\boldsymbol{y} - \boldsymbol{v}_{d_j}) & \text{otherwise,} \end{cases} \quad (6.3)$$

where $\boldsymbol{v}_{d_j}$ is a suitable center of class d_j, that is commonly called *centroid* as it is biased by the cut-off effect of the cluster border,

inducing a dual minimization problem.

- the solution of (6.1) is:

$$D(\boldsymbol{y}) = \arg\min_j \left\{ (\boldsymbol{y} - \boldsymbol{v}_{d_j})^T(\boldsymbol{y} - \boldsymbol{v}_{d_j}) \right\}. \quad (6.4)$$

To operationally complete the rule we must define the centroids as well – a task requiring statistical intervention. Our intent to maximize (6.1) with loss function (6.3) implies the dual problem of individually minimizing the sum of the Euclidean distance of each item w.r.t. the centroid of the class it has been assigned to. For whatever assignment of items to clusters, the latter problem reads:

$$\boldsymbol{v}_{d_j} = \arg\min_{\boldsymbol{v}} \left\{ \sum_{\boldsymbol{y} \in d_j} (\boldsymbol{y} - \boldsymbol{v})^T(\boldsymbol{y} - \boldsymbol{v}) \right\}, \quad (6.5)$$

for which the solution comes in the form:

$$\boldsymbol{v}_{d_j} = \frac{1}{m_j} \sum_{\boldsymbol{y} \in d_j} \boldsymbol{y}, \quad (6.6)$$

where m_j is the number of patterns belonging to j-th cluster. Note that this solution comes also from the companion optimization problem (6.2) when we assume the loss function

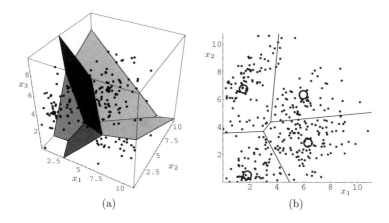

Fig. 6.3. Voronoi tessellation delimiting four clusters of: (a) a set of points in $\mathbb{R}^3$ and (b) one of their projections in $\mathbb{R}^2$. Bold points in (a) and circles in (b) identify the centroids of the Voronoi cells.

to be $\lambda[D(\boldsymbol{y})|d_i] = 0$ if the true label of $\boldsymbol{y}$ is i, 1 otherwise, and the distribution law of the points around the cluster centroid to be a multidimensional Gaussian distribution law [47], as $\boldsymbol{v}_{d_j}$ is a minimum variance weakly unbiased estimator of $\mathrm{E}[\boldsymbol{Y}_j]$, with $\boldsymbol{Y}_j$ denoting the random pattern describing points attributed to d_j. With our approach we avoid mentioning the $\boldsymbol{Y}_j$. Rather, we may suppose that Gaussian distribution law is a good model for dealing with future items.

Example 6.1. Coming back to Fig. 6.2, we adopted the metric (6.3) to obtain the pattern profiles shown in the picture.

Example 6.2. To better understand the output of decision rule (6.4), let us consider the clusters determined with this rule on the set of points of the artificial case study as illustrated in Fig. 6.3. It consists of a sample drawn from a mixture of Gaussian variables (say objects' families like books, shoes, sporting implements or boxes) in a three-dimensional space (e.g. considered in terms of weight, size and maintenance effort). We devise the location of the center and the covariance matrix of the single variables and check how these parameters influence the performance of a clustering algorithm, having fixed a number of initial centroids equal to 4.

The delimiters of the clusters form a *Voronoi tessellation* [58] whose convex polytope is the union of the edges of Voronoi cells. Namely, denoting with $\mathfrak{V}$ the set of centroids identified according to (6.5), a *Voronoi cell* for $v_j \in \mathfrak{V}$ is the set of all points $y \in \mathbb{R}^n$ (n equals 3 in Fig. 6.3(a), 2 in Fig. 6.3(b)) closer to v_j than to any other point of $\mathfrak{V}$.

<small>A variety of algorithms for exploiting the metric</small>

Rules. The implementation of rule (6.4) or similar ones deriving from other loss functions constitutes the core of a wide family of clustering algorithms that we will discuss in Section 6.3. Thus, implementing (6.4) we are implicitly adopting an *agglomerative rule* matching a general model where the granules of a same cluster are assumed to be dispersed around a central value representing the key characteristic of a local population of data. At the opposite side we have *aggregative rules* matching models where this property is distributed on the cluster items so that it emerges exactly from their aggregation. This induces techniques like Nearest Neighbor [12] where the distance is computed between the item to be clustered and those among the items already attributed to a cluster which are close to it. The label of the closest item decrees the assignment. Thus, a companion of Example 6.2 for aggregative rule using the same Euclidean distance for metric is represented by the following case study.

Example 6.3. Look at the 2-spiral dataset shown in Fig. 6.4(a). It is obtained from a base function plus a noise term ε, so that the i-th point reads in polar coordinates (radius r, angle θ):

$$r_i = a_i; \quad \theta_i = a_i + \varepsilon_i + \alpha, \qquad (6.7)$$

where a_i and ε_i are randomly sampled, the former uniformly in $[0, 4\pi]$, the latter from a Gaussian variable with $\mu = 0$ and $\sigma = 0.5$. We have 127 gray points and an equal number of black ones, where the difference stands in the initial phase α set to either 0 or π. Figs. 6.4(b) and (c) represent a successful implementation and an unsuccessful one of the Nearest Neighbor method. The gray tunes refer to the labels attributed by the clustering and the different fate depends on the choice of the initial aggregation points: the two highest distance points lead the algorithm to the original clusters, two random points to new different ones confusing 39 points. However, all that glitters isn't gold. Indeed, Fig. 6.4(d) shows a bad result when the algorithm is applied with the first option but to a more noisy dataset, where ε_i is drawn from a Gassian variable with $\sigma = 1.2$. In this case we have 23 labeling mistakes.

Actually we have a variety of rules exploiting metrics for deciding to which cluster an item should be attributed to. We will give a panoramic of them in Section 6.3.3.

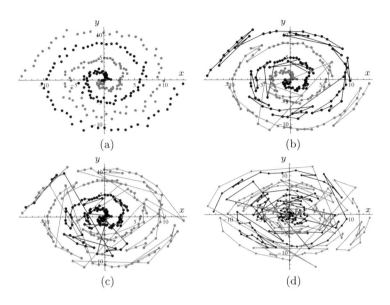

Fig. 6.4. Nearest Neighbor algorithm applied to a 2-spiral dataset. (a) small noise dataset; gray tunes distinguish the two original labels of the points. (b) successful clustering; gray tunes distinguish the labels computed by the clustering algorithm; segments track the aggregation sequence. (c) unsuccessful clustering, same notation. (d) unsuccessful clustering obtained by same algorithm as in (b) but applied to a dataset more noisy than in (a).

Representation. A third direction of this taxonomy concerns the representation of the data. In the same way as if you bring your blood sample to a clinical laboratory to be examined to understand whether you are healthy or ill, the operator will ask you what you want to find from the exam; also for clustering objects of your attic you must ask yourself which features you see in these objects. Maybe you want to consider their age, utility, conservation state, and so on. These are features that you have determined by yourself, in the prejudice that they are really useful to suitably clustering your objects. In many cases, however, you have a lot of characteristics of the objects but do not rely on them as patterns for classifying the objects. For instance, as in the case of many large databases, you have a long record of data (features in terms of variables' assignments) associated to the single item, but the similarity measure between two items based on the first 50 features is quite different from the measure based on the subsequent 50 features, and you ask yourself which one of the two measures is more reliable. Rather, we may look for

But, what do you want to observe of a record?

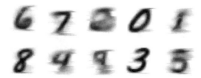

Fig. 6.5. Average feature profile of handwritten digits when we use the first 30 principal components out of the 784 total attributes.

suitable functions, for instance linear, of the original records, obtaining $s = W^T y$, with W denoting a linear operator mapping each patterns into either shorter records s reporting the true important characteristics of the objects or longer records s including necessary hidden variables allowing for their efficient classification. As the new records will be given in input, as intermediate patterns, to a loss function ℓ in (6.1) to accomplish your tidying up task, the whole accounts for identifying suitable families $\{\ell'\}$ of loss functions on the basis of special properties we may require to the intermediate records. It is a subtle question whether we are managing a new metric or a new representation through an old metric. We put the divide in the generality of the representation. If the properties we require to the new records are independent of the loss function they will feed, i.e. they represent a goal *per se*, we speak of new representation, like it happens with famous PCA [44, 30] or ICA [27] data processing. Otherwise we assume to manage a suitable preprocessing of the data as a part of loss function implementation.

A soft divide between metrics and representation

Example 6.4. Continuing Example 6.1, Fig. 6.5 denotes the improvement of our clustering if we linearly compress the initial 28×28 attributes of the pattern vector into 30 new features according to the PCA strategy as it will be discussed in Section 7.1.

Of course, since the Voronoi tessellation generalizes to any metric defined on $\mathbb{R}^n$, provided the centroid set $\mathfrak{V} \subseteq \mathbb{R}^n$ is discrete, all what you may do with a linear transform and a metric like (6.3) is to separate with a simple surface, such as a parabola, and in general with smooth convex surfaces, i.e. linear surfaces in a higher dimensional spaces, points that are not shuffled each other like in a salt-pepper background. Otherwise you need a new thorough description of the objects through more separable features; and this cannot happen through simple linear transforms of the original features. You may need logical operators such as hierarchical procedures, or, on the contrary, entirely subsymbolical operators, such as neural networks. We will consider some examples of these methods in Chapter 7 as for symbolic methods and in Chapter 8 for subsymbolic and hybrid methods.

6.2 Genotype Value of Metrics

A leitmotiv of this book is that we learn relations between a set of data in order to apply them to future elements of the same family of data. Also, as in the attic case, the reason why we tidied up it is, apart to make your parent pleased of you, that if a new item comes in the attic you immediately know where to put it or, *vice versa*, in search of an item with some characteristics, you promptly know which shelf may be inspected. Both goals may be resumed by the minimization of the probability of addressing a wrong shelf in the future. Namely, synthesize the history of future shelving inquiries through the distribution law of the random inquiry Y; associate to the exits of the inquiries a Bernoullian variable X being 1 in case of wrong addressing and 0 elsewhere; you are interested in the expected value $E[X]$. This in principle makes the goal in our perspective coinciding with the one in the *aprioristic* paradigm as mentioned before. Indeed, since decision rules for attributing items to shelves have been learnt from previous data, you obtain $P(X = 1)$ (hence $E[X]$) from the distribution law of the random variable Y. Moreover, we learnt from Chapters 1 and 3 that the c.d.f. F_Y of Y coincides with the limit of its empirical c.d.f. $\widehat{F}_Y$ with the increase of the sample size. The peculiarity of the present problem in our approach, however, is that the ordering of the items at the basis of the empirical c.d.f. is exactly the goal of our clustering (remember that $\widehat{F}_Y(t)$ is the frequency of items in a sample preceeding t in a given order). We pursue it by fixing a structure between data, and do it by devising a loss function.

> You have a metric because you assume a structure under the data.

As we mentioned before, the classical way of introducing clustering starts from a probability model of the data. In turn probability is rooted on measure theory through which we deal with the splitting hairs problem of the asymptotic measure of points. In our framework, where points unavoidably broad into blobs, we look for more elementary structures based on the notion of distance in a feature space that is inherently discrete, having the continuous extensions as suitable approximations when there is the need. We will base our clustering exactly on the distance notions we are able to state between objects, where the optimality direction to follow in order to improve these notions is drawn from the Levin inequality enunciated in Chapter 2 in terms of

Kolmogorov complexity. Namely, with reference to $x, z \in \mathfrak{X}$ as in Definition 2.5, we have:

$$f_{X|z}(x) \leq 2^{-K_z(x)}, \qquad (6.8)$$

where $K_z(x)$ is the length of the shortest program computing x on a universal computing machine having z in input. It essentially tells us what we may expect to describe of the future objects under clustering after the mutual structure we have stated about them. It is customary to fill up the inequality with two complementary actions as follows. On one side, we decrease the right member of (6.8) by substituting $K_z(x)$ with $d(x, z)$. Since Kolmogorov complexity is not computable in general (by definition), we intend with the latter the shortest length we may achieve with the current knowledge about the data, hence the best computation we may set up to generate them. On the other, we change inequality into equality, thus looking for the *maximum likelihood* of the data.

The mother of all metrics: Kolmogorov complexity.

In conclusion, moving to normed spaces, we approximately close the loop between data structure and their distribution law with the following relation:

You say a metric and you get a probability density.

$$f_{Y-v}(y) \propto e^{-\nu(y-v)} = \kappa e^{-\nu(y-v)}, \qquad (6.9)$$

for a suitable *normalization* constant $\kappa = 1/\int_{\mathfrak{Y}} e^{-\nu(y-v)} dy$, where $\nu(y - v)$ denotes the norm (in its weakest acceptation – see Appendix A) at the basis of the loss function ℓ for y now depending on the parameters of the cluster which the item y is attributed to. This may prove a satisfactory relationship since we are not requested to identify the exact dependence, as we are looking for a decision rule that optimizes the risk of future mistakes, not for exactly determining this risk. In conclusion, with the metric we use in our clustering problem we are exploiting all the structural knowledge we have about data. This will turn useful to us both to cluster present data and to figure what will happen in the future. In the favorite idea of conventional approaches that the items really belong to clusters and we want to correctly discover these memberships, the optimization rule (6.1) typically points at a direction for identifying the minimum entropy perspective with which to observe the data [28].

Coming back to Section 6.1, the solution (6.4) is based on a l_2 metric and attributes a distance $(y - v_{d_j})^T(y - v_{d_j})$ to points assigned to the j-th cluster from its centroid v_{d_j} for each j. With a density function $f_Y(y) \propto e^{-(y-v_{d_j})^T(y-v_{d_j})}$ this solution determines cluster assignments that minimize the sample entropy of this data model $\widehat{H} = 1/m \sum_{i=1}^{m} -\log f_Y(y_i) = 1/m \sum_{i=1}^{m}$

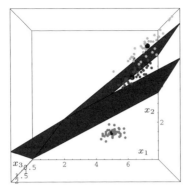

Fig. 6.6. Voronoi tessellation after k-means algorithm applied to Iris dataset using the correlation coefficient as similarity metric; same notation as in Fig. 6.3.

$(\boldsymbol{y} - \boldsymbol{v}_{D(\boldsymbol{y})})^T(\boldsymbol{y} - \boldsymbol{v}_{D(\boldsymbol{y})}) + c$ for a suitable c. In any case, maintaining our agnostic approach we have that **if:**

$$\ell[D(\boldsymbol{y}), d_j] = \begin{cases} \nu(\boldsymbol{y} - \boldsymbol{v}_j) & \text{if } D(\boldsymbol{y}) \neq d_j \\ 0 & \text{otherwise,} \end{cases} \quad (6.10)$$

defining:

$$\ell_0[D(\boldsymbol{y}), d_j] = \nu(\boldsymbol{y} - \boldsymbol{v}_j) \quad \text{for all } j, \quad (6.11)$$

then the maximization problem (6.1) translates into the minimization problem:

$$D = \arg\min_{\widetilde{D}} \left\{ \sum_{\boldsymbol{y} \in \mathfrak{Y}} \ell_0 \left[\widetilde{D}(\boldsymbol{y}), \widetilde{D}(\boldsymbol{y}) \right] \right\}, \quad (6.12)$$

which reads a sample version of:

$$\min_{\widetilde{D}} \left\{ \mathrm{E}_Y \left[\ell_0 \left[\widetilde{D}(\boldsymbol{y}), \widetilde{D}(\boldsymbol{y}) \right] \right] \right\}, \quad (6.13)$$

representing the target of the classification subsequent to the clustering. As an extension of (6.5), the problem (6.12) has solution:

$$D(\boldsymbol{y}) = \arg\min_{j} \nu(\boldsymbol{y} - \boldsymbol{v}_j). \quad (6.14)$$

Put it simply, we are managing for minimizing the above sum as an average value of the conditional complexity in the sample we are dealing with, which stands for hitting at its most concise description.

<small>Conciseness is the common quality we want of a metric.</small>

The efficacy of these distances obviously depends on the dataset they are applied to.

204   The Clustering Problem

Example 6.5. With reference to the Iris dataset using the correlation coefficient metric we develop a classifier performing with an error rate of 4%, nearly of the same order than the best classifiers reported in the literature [5], in spite of the constraint introduced by this metric of having the intersection of the clusters corresponding to the bisector of the positive orthant (see Fig. 6.6).

All metrics in Tables A.1 and A.2 share the feature of being (possibly piecewise) continuous functions of their argument, so that we may appreciate the sensitivity $\partial \nu / \partial \boldsymbol{y}$ of the norm ν to the single coordinates or suitable functions of them as we will see in Section 7.1.

<small>Metric may arise also from logical checks on pairs of data.</small> With categorical data we mean data assuming only a small number of values, each corresponding to a specific category value or label. By definition, a continuous metering of these data is not appropriate. Instead, a rich family of distances are considered in Table A.3 mainly based on *logical* properties the dataset owns.

6.3 Clustering Algorithms

The activity of grouping objects on the basis of their characteristics is as old as the existence of attics to be tidied up, societies to be ordered, etc. Its formalization into automatic procedures is a typical product of crossing of the technological facility offered by the computers and the methodological need raised by the users such as service companies, drug researchers and so on. Going back in the literature, the root computations for clustering are commonly attributed to McQueen [37], while the systematic discussion of the clustering problem is found in a pair of books in the seventies [17, 24]. The common practice, identified in that time

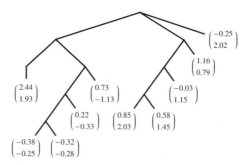

Fig. 6.7. A typical tree graph visualizing agglomerations of points drawn from a synthetic dataset.

clustering algorithms with k-means (also known as Lloyd's algorithm [36]), was in the thread of agglomerative algorithms. You start from a set of items that you tentatively assume as representative of a cluster and increase clusters by attributing each new element to the closer representative, according to a given metric. The subsequent iteration starts from updating the representatives on the basis of the last attribution and proceed with a new attribution. Looking at the tree graph like the one in Fig. 6.7 where in the roots we put the labels of the clusters and on the leaves the elements to be clustered, we may say that the above procedure is top-down. You start from the cluster representatives, even though not perfectly defined and subjected to be refined during the procedure, then you go down in order to state a single connection between clusters and items. On the contrary you may start from the bottom, just considering subsets of elements in order to check their mutual similarity/dissimilarity, and in such a way re-ascend to the root, thus finding the cluster label to be attributed to the element. Both procedures may be hierarchical, i.e. may proceed by partial partitioning/aggregation as we will see in a moment. Moreover we may have intermediate algorithms where the procedure is intrinsically top-down, but the number of clusters depends on a bottom-up behavior of the elements. Finally, as mentioned above, a third factor of this taxonomy concerns the way of characterizing the clusters: through either points or distributed properties (see Fig. 6.7). In this section, we will show some exemplars of the algorithms filling up the cells of the taxonomy, where many cells are empty and many other are overcrowded, expressing nuances of the researchers and peculiarity of the clustering problems they are called to solve. We describe the algorithms mainly in terms of the criteria through which they manage distances, paying less attention to the operational strategies to optimize the specific cost functions. These strategies constitute a relevant topic *per se*. However we may assume them already consolidated in the software available on the network, leaving further algorithmic sophistication to other books.

Either fix clusters and group data inside,

or look at data and leave clusters to emerge from them.

Intermediate strategies introduce hierarchy in the process.

Grouping may be based either on point or distributed features.

6.3.1 A Template Procedure

Let us start our excursus over the algorithms with the four steps in Algorithm 6.1 through which an archetypal k-means procedure may be described.

The magic four

Example 6.6. Fig. 6.8 displays a typical output of the basic version of this algorithm applied to some synthetic data using the Euclidean distance. It is analogous to the case shown in Fig. 6.3(a), yet restricted to a two-dimensional space. Data comes from

Graphic and

Algorithm 6.1. k-means algorithm

Given a set of items $\{s_1, \ldots, s_m\}$, with $s_i \in \mathfrak{S}$, a set of patterns $\{\boldsymbol{y}_1, \ldots, \boldsymbol{y}_m\}$ identifying the items within $\mathfrak{Y} \subseteq \mathbb{R}^n$, and a norm ν on the patterns,

1. choose k centroids $\boldsymbol{v}_i$ of clusters in $\mathbb{R}^n$ in a reasonable way (*initialization*)
2. attribute each point representing an item to the closest centroid according to $\nu(\boldsymbol{y} - \boldsymbol{v})$ (*assignment step*)
3. recompute the cluster centroids as the barycenters of the annexed points (*update step*)
4. loop steps 2 − 3 up to fill a *stop criterion*.

bivariate Gaussian variables, for which we determined the location of the center and the covariance matrix in order to check how these parameters influence the performance of the clustering algorithm. In particular, starting with $k = 4$ points randomly chosen from the whole dataset as initial centroids, the algorithm converges after a few iterations.

numerical diagnostic tools

While clustering algorithms fall in the class of unsupervised learning methods, since the label of the objects is not given in advance, the assessment of their efficiency is typically realized in a supervised mode. We are used to jointly appreciate the performance of the algorithm through the above graphs and the *confusion matrices* (see Table 6.1). Thanks to the gray tune differentiation in the pictures, with the former we realize how points generated by one of the Gaussian distributions, say a class, in the mixture are separated by the others in the tessellation. With the latter you have in each element (i, j) of the matrix the number of items belonging to class i and attributed to cluster j. If you index clusters with the same indices of the class mostly labeling its items, apart from rare pathologies inducing the same index on more than one cluster, you say that cluster i approximates class j where their intersection represents the set of "correctly classified items" (see Table 6.1(a)). Moreover, having normalized these numbers over the rows, you obtain frequencies whose sum of off-diagonal terms on row i represents the percentage of negative errors for class i, i.e. the percentage of points belonging to class i and attributed to differently indexed clusters (see Table 6.1(b)). Analogously, having normalized these numbers over the columns, you obtain frequencies whose sum of off-diagonal terms on column j represents the percentage of negative errors for cluster i, i.e. the percentage of points attributed

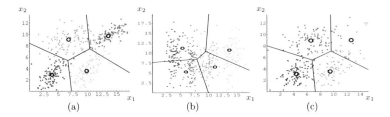

Fig. 6.8. k-means algorithm applied to three datasets generated as a mixture of four Gaussian distributions with different means and standard deviations. Points with the same gray level belong to the same population. Lines: polytopes delimiting each cluster; bold circles: cluster centroids. (a) equilibrated dataset (100 items from each population) and initial cluster centroids randomly picked within the whole dataset, (b) same as (a) with elongated population's shape; and (c) non equilibrated dataset (20 points from one population and 100 from each other one).

Table 6.1. Confusion matrix for the three examples shown in Fig. 6.8 normalized by rows.

	1	2	3	4
1	0.97	0	0.01	0.02
2	0	0.99	0.01	0
3	0.03	0.13	0.83	0.01
4	0.02	0.1	0.01	0.87

(a)

	1	2	3	4
1	0.66	0.34	0	0
2	0.4	0.6	0	0
3	0.09	0.06	0.69	0.16
4	0	0	0.27	0.73

(b)

	1	2	3	4
1	0.96	0.01	0.03	0
2	0.02	0.86	0	0.12
3	0.05	0	0.7	0.25
4	0	0	0	1

(c)

to cluster j and coming from differently indexed classes (see Table 6.1(c)).

Example 6.7. Continuing Example 6.6 we see that the clustering algorithm provides a sound classification performance in the first example (Fig. 6.8(a)), while suffers a drastic degratation in case (b) and (c), due to the elongated shape of the source distributions in the first case, and to the unbalancing of the classes into the dataset in the second one (the small cluster in the upper-right corner contains 20 points, compared to the 100 points of the other clusters).

As a matter of fact, we are enabled through this algorithm to draw linear separators between classes, that become planes as illustrated in Fig. 6.9 and hyperplanes in general.

A (possibly) linear tessellation,

Example 6.8. Coming back to the Iris dataset as discussed in Example 6.5, Fig. 6.10(a) illustrates the output of k-means algorithm

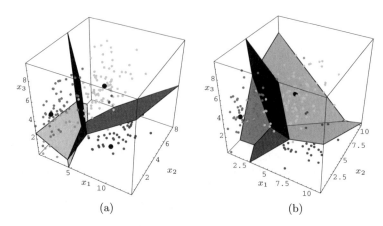

Fig. 6.9. k-means algorithm applied to three equilibrated datasets in $\mathbb{R}^3$ generated as a mixture of four Gaussian distributions with different means and standard deviations, where: (a) Euclidean distance, and (b) correlation coefficient is used as a metric. Initial cluster centroids are randomly picked within the whole dataset. Polygons: Voronoi tessellation; bold points: cluster centroids.

when starting with centroids randomly picked within the whole dataset and considering only first, third and fourth feature (hence, sepal length, petal length and petal width) for rendering convenience, discarding the component with less variance (see later on the PCA criterion supporting this choice). Only a fraction 0.093 of points are misclassified, corresponding to the 14 points lying on the boundary of the two non-separable classes. Note that the best we did by considering all the four features still brings to the same error of 0.093. However, the addition of the last feature introduces instability to the process, which converges often toward bad solutions.

Example 6.9. With reference to Example 6.4, things go worse for the MNIST dataset, where the high numbers of features, that not always classes and examples together with the high variability in the is adequate. writing style prevent the clustering algorithm to perform well. Figs. 6.10(b-c) show respectively the ten final centroids obtained by applying Algorithm 6.1 to the first 1000 examples in the MNIST dataset starting from: (b) $k = 10$ initial centroids, each randomly selected from the subset of profiles corresponding to a given digit, and (c) $k = 10$ initial centroids chosen as linear combinations of 5 digits uniformly picked from the whole dataset. Even though from the former picture we can clearly identify the different digits, the misclassification error attested around 0.416, while it moves beyond 0.88 in the last case as can be seen from

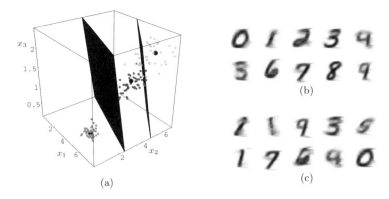

Fig. 6.10. k-means algorithm applied to (a) Iris dataset and (b-c) MNIST handwritten digits. (a) Same notation as in Fig. 6.9, with initial centroids randomly picked from the whole database; (b) final centroids obtained by starting with initial centers selected from the true clusters, and (c) same as (b) with initial centers generated as linear combination of randomly drawn objects from the whole population.

the more blurred shapes of the centroids and the repetition of digit 1.

The main advantages of k-means algorithm are its simplicity and *observed* speed, which allow it to run on large datasets. As for the latter, even if a recent result [2] shows a lower bound construction of a family of k-means instances for which the running time is *superpolynomial* (i.e. $2^{O\sqrt{m}}$ with m number of input patterns), "in practice the number of iterations is generally much less than the number of points" [18]. The above discrepancy has been partly circumvented by means of probabilistic reasoning (e.g. smoothed analysis [52]) showing a polynomial running time of the algorithm in smoothed high-dimensional settings with high probability [23]. Yet it does not systematically yield the same result with each run of the algorithm. Moreover it is highly sensitive to its free parameters.

A first aid to the algorithm has been supplied by many researchers and practitioners in terms of smart exploitation of the choices left open by the procedure: **Looking for improvements**

a) *number k and values of initial centroids*: how do you determine k and attribute initial values to the centers?
b) *aggregation vs. agglomeration strategy*: where to localize the distinctive characters of a cluster: either concentrated in a single point or distributed on the whole set of points grouped into a cluster?

c) *stopping rule*: when do you decide that the items' grouping is satisfactory?

d) *sequential vs. parallel implementation*: do you update the cluster center after each annexation or after all items have been processed?

Answering these questions gives rise to a large variety of algorithms that really work differently, so that we have the problem of which algorithm to choose for which data and why. This is the vein along which we will try to solve the puzzle of granular computing in the next sections.

6.3.2 Initial Numbers and Values of Centroids

The initial choice of the number of clusters plays a decisive role in k-means-like algorithms. In the previous experiments we set it exactly to be equal to the number of classes that we knew generated the dataset; in general this number is unknown. One common *rule of thumb* consists in choosing a number of clusters so that adding another cluster does not add tangible information. According to the general wisdom: "look for well separated clusters of tightly grouped items", we are used to split the overall dispersion Δ_y of the object features around their general mean v into a *between clusters* component Δ_b and a *within class* component Δ_w as follows, denoting with $d(y_i, y_j)$ the square of the Euclidean distance d_{ij} in Table A.2, and v_j a v_{d_j} as in (6.6):

> A saddle point with maximal between-cluster distances and minimal within-cluster distances

$$\Delta_y(k) = \frac{1}{k}\sum_{j=1}^{k}\frac{1}{m_j}\sum_{i\in d_j} d(y_i, v);$$

$$\Delta_w(k) = \frac{1}{k}\sum_{j=1}^{k}\frac{1}{m_j}\sum_{i\in d_j} d(y_i, v_j); \quad \Delta_b(k) = \frac{1}{k}\sum_{j=1}^{k} d(v_j, v).$$
(6.15)

By simple algebra we realize that $\Delta_y = \Delta_b + \Delta_w$. This constitutes the master equation of a vast vein of methods rooten on Analysis of Variance (ANOVA) [55] to check the significance of the split we operate of a sample into subsamples/clusters.

For instance, the family of heuristics resumed by the so-called *elbow criterion* [38] is:

> graph the clustering information, resumed through an index variously connected to the above dispersion indices, against the number of clusters and choose the value of k where the marginal information gain will drop, i.e. the information increase flattens remarkably generating an angle (the elbow) in the graph.

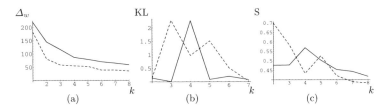

Fig. 6.11. The elbow criterion and two ways of formalizing it for the Iris dataset (dashed line) and the synthetic dataset (plain line) reported in Fig. 6.8(a). (a) Within cluster dispersion $\Delta_w(k)$ versus number of clusters k; (b) trend of KL index (6.16) and (c) Silhouette statistic (6.17) versus k.

Typical information indices formalize the above heuristic by maximizing suitable functions of the above Δs, as the Krzanowski-Lai (KL) [34] and Calinski-Harabasz (CH) [9] index, defined as:

> You can use several indices

$$\mathrm{KL}(k) = \left| \frac{\psi(k)}{\psi(k-1)} \right|; \quad \mathrm{CH}(k) = \frac{\Delta_b/(k-1)}{\Delta_w/(m-k)}, \quad (6.16)$$

where $\psi(k) = (k-1)^{2/n}\Delta_w(k-1) - k^{2/n}\Delta_w(k)$, with n denoting the dimension of the feature space.

Alternatively, with *Silhouette* statistic S [31] we measure how well matched is an object to the other objects in its own cluster versus how well matched it would be if it was moved to another cluster. Formally it is defined as:

$$\mathrm{S} = \sum_{i=1}^{m} \frac{1}{m} \frac{\bar{c}_j - \bar{d}_j}{\max\{\bar{d}_j, \bar{c}_j\}}, \quad (6.17)$$

where $\bar{d}_j$ is the average of all the distances within the cluster $\boldsymbol{y}_j$ belongs to, and $\bar{c}_j$ is the minimum of the mean distances between $\boldsymbol{y}_j$ and all the points in the closest clusters.

Example 6.10. Fig. 6.11(a) shows the course of $\Delta_w(k)$ versus the clusters number k for the Iris dataset (dashed line) and the synthetic dataset reported in Fig. 6.8(a) (plain line), denoting a marked change in the trend of the dispersion when $k = 3$ for the first example and $k = 4$ for the second one. Actually, in both cases an earlier angle appears in $k = 2$, but the flattening of Δ_w is not too much pronounced). Figs. 6.11(b-c) show respectively the plot of the KL index and Silhouette statistic versus k. In particular the former provides a clear and correct solution to the problem with the processed datasets.

The reader may imagine the huge set of variants of these generic methods (see for instance the GAP statistic [54] and the H index

[24]) that have been proposed in the literature. This number still increases if we consider the options, besides random selection, for initializing the centroids. For instance, in GNAT [8] a random sample is drawn from a dataset and the initial k centers are picked from this subsample in such a way that, after selecting a sample element, the choice falls upon the next object that is farthest away from the previously selected ones. An exhaustive search based on the same principle may prove more accurate, though more time consuming [62].

6.3.3 Aggregation vs. Agglomeration

The above discussion dealing with metrics and methods has been dealt with the subtending idea that the representative of a cluster is a wisely selected point constituting its centroid, where the way of contrasting it with candidate members of the cluster is demanded to the metric. On the contrary of this top-down procedure, we may start from the objects and compare them reciprocally in order to induce some regularity to emerge. This stands for having the representative of a cluster smeared on all its members, sharing ensemble characteristics like texture or fractal properties or even less structured ones. A template of bottom-up cluster algorithms is the *hierarchical clustering* [29, 31]. In this case we start from as many clusters as the number of elements in the dataset and proceed by subsequent grouping of items according to a given merging rule between current clusters. The typical outputs of this procedure are *dendrograms* or *profiles graphs* like in Figs. 6.12(a-b) respectively, whose layout intuitively depends on the grouping rule.

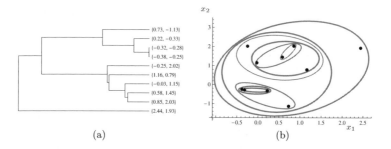

Fig. 6.12. Same toy example as the one in Fig. 6.7 showing: (a) a typical dendrogram aggregating clusters through single-link method and assuming Euclidean distance between pair of points, and (b) corresponding profiles graph showing the incremental clustering of points. Leaves and point labels: items to be clustered.

Table 6.2. Most common linkage methods together with the induced intercluster dissimilarity.

Linkage method	Intercluster dissimilarity
single	smallest intercluster dissimilarity
average	average intercluster dissimilarity
complete	largest intercluster dissimilarity
weighted	weighted average intercluster dissimilarity
centroid	distance between cluster centroids
median	distance between cluster medians
variance	the sum of intracluster variance
Hausdorff	the Hausdorff distance between clusters (see (5.10))
Ward	the increase in variance that would occur if two clusters were fused

Algorithm 6.2. Hierarchical Clustering

With the same notation of Algorithm 6.1, for a given linkage method representing dissimilarity between two clusters, and a threshold ϑ,

1. start from m clusters containing each a single element in $\mathfrak{S}$ (*initialization*)
2. compute the pairwise dissimilarity between each couple of clusters (*pairwise matrix*)
3. merge all the clusters whose distance does not exceed ϑ (*merge step*)
4. loop steps 2 − 3 up to fill a *stop criterion*.

Merges are based on a given ensemble feature of the clusters that is updated at each production of a new cluster. In Table 6.2 we characterize the most common features through the method for computing them and the merging criterion they realize. With *single* and *complete* methods each item in a cluster maintains its individuality and the features are computed in relation to a contrasting cluster. Namely, with the former method we consider the smallest distance, in the profiles space, between pair of elements belonging to the two clusters, with the latter the greatest. With the other methods we substitute the actually clustered elements with a virtual one having a *central profile* represented for instance by the average, median or other statistics of the single profiles. Each of these features is used to compute a specific distance between clusters, such as the distance between cluster centroids, treated as an index of their dissimilarity. Note however

Hierarchies in many respects,

that with the Ward method [59] the dissimilarity index does not satisfy the properties presented in Definition A.1.

The merging criterion is based on one of the above dissimilarity indices or similar. We distinguish two ways of implementing the criterion. Namely, we may have a binary tree if we aggregate the closest pair of clusters at each step. Alternatively we may set a threshold, so that at each step all clusters merge whose dissimilarity index is less than the specified threshold. In both cases the algorithm iterates either until a single cluster remains, or when a proximity threshold has been exceeded. A way of generating the dendrogram in Fig. 6.12 is shown in Algorithm 6.2.

Example 6.11. Fig. 6.13 reports the dendrogram obtained on the Iris dataset. At each iteration we grouped the closest clusters according to the *single linkage* method. This gives rise to a special tree, a dendrogram indeed, where, independently from the iteration number, subsequent merges determine in parallel subsequent ramification levels. Having the root at the level 0 and the leaves at the highest level, our graph is truncated in the back way at level 14 and starts from level 2, for the sake of readability. We used Euclidean distance between data items in the patterns space. Plain labels denote the item's position within the dataset; alternatively when the label does not refer to a leaf, boxed labels specify the cardinality of the clusters they characterize, i.e. the number of leaves of the subtree they represent. From the dendrogram we recognize that the set of patterns belonging to the class *Iris Setosa*, constituting the cluster linearly separable from the others, completely split at level 3 (lowest group of four lines). By contrast, the two remaining huge boxes (with 38 and 39 leafs) separately belonging to the two remaining classes split only at a higher level. The other points represent small clusters that cannot be classified as belonging to a specific one of the three main classes. For the sake of completeness, Fig. 6.14 shows the smoothed convex hull of the first, second and 14-th level of profile graph referred to the two coordinates with highest variance of the Iris dataset, confirming the considerations on the dendrogram in Fig. 6.13.

Clustering hierarchy may be implemented in a dual mode as well. With *divisive* algorithm such as DIANA [31] we start from a single cluster, then iteratively split it in more groups. This types of algorithms, however, finds rare applications.

At an intermediate position you have methods that work as a k-means basically, but negotiate k – i.e. the number of clusters – at run-time. For instance QT Clust algorithm [26] uses a maximum intra-cluster diameter $d_{\max}$ for this negotiation: once $d_{\max}$

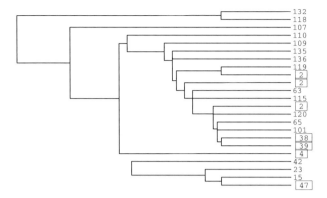

Fig. 6.13. Iris dataset dendrogram, produced with Euclidean distance between pairs of points and single-linkage method for intercluster dissimilarity. The graph is visualized from the second to the 14-th level. Plain labels: item index within the dataset; boxed labels: cardinality of the cluster they represent.

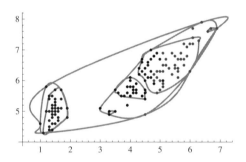

Fig. 6.14. First, second and 14-th level of the profile graph referred to the Iris dataset when projected on the two most dispersed coordinates. Points' gray levels: original labels.

has been selected, a candidate cluster with this diameter is centered on each point in the dataset, but only the one containing the majority of points wins the first tournament. This cluster is saved, its points deleted and the algorithm starts a new cycle until no points remain.

The most popular algorithm of this category is ISODATA [3] (Iterative Self-Organizing Data Analysis Techniques) which has gained visibility thanks to its sophisticated management of the centroids that are automatically adjusted during the iteration. It consists of: i) merging similar clusters, i.e. clusters whose distance between centroids goes below a certain threshold, ii) splitting clusters with large standard deviations, and iii) erasing

ISODATA negotiations

Algorithm 6.3. ISODATA algorithm

With the same notation of Algorithm 6.1,

---

1. fix: (*initialization*)
   - thresholds $\vartheta_P, \vartheta_N, \vartheta_S$, and ϑ_D,
   - a constant α,
   - k centroids $\boldsymbol{v}_i$ of clusters in $\mathbb{R}^n$ in a reasonable way.
2. assign each item to the closest centroid according to the fixed metric chosen (*assignment step*)
3. discard clusters with fewer than ϑ_N elements (*erasing step*)
4. update the cluster centroids (*update step*)
5. (*split step*):
   - compute the standard deviation vector
   $$\Sigma_j = (\sigma_1^{(j)}, \ldots, \sigma_i^{(j)}, \ldots, \sigma_t^{(j)}), \quad \text{for } j = 1, \ldots, k,$$
   where $\sigma_i^{(j)}$ denotes the standard deviation of the i-th component within the cluster j
   - compute the maximum component $\sigma_{\max}^{(j)}$ of each Σ_j
   - if $\sigma_{\max}^{(j)} > \vartheta_S$, split the j-th cluster into two new clusters by adding $\pm \delta$ to the old centroid $\boldsymbol{v}_j$, where $\delta = \alpha \sigma_{\max}^{(j)}$ for a suitable $\alpha > 0$, obtaining $\boldsymbol{v}_j + \delta$ and $\boldsymbol{v}_j - \delta$
6. (*merge step*)
   - compute the pairwise distance r_{ij} between the centroids $\boldsymbol{v}_i$ and $\boldsymbol{v}_j$
   - select at most ϑ_P smallest r_{ij}s which are smaller than ϑ_D
   - join the selected centroids by replacing $\boldsymbol{v}_i$ and $\boldsymbol{v}_j$ with their mean, weighted by the number of items in each cluster, and goto step 2
7. loop steps $2 - 5$ up to fill a *stop criterion*.

---

small clusters. One way of implementing the above facilities is sketched in Algorithm 6.3. Some versions of the ISODATA algorithm specialize the conditions in both merge and split step by considering both the number of desired clusters and the within/between clusters variance. As a matter of fact, procedures like this are surely more flexible than the basic k-means, but the user has to choose empirically many more threshold parameters whose values could substantially impact the quality of results.

Example 6.12. Fig. 6.15 shows an initial configuration and final output of a simple version of the ISODATA algorithm applied to the synthetic dataset introduced in Example 6.6. We start giving 10 random centers (see Fig. 6.15(a)) and at each iteration we are allowed to: i) merge two clusters whose centers have a distance less than a threshold set to 4, allowing only 2 merges per iteration, and

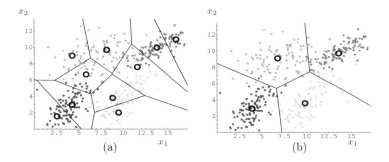

Fig. 6.15. ISODATA algorithm applied to synthetic data: (a) initial clustering starting from 10 centers; (b) final output of the algorithm.

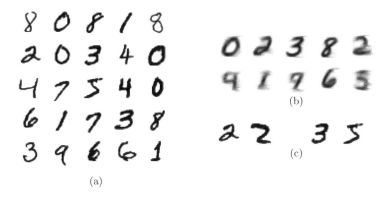

Fig. 6.16. ISODATA algorithm applied to handwritten digits: (a) initial centroids; (b) final output of the algorithm; and (c) two couples of digits: the distance between the former (having the same label) is greater than the distance between the latter (belonging to two different classes).

ii) delete clusters containing at most 5 points. No splitting facility is used.

Example 6.13. Things go worse with the MNIST dataset. Here we started from the 25 digits shown in Fig. 6.16(a) but what we obtained was a set of 10 clusters (whose representatives are shown in Fig. 6.16(b)) misclassifying a fraction 0.43 of examples. The key reason behind such a poor performance is that the complexity of geometry of data is not well represented by the Euclidean distance. For instance, as shown in Fig. 6.16(c), a pair of samples of 3 and 5 have a distance of 8.5, while a pair of not particularly distorted 2s have a distance of 11.3. Thus, any merging/splitting criterion based on this distance is not efficient.

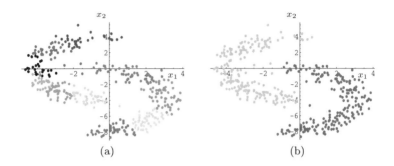

Fig. 6.17. Agglomerative clustering applied to a synthetic dataset halfring shaped: (a) clusters obtained by k-means algorithm where k is set to 10; and (b) two clusters obtained by aggregating those clusters obtained in (a) by minimizing the average intercluster dissimilarity.

Merge clusters but not centroids. This highlights the room for another aggregation strategy, where we merge clusters but not their centroids. We start from a partitioning algorithm (like k-means) obtaining an initial set of clusters, then we proceed by merging the indices of two clusters when we recognize they represent the same class, without merging their centroids. Namely, we choose one of the inter-clusters dissimilarities reported in Table 6.2 or some other functions we may decide to invent (usually denoted as *consensus functions* [39]) and, following the hierarchical paradigm, we merge clusters whenever their distance drop below a given threshold. By this approach we will attribute the same meaning to the symbols grouped in the two clusters, but we maintain centroids separated, in order to avoid collapsing them into a new point that is not representative of the elements of the two clusters.

Example 6.14. Even though with this strategy we gain just a few percentage of correctness in the MNIST dataset, with datasets as in Fig. 6.17 we are able to isolate the two main groups in spite of their non convex envelopes. In particular, Fig. 6.17(a) shows the 10 clusters obtained with naive k-means algorithm and Fig. 6.17(b) highlights that the two halfrings groups are definitely discovered by aggregating the 10 clusters through the minimization of the average intercluster dissimilarity (see Table 6.2). The minimization is achieved as a recursive post-processing of the 10 clusters in Fig. 6.17(a). Namely, we iteratively merge in a unique cluster the couple of clusters having minimum dissimilarity below a given threshold. The procedure stops with only 2 of them above the threshold.

6.3.4 Implementation Mode and Stopping Rule

As we premised, the technical aspects of the clustering algorithms are out of the scope of this book. Thus, in this paragraph we only fix some stakes delimiting these aspects from a computer science perspective. First of all let us highlight that until now we have interpreted suitability in Definition 6.1 in terms of optimality. In this respect the maximization of the cost function (6.1) represents an NP-hard problem [19]. This means that we have no guarantee of reaching an optimal solution in reasonable time. In fact, k-means is a local optimization algorithm generally reaching a fixed point in a few iterations. As it is, it suffers of the typical drawback of greedy algorithms [4] such as the risk of entering limit cycles, the habit of getting stuck into a local optimum far from the global one, the dependence on the initialization of the algorithm. Many recipes have been implemented to overcome these kinds of drawbacks that are common to a lot of optimization problems. The main directions that have been stressed concern:

We solve approximately a hard optimization problem, trying many ways to escape local optima.

1. partially exhaustive exploration of the search space. The idea is to systematically explore a part of this space involving a polynomial number of trials. An algorithm guaranteeing a fast convergence to a minimum in probability is based on the *exact relocation test* [61]: an object is assigned to a new cluster when the overall value of the specific loss $\sum_{i=1}^{m} \ell_0[D(\boldsymbol{y}_i), D(\boldsymbol{y}_i)]$ effectively diminishes;
2. branching the above exploration while pruning not promising branches and avoiding to repeat a same exploration (in tabu search fashion) [22];
3. injecting randomness in the search path, allowing to overcome the local minima traps. A wide family of simulated annealing algorithms have been implemented in the last decades with this aim [32];
4. multistarting. A way of reducing the influence of the initial configuration is to have many different initial conditions. This may be pursued either by a parallel algorithm jointly driving more search paths, or by a sequential algorithm jumping to different sites of the optimization landscape when a search path looks having reached a fixed point;
5. evolutionary algorithms. A way of efficiently exploiting the last two points is represented today by evolutionary algorithms, such as genetic algorithms and artificial immune systems that we will discuss in detail in Chapters 8 and 9.

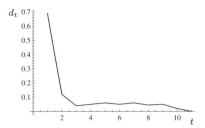

Fig. 6.18. Graph of the total displacement d_t of all centroids w.r.t. the number of iterations t for the Iris dataset.

We generally implement incremental clustering without knowing when to stop. All variants along these directions give rise to incremental algorithms, so that they suffer of the general weakness of not having a clear indicator of when the computation has to be assumed as concluded. An implementation may choose to stop the algorithm either after a certain number of iterations, or when the changes in the means have become small, or when only a few items cross the clusters boundaries. A typical rule used in our experiments is: "if all clusters remain unchanged, then stop". For instance, in Fig. 6.18 we show the trend of the total variations in all centroids position from an iteration to the next one. Therein the algorithm was stopped when this value dropped below a fixed threshold (here equal to 0.001).

6.4 The Weighted Sum Option

As considered in Section 6.2, in case the cost function is based on a norm $\nu(\boldsymbol{y} - \boldsymbol{v}_j)$, with $\boldsymbol{v}_j$ the centroid of cluster j, we may move from (6.1) to its dual formulation (6.12) that may be interpreted as a sample version of the classification goal $\min_{\widetilde{D}}$ $\left\{ \sum_{\boldsymbol{y} \in \mathfrak{Y}} \ell_0 \left[\widetilde{D}(\boldsymbol{y}), \widetilde{D}(\boldsymbol{y}) \right] \right\}$ referred to the whole pattern space $\mathfrak{Y}$. **In case the assignments are not univocal,** This in case of sharp classification. When we cannot univocally assign a pattern to a single cluster, we may consider $\mathfrak{Y}$ to be the projection of an extended space $\mathfrak{H}$ partitioned by the questioned clusters. Their projections in $\mathfrak{Y}$ overlap, so that we may wonder **you may introduce probabilities and minimize entropies,** around a probability $f_{C|\boldsymbol{y}}(c)$ that we have met a cluster c given that we have observed a feature vector $\boldsymbol{y}$ or, *vice versa* a probability density $f_{\boldsymbol{Y}|C=c}(\boldsymbol{y})$ that $\boldsymbol{Y} = \boldsymbol{y}$ within a cluster c. Then the dual formulation (6.4) of the clustering problem would read:

$$D = \arg\min_{\widetilde{D}} \left\{ \sum_{c=1}^{k} f_C(c) \ell_0 \left[\widetilde{D}(\boldsymbol{y}), d_c \right] \right\}. \tag{6.18}$$

In this framework, in view of (6.9), target (6.18) looks for minimizing the conditional entropy $H[\boldsymbol{Y}|C]$, defined as the mean value of the conditional entropies of $\boldsymbol{Y}$ given a specific cluster it belongs to, namely:

$$H[\boldsymbol{Y}|C] = -\sum_{c=1}^{k} f_C(c) \int_{\boldsymbol{y}\in\mathfrak{Y}} f_{\boldsymbol{Y}|C=c}(\boldsymbol{y}) \log f_{\boldsymbol{Y}|C=c}(\boldsymbol{y}) \mathrm{d}\boldsymbol{y}. \tag{6.19}$$

In force of (6.9) we read (6.19) as:

$$H[\boldsymbol{Y}|C] = \sum_{c=1}^{k} \int_{\boldsymbol{y}\in\mathfrak{Y}} f_C(c) f_{\boldsymbol{Y}|C=c}(\boldsymbol{y}) \nu(\boldsymbol{y} - \boldsymbol{v}_c) \mathrm{d}\boldsymbol{y} + a =$$

$$\int_{\boldsymbol{y}\in\mathfrak{Y}} \sum_{c=1}^{k} f_{C|\boldsymbol{y}}(c) f_{\boldsymbol{Y}}(\boldsymbol{y}) \nu(\boldsymbol{y} - \boldsymbol{v}_c) \mathrm{d}\boldsymbol{y} + a, \tag{6.20}$$

where a is a suitable constant. Willing to minimize $H[\boldsymbol{Y}|C]$, the distribution $f_{C|\boldsymbol{y}}$ represents the true goal of our inference. Thus we are no more looking for a sharp decision D about item assignments, rather we search for the probability that the item belongs to any of the candidate clusters. This implies $\sum_c f_{C|\boldsymbol{y}}(c) = 1$ for each $\boldsymbol{y}$. We don't know, however, $f_{\boldsymbol{Y}}$; thus, we come to a fuzzy framework in place of a probabilistic one. Namely, we sterilize $f_{\boldsymbol{Y}}$ by setting it to 1, denote $f_{C|\boldsymbol{y}}$ with $u_{c\boldsymbol{y}}$ and focus on the goal:

[margin: or at least membership functions,]

$$\boldsymbol{u} = \arg\min_{\widetilde{\boldsymbol{u}}} \left\{ \sum_{c=1}^{k} \int_{\boldsymbol{y}\in\mathfrak{Y}} u_{c\boldsymbol{y}} \ell_0 \left[\widetilde{D}(\boldsymbol{y}), d_c \right] \right\}, \tag{6.21}$$

which reads:

$$\boldsymbol{u} = \arg\min_{\widetilde{\boldsymbol{u}}} \left\{ \int_{\boldsymbol{y}\in\mathfrak{Y}} \sum_{c=1}^{k} u_{c\boldsymbol{y}} \nu(\boldsymbol{y} - \boldsymbol{v}_c) \right\}, \tag{6.22}$$

where $u_{c\boldsymbol{y}}$ is a fuzziness measure, as it comes from the ratio $f_{\boldsymbol{Y}|C=c}(\boldsymbol{y})/f_{\boldsymbol{Y}}(\boldsymbol{y})$. Moreover, in order to render the solution far from the extremes $\{0,1\}$ on each $\boldsymbol{y}$ we should add more constraints. The one used by the Fuzzy C-Means algorithm (for short FCM) concerns the sum of powers of $u_{c\boldsymbol{y}}$. Namely we write as $f_{C|\boldsymbol{y}} = u_{c\boldsymbol{y}}^r$, so that the normalization constraint is $\sum_c u_{c\boldsymbol{y}}^r = 1$. Then as a further constraint it is required that $\sum_c f_{C|\boldsymbol{y}}^{-r}(c) = 1$, i.e. $\sum_c u_{c\boldsymbol{y}} = 1$. Let us go through this algorithm which proved very suitable in many conceptual tasks and applications.

[margin: with consistency constraints.]

6.4.1 The Fuzzy C-Means Algorithm

As mentioned before we move from the identification of a decision rule D as in (6.18) to the design of membership functions specific of each fuzzy set. With the fuzzy sets notation, the latter is carried out through the minimization of the following weighted sum:

A weighted sum minimization,

$$C = \sum_{i=1}^{m} \sum_{c=1}^{k} u_{ci}^r \nu(\boldsymbol{y}_i - \boldsymbol{v}_c), \qquad (6.23)$$

where $\boldsymbol{v}_c$s are n-dimensional prototypes of the clusters and $U = [u_{ci}]$ achieves the additional functionality of a partition matrix expressing a way of allocation of the data to the corresponding clusters; u_{ci} is the membership degree of data $\boldsymbol{y}_i$ in the c-th cluster. The distance between the data $\boldsymbol{y}_i$ and prototype $\boldsymbol{v}_c$ is specified in terms of a suitable norm ν. The fuzzification coefficient $r > 1$ expresses the impact of the membership grades on the individual clusters.

with constrained weights,

The partitioning functionality adds the further requirement on u_{ci} of smearing a unitary membership mass of each pattern on the candidate clusters. Hence, we obtain the characterization of the partition matrix through the following two properties:

$$0 < \sum_{i=1}^{m} u_{ci} < m, \quad c = 1, 2, \ldots, k, \qquad (6.24)$$

$$\sum_{c=1}^{k} u_{ci} = 1, \quad i = 1, 2, \ldots, m. \qquad (6.25)$$

Let us denote by U a matrix satisfying (6.24-6.25). The first requirement states that each cluster has to be nonempty and different from the entire set. The second requirement states that the sum of the membership grades should be confined to 1.

through a iterative two steps procedure.

The minimization of C must be completed w.r.t. U and the prototypes $\boldsymbol{v}_c$ of the clusters. As for the latter, for ν coinciding with l^2 norm, from simple algebra we know that the optimal centroid/prototype of cluster c is represented by the center of mass of the $\boldsymbol{y}_i$s each affected by a mass u_{ci}, whatever c and u assignment inside it. Namely:

$$v_{cj} = \frac{\sum_{i=1}^{m} u_{ci}^r y_{ij}}{\sum_{i=1}^{m} u_{ci}^r}, \qquad (6.26)$$

with c ranging on the k centroids and j on the dimensionality n of the space $\mathcal{Y}$. A bit sophisticated computation shows that the optimal u assignment is given by:

$$u_{ci} = \frac{1}{\sum_{\kappa=1}^{k} \left(\frac{\nu(\boldsymbol{y}_i - \boldsymbol{v}_c)}{\nu(\boldsymbol{y}_i - \boldsymbol{v}_\kappa)}\right)^{\frac{1}{r-1}}}. \quad (6.27)$$

Overall, the FCM clustering is completed through a sequence of iterations where we start from some random allocation of data (a certain randomly initialized partition matrix) and carry out the following updates by adjusting the values of the partition matrix and the prototypes. Thus this algorithm complies with the iterative procedure sketched by the Algorithm 6.1 provided that step 2 is interpreted in a fuzzy fashion, where the attribution of a cluster to a centroid is intended as the determination of the membership function of the point to the centroid.

The key features of the FCM algorithm are enumerated as in Table 6.3. In particular:

The FCM key features:

- The termination condition is quantified by looking at the changes in the membership values of the successive partition matrices. Denote by $U(t)$ and $U(t+1)$ the two partition matrices produced in two consecutive iterations of the algorithm. If the distance $|U(t+1) - U(t)|$ is less than a small predefined threshold ε then we terminate the algorithm. Typically, one considers the Chebyshev distance between the partition matrices (refer to Table A.2) meaning that we look at the maximal values of absolute differences between the membership grades of the partition matrices obtained in two consecutive iterations.

stop when nothing new happens,

- The fuzzification coefficient exhibits a direct impact on the geometry of fuzzy sets generated by the algorithm. Typically the value of r is assumed to be equal to 2.0 (with a vague reminiscence of quantum mechanics [56]). Lower values of r (that are closer to 1) yield membership functions that start resembling characteristic functions of sets; most of the membership values becomes localized around 1 or 0. The increase of the fuzzification coefficient ($r = 3, 4$, etc.) produces spiky membership functions with the membership grades equal to 1 at the prototypes and a fast decline of the values when moving away from the prototypes. Several illustrative examples of the membership functions are shown in Fig. 6.19. In addition to the varying shape of the membership functions, observe that the requirement put on the sum of membership grades imposed on the fuzzy sets yields some rippling effect: the membership functions are not unimodal but may exhibit some ripples whose intensity depends upon the distribution of the prototypes and the values of the fuzzification coefficient.

a coefficient for shaping the membership functions,

Table 6.3. The main features of the FCM clustering algorithm.

Feature of the FCM algorithm	Representation and optimization aspects
Number of clusters k	Structure in the dataset and the number of fuzzy sets estimated by the method; the increase in the number of clusters produces lower values of the objective function however given the semantics of fuzzy sets one should maintain this number quite low (5-9 information granules)
Objective function C	Develops the structure aimed at the minimization of C; iterative process supports the determination of the local minimum of C
Norm ν	Reflects (or imposes) a geometry of the clusters one is looking for; essential design parameter affecting the shape of the membership functions
Fuzzification coefficient r	Implies a certain shape of membership functions present in the partition matrix; essential design parameter. Low values of r (being close to 1.0) induce characteristic function. The values higher than 2.0 yield spiky membership functions
Clustering depth	Defines the grouping hierarchy level. We may proceed top-down like with ISODATA scheme taking track of the intermediate groups whenever they have a semantic meaning
Termination criterion	Distance between partition matrices in two successive iterations; the algorithm terminated once the distance drop below some assumed positive threshold ε, that is $|U(t+1) - U(t)| < \varepsilon$

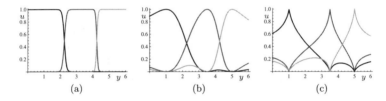

Fig. 6.19. Examples of membership functions u of fuzzy sets when computed w.r.t. different prototypes $v = 1$ (black curves), $v = 3.5$ (dark gray curves) and $v = 5$ (light gray curves). The fuzzification coefficient r assumes values of 1.2 (a), 2.0 (b) and 3.5 (c). The intensity of the rippling effect is affected by the values of r, increasing with higher values of the latter.

- Membership functions offer an interesting feature of evaluating an extent to which a certain pattern is shared between different clusters and in this sense become difficult to allocate to a single cluster (fuzzy set). Let us introduce the following index which serves as a certain separation measure:

an index for appreciating how much clusters are separated,

$$v(u_1, u_2, \ldots, u_k) = 1 - k^k \prod_{c=1}^{k} u_c, \qquad (6.28)$$

where $u_1, u_2, \ldots, u_k$ are the membership degrees for some pattern (here we dropped the index of patterns for the sake of readability). If only one of membership degrees, say $u_c = 1$, and the remaining are equal to zero, then the separation index attains its maximum equal to 1. On the other extreme, when the pattern is shared by all clusters to the same degree equal to $1/k$, then the value of the index drops down to zero. This means that there is no separation between the clusters as reported for this specific item. In particular, for $k = 2$ the above expression relates directly to the entropy of fuzzy sets, in that $v(\boldsymbol{u}) = v(u, 1 - u) = v(u, 1 - u) = 1 - 4u(1 - u) = 1 - H(\boldsymbol{u})$.

- While the number of clusters is typically limited to a few information granules, we can easily proceed top-down with successive refinements of fuzzy sets. This can be done by splitting those fuzzy clusters with the highest heterogeneity. Let us assume that we have already constructed κ fuzzy clusters. Each of them can be characterized by the performance index: another for appreciating how the single cluster is dispersed.

$$V_c = \sum_{i=1}^{m} u_{ci}^r \nu(\boldsymbol{y}_i - \boldsymbol{v}_c)^2, \quad (6.29)$$

for $c = 1, 2, \ldots, k$. The higher the value of v_c the more heterogeneous the c-th cluster. The one with the highest value of v_c, that is the one for which we have $c_{\max} = \arg\max_c v_c$, is refined by being split into two clusters [1]. Denote the set of data associated with the $c_{\max}$-th cluster by $\mathfrak{Y}(c_{\max})$, i.e.

$$\mathfrak{Y}(c_{\max}) = \{\boldsymbol{y}_i \in \mathfrak{Y}, u_{c_{max}i} = \max_c u_{ci}\}. \quad (6.30)$$

We group the elements in $\mathfrak{Y}(c_{\max})$ by forming two clusters which lead to two more specific (detailed) fuzzy sets. This gives rise to a hierarchical structure of the family of fuzzy sets as illustrated in Fig. 6.20. The relevance of this construct in the setting of fuzzy sets is that it emphasizes the essence of forming a hierarchy of fuzzy sets rather than working with a single level structure of a large number of components whose semantics could not be retained. The process of further refinements is realized in the same way by picking up the cluster with highest heterogeneity and splitting it into two consecutive clusters.

It is worth emphasizing that the FCM algorithm is a highly representative method of membership estimation that profoundly dwells on the use of experimental data. In contrast to some other techniques presented so far that are also data-driven, FCM can easily cope with multivariable experimental data.

---

[1] Note the analogy with the ISODATA clustering Algorithm 6.3.

226   The Clustering Problem

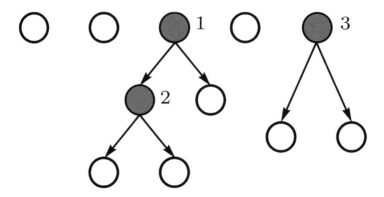

Fig. 6.20. Successive refinements of fuzzy sets through fuzzy clustering applied to the clusters with highest heterogeneity. The numbers indicate the order of the splits.

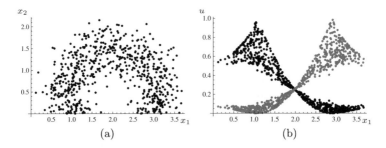

Fig. 6.21. (a) Two-dimensional synthetic dataset. (b) Membership grades of both clusters developed by the clustering algorithm (respectively black and gray points).

Example 6.15. Shadowed sets are instrumental in fuzzy cluster analysis, especially when used to describe results produced there. Consider the dataset shown in Fig. 6.21(a). The idea is to transform into a shadowed set the membership grades of the partition matrix returned by standard FCM algorithm run with $k = 2$ clusters. The prototypes are equal to $v_1 = (1.08, 0.81)$, $v_2 = (2.91, 0.83)$ and reflect the structure of the dataset. The membership functions of both clusters (fuzzy relation) are visualized in Fig. 6.21(b) with different colors. The determination of the threshold level α_{opt} as in (4.14) inducing a consistent shadowed set is completed through (4.16) by simple enumeration. Following the graphs in Fig. 6.22(a), we obtain $\alpha = 0.41$, which highlights several patterns to be treated as potential candidates for further thorough analysis. When we complete clustering with $k = 6$ the

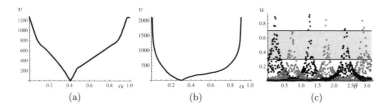

Fig. 6.22. Index v (computed from (4.13)) viewed as a function of α for a number of clusters $k = 2$ (a) and $k = 6$ (b). (c) Distribution of membership grades of the six clusters as computed by FCM algorithm and shadow region determined by $\alpha_{\text{opt}} = 0.3$. For the sake of visualization the membership grades in the last picture are drawn w.r.t. the angle θ each point forms with x_1 when centered in $(2,0)$ (see Fig. 6.21(a)).

results become quite different: as shown in Fig. 6.22(b) the optimal value of α, here equal to 0.3, induces a shadowed set as in picture (c) reducing the overlap between the clusters quite visibly.

6.5 Discriminant Analysis

If we know to which class to attribute the items, then the problem is concerned with finding metric and features supporting the classification – a problem positioned in the realm of *discriminant analysis*. We absolutely abandon any attempt to deduce it from a prejudicial meaning of the data. On the contrary we rely on a loss function recalling random mechanisms to identify these ingredients under the sole constraint that it does not contradict the subsequent classification. Paradoxically, this represents a way to deduce meaning to the features as a synthesis of the original data that statistically proves useful for classification tasks, hence worth to be declared as a property of the items we are clustering (remember our previous consideration about the meaning of word "rice"). Leaving the representation problem for granted, since we will discuss it in the next chapter, the selection of the metric is typically up to us, like in the cluster analysis. In these conditions the analysis is mainly of functional kind in the conventional approaches, whilst it remains of statistic kind with ours. As mentioned in Section 6.1, having the probabilistic models of the populations representing the classes the object must be assigned to, the problem stands in identifying the correct decision function minimizing the risk R. In case of free parameters, these may be estimated *independently* from the decision function since we know the labels of the sample points. *Vice versa*,

You know the class of each data item, but you do not know why.

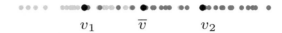

Fig. 6.23. One-dimensional points belonging to two classes (light and dark gray points, respectively). External black points: centroids v_1 and v_2; central black point: mean value $\bar{v}$ between them.

Still in search of compatible models. in our framework we consider a parameters' population which is compatible with the sample points and the adopted decision rule. As a result we obtain probabilitstic appreciations on the efficiency of the rule. Let us consider the sample in Fig. 6.23 drawn from a mixture of two Gaussian variables Y_1 and Y_2 with means μ_1 and μ_2, respectively, and same standard deviation σ. The minimization of R as in (6.2) implies the rule:

$$\text{assign} \begin{cases} y \text{ to } d_1 & \text{if } |y - \mu_1| \leq |y - \mu_2|, \\ y \text{ to } d_2 & \text{otherwise.} \end{cases} \quad (6.31)$$

In case the means are not known we substitute them with their estimates $v_c = \sum_{y_i \in d_c} y_i / m_c$ with $c = 1, 2$ and m_c denoting the number of sample elements coming from class d_c.

Coming to our framework, we know that the metric with which to deal with the sample is the l_2 norm, having Gaussian distribution as companion. First of all we look for the parameter distributions compatible with the sample labeling. This may represent a discrete variant of the regression problem we treated in Section 3.2.1. Thus,

- since we adopted a quadratic loss function (for instance), we refer to a pair of Gaussian variables Y_1 and Y_2 whose specifications we may derive from specifications of a Normal distribution Z as follows:

Hence sampling mechanism,

$$y_{ci} = \mu_c + \sigma z_i, \quad c = 1, 2, \quad (6.32)$$

where σ is the standard deviation that is common to the two populations and we assume known for the moment, and μ_c characterizes the c-th cluster;
- since the centroids (sample mean) v_1 and v_2 of the clusters are the sole statistics drawn from data (hence the sole relevant properties of them), we would use the same threshold $\bar{v} = (v_1 + v_2)/2$ for any pair of populations giving rise to the same statistics v_1 and v_2;

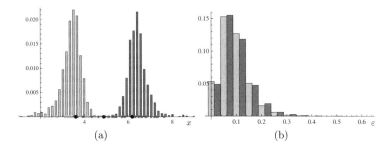

Fig. 6.24. (a) M_1 and M_2 centroid populations (black and gray bars respectively); (b) corresponding E_1 and E_2 populations of the misclassification probabilities (same colors as in (a)).

- according to our model (6.32), the centers μ_i^*s of the populations underlying the samples are the solution of equations:

hence master equations,

$$v_1 = \mu_1^* + \frac{\sigma}{m_1^*} \sum_{i=1}^{m_1^*} z_i \qquad (6.33)$$

$$v_2 = \mu_2^* + \frac{\sigma}{m_2^*} \sum_{j=m_1^*+1}^{m_1^*+m_2^*} z_j, \qquad (6.34)$$

for any set $z = \{z_1, \ldots, z_{m_1^*}, z_{m_1^*+1}, \ldots, z_{m_1^*+m_2^*}\}$ drawn from Z, such that the related ys as in (6.32) fall inside the respective clusters with centroids v_1 and v_2. This means that we sample items from Z in number of $m_c \simeq \gamma(m_1 + m_2)/2$ with γ somehow greater than 1 and iteratively for each cluster c we: 1) compute center μ_c^* from (6.33-6.34), 2) remove the farthest y trespassing the threshold from the cluster μ_c^* it belongs to and go back to 1) to update the center, until all points fall inside the cluster.

By iterating (6.33) and (6.34) in a bootstrap mode with a huge number of z replicas, we obtain M_1 and M_2 populations as in Fig. 6.24(a), and corresponding E_1 and E_2 populations of the misclassification probabilities, i.e. probabilities $\varepsilon_1 = 1 - F_Z((\overline{v} - \mu_1^*)/\sigma)$ of points from population 1 to fall out of cluster 1, and analogous probability $\varepsilon_2 = F_Z((\overline{v} - \mu_2^*)/\sigma)$ for population 2 (Fig. 6.24(b)). The latter distributions are effective indicators of the discriminant ability of the algorithm. In particular, independently from the means of the two Gaussian variables from which we sample points as in Fig. 6.23, since these variables have equal variance, histograms of M_1 and M_2 are almost specular, and E_1 and E_2 almost the same. Moreover, in view of the mentioned border biasing effect, we prefer focusing on the modes of

in order to obtain parameter distributions.

the two distributions that are 3.51 and 6.41. Due to the peculiarities of the sample elements, the M and E distributions are not symmetric around the mean however, with the effect of having $\mathrm{E}[E_1] = 0.085$ and $\mathrm{E}[E_2] = 0.086$, summing up to a percentage of misclassification equal to 17.1%, that is smaller than the percentage induced by the threshold on the original populations (19+13%) but greater than the sum of frequencies of the actually misclassified sample items $(6 + 8\%)$.

6.6 Conclusions

In this chapter we face the same topic of Chapter 1, but considering many frameworks for the same data contemporarily. You may image Hansel following customary k paths starting from the wood, say leading to the pasture land, to the river or home, so that looking at a pebble he wonders which path is indicator of. Maybe you may have got a similar experience with ski tracks in a piste. In essence you need a data ordering function with each framework. On the base of this ordering you attribute, partly or completely, the single data item to a framework (the core of clustering); in turn the ordering is tuned on the data you have attributed to the framework (the skill of clustering). With the goal (6.1) the solution of our problem is definitely linked to the information we may draw from data rather than on a preexisting model of them. Having the shortest program $\pi(y)$ determining the Kolmogorov complexity measure $K(y)$ as the template of the information underlying a data item, a pattern y in this context, we move to distances as less efficient but computable complexity measures, that become ordering functions of this multiframework scenario. Distances $d(y, v)$ are functions of two arguments, one is the pattern y, the other is a prototype v, for instance the cluster centroid, representing the actual reference of the current framework. Substituting $K_v(y) - c$ with $d(y, v)$ in (2.21) and renaming $d(y, v)$ with the norm $\nu(y - v)$ you obtain (6.9), in the maximum likelihood reading of (2.21). Hence probabilistic models come as a corollary of clustering. You don't need a true metric to draw these models, but only a weak norm functionality as depicted in Definition A.4 allowing the pattern positioning w.r.t. the prototypes.

This is why we assume norms as basic constituents of a clustering algorithm and focus on how they are exploited both from information management and from algorithmic perspective. We distinguish between crisp norms and fuzzy norms, where with the latter we manage fuzzy clustering algorithms. As mentioned in Appendix A.3, the latter miss the *positive definiteness*, which renders the contour lines of the norm vague. In FCM algorithm we obtain this effects adding a weight to the norms of the patterns with the role of partitioning function. We based our discussion on rather conventional norms, avoiding tossing more *exotic* norms or pseudo-norms such as textures [48], pattern likelihood under candidate Hidden Markov models [51], or algorithmic steps to compute one pattern from prototypes [46], and so on.

The information management is highly connected both to the data representation, a topic that we will treat in the next chapter, and to the structure of the clustering algorithm. As for the latter, we distinguish notion of horizontal structure entailing strategies of nearest neighboring, texture, centroids, etc., and vertical structures allowing a progression in the aggregation or splitting of the clusters. The general feeling we should capture from the examples we show is that there is not an optimal metric or an optimal algorithm in absolute. Rather both depend on the data under question, thus leaving us with the unavoidable charge of deciding both, possibly after either a deep numerical trial, or drawing from analogous instances.

6.7 Exercises

1. Discuss the difference between metric and representation in a clustering problem.
2. Show that the maximization problem (6.1) is equivalent to (6.4) for ℓ related to the Euclidean distance, as in (6.3).
3. Repeat Exercise 2 using a different distance chosen among the ones in Table A.2.
4. Apply the k-means algorithm to the Iris dataset alternatively using: the first 2 and all pattern features, l_2 and $l_{1,\varepsilon}$ norms for a suitable ε (look at the ε-insensitive loss function in Table A.1), $k = 3$ and $k = 5$, and a stop criterion decided by yourself. Compare the results. With all options attribute labels and derive confusion tables both in terms of absolute values and percentages. Using only two features per pattern, draw the Voronoi tessellations entailed by the solutions.
5. Use the techniques described in Section 6.3.2 in order to find an appropriate number of clusters for the Iris dataset, and compare the result with the number of classes in the dataset.
6. Apply the hierarchical clustering Algorithm 6.2 to the Cox dataset, using the *complete* linkage method as enunciated in Table 6.2 and θ equal to $d_{\min}l$, with $d_{\min}$ the minimal distance between each pair of patterns and l the clustering level. Draw the last 7 levels of the dendrogram.

7. Apply the ISODATA Algorithm 6.3 to the patterns referring to 4 provinces in the Swiss dataset, choosing parameters that guarantee a good separability of the provinces.
8. Repeat the experiment in Exercise 7 by using the Nearest Neighborhood algorithm shown in Example 6.3.
9. Select three digits from the MNIST dataset and apply the Fuzzy C-Means algorithm with fuzzification factor $r = 3.2$ to the corresponding patterns, deciding an appropriate number of centroids. Afterwards, check whether or not patterns assigned to the same digits belong to a same cluster.
10. Apply discriminant analysis to the two classes *normal, carriers* of the Cox dataset. Use the assignment rule (6.31) with l_1 substituted by the angular distance defined in Table A.2 between the class centroid and the current pattern. Compare the results you obtain by considering separately the blood measures or directly their mean after having normalized each measure in the $[0, 1]$ interval (for a smarter normalization see Example 12.5). In the last case, draw the Voronoi tessellation induced by the classification rule.

7 Suitably Representing Data

We may organize the search for the metric d at the basis of the cost function ℓ into two main steps: search for a *sound representation* of the data and use of a metric appropriate to the representation. The term sound stands for a representation allowing to better understanding the data, for instance by decoupling original signals, removing noise, discarding meaningless details, and alike. The result of the splitting could prove less efficient than the direct metric, but more manageable in most cases. Essentially, we are looking for rewriting the metric instances $d(\boldsymbol{y}_i, \boldsymbol{y}_j)$ as a composition $d'(g(\boldsymbol{y}_i), g(\boldsymbol{y}_j))$, with g optimizing the cost function C, namely:

Subproblem: smartly represent the data,

$$g = \arg\max_{\widetilde{g}} \mathsf{C}[\widetilde{g}] = \arg\max_{\widetilde{g}} \left\{ \sum_{\boldsymbol{y} \in \mathfrak{Y}} \sum_{j=1}^{k} \ell\left[D(\widetilde{g}(\boldsymbol{y})), d_j\right] \right\}. \quad (7.1)$$

In principle, this maximization is a part of the overall goal of optimizing the clustering procedure, hence depends on both metric and assignment decision rule. However, we will treat it separately, possibly as a step of an incremental procedure. We are used to:

- solve the problem independently of the aggregative/agglomerative option and of the centroid selection in the second case. Hence we will conventionally fix the second argument of d at the origin $\mathbf{0}$ of the feature space;

  independently of the specific clustering,

- work with a linear g, so that the target becomes:

  possibly through a linear transform

$$W = \arg\max_{\widetilde{W}} \mathsf{C}[\widetilde{W}] = \arg\max_{\widetilde{W}} \left\{ \sum_{\boldsymbol{y} \in \mathfrak{Y}} \sum_{j=1}^{k} \ell\left[D(\widetilde{W}^T \boldsymbol{y}), d_j\right] \right\}, \quad (7.2)$$

where by $\widetilde{W}^T$ we denote the transpose of the matrix $\widetilde{W}$.

As mentioned before, very often this kind of maximization problem finds its dual in the minimization of an analogous expression related to the single clusters like in (6.4). Thus a very general template of the problem is the following:

$$W = \arg\min_{\widetilde{W}} \check{C}[\widetilde{W}] = \arg\min_{\widetilde{W}} \left\{ \sum_{y \in \mathcal{Y}'} \nu(\widetilde{W}^T y) \right\}, \quad (7.3)$$

since $\nu(y) = d(y, \mathbf{0})$, with $\mathcal{Y}'$ forming a suitable subset of $\mathcal{Y}$. Of the nonlinear option we will discuss later on. The evident benefit of the linear option is associated with the simplicity of the involved computations, such as the derivatives used to pursue the optimum points. Actually, within our perspective we realize that most functions of common use are linear. The sophistication of their coefficients identification however comes with the availability of the computer resources. At the beginning we based them on the sole first moment of the data under question, the sample mean vector indeed. Then we moved to the second order moments, and most recently to some higher order moment. We will discuss the last two options in a while, leaving the former as being very much straightforward.

<small>with smart coefficients.</small>

The general understatement is that if you understand the feature you are processing then you may cluster well the related objects. Thus we expect that a suitable W well meets both objectives. Focusing on the former, as a more general one, hence disputable per se, the rule of thumb is that features are as more understandable as they are separable. In the recent years researchers are looking for a structural separability and a semantic issue. The former states that convenient features simply add in the metric of an object, such as it happens for the norms we saw in the previous chapter. The probabilistic counterpart via (6.9) of this property is the independence of the features that is tossed in various ways in the literature. Hence, on the one hand having separable features to order the objects heaped in your attic, such as the weight of the objects and their usability, offers the analytical commodity of considering each feature separately. On the other one, the features simply combine through Cartesian product, so that you may arrange a grid of shelves to store the objects with the expectation that no any of them will be left drastically empty; rather the relative filling rates between cells in a row remain constant with columns, and *vice versa*.

<small>Each feature with an its own meaning</small>

At a semantic level separability stands for sparseness. Namely, only one or few features are relevant in the single objects, so that the meaningful features may be identified with the directions along which the objects lie in the feature space. Of course sparseness helps independence, but with very rare occupations

<small>Either a grid of classification cells,</small>

of the mentioned Cartesian product cells. Rather, an ideal condition with your tiding up activity is that for each object category you have an almost exclusive feature that both characterizes the object and differentiates the specific items. Thus, on one side you manage a limited number, possibly one, of features to cluster each object. On the other, you may arrange the shelves one line for each combination of the above limited number of features, being sure that no ambiguities will arise in locating the single objects.

<small>or simply parallel lines.</small>

7.1 Principal Components Analysis

Second order moments are directly involved by l^2 metric as follows. Denoting $W^T \boldsymbol{y}$ with $\boldsymbol{a}$, where we assume the i-th component a_i of $\boldsymbol{a}$ as being generated by the dot product $\boldsymbol{w}_i \cdot \boldsymbol{y}$, with $\boldsymbol{w}_i$ the i-th column vector of W, $\nu(\boldsymbol{a})$ becomes:

$$\nu(\boldsymbol{a}) = \boldsymbol{a}^T \boldsymbol{a} \qquad (7.4)$$

and the cost function reads:

$$\check{C} = \sum_{i=1}^{m} (W^T \boldsymbol{y}_i)^T (W^T \boldsymbol{y}_i) = \sum_{i=1}^{m} \nu(\boldsymbol{a}_i), \qquad (7.5)$$

for a suitable m denoting the cardinality of a subset of $\mathfrak{Y}$. With this choice for ℓ, the derivative of the single term in the sum takes on the form:

$$\phi(\boldsymbol{a}) = \partial \ell(\boldsymbol{a})/\partial \boldsymbol{a} = 2\boldsymbol{a}, \qquad (7.6)$$

stressing a linear dependence in its argument. Now, writing $\check{C}$ as:

$$\check{C} = \sum_{i,j=1}^{m} \boldsymbol{w}_i^T \Sigma_{\boldsymbol{Y}} \boldsymbol{w}_j \qquad (7.7)$$

where $\Sigma_{\boldsymbol{Y}}$ is the covariance matrix of $\boldsymbol{Y}$, the landscape with W has infinite optima constituted by the quotient set of all orthogonal linear transforms, i.e. such that $\boldsymbol{w}_i \cdot \boldsymbol{w}_j = 0$, $\forall i \neq j$, for non degenerate, i.e. finite and non null, $\boldsymbol{w}_i$s. We obtain these optima as the solution of the problem of maximizing C under the constraint that all $\nu(\boldsymbol{w}_j) \leq 1$ or minimizing $\check{C}$ under the constraint that all $\nu(\boldsymbol{w}_j) \geq 1$. As a matter of fact we obtain a rotation matrix W of the $\boldsymbol{Y}$s. Hence $\check{C}$ represents the sum of the quadratic distances of the m objects from the axes origin, whose value does not change with any rotation of the framework. This is why any orthogonal transform with $\nu(\boldsymbol{w}_j) =$

<small>With a l^2 metric</small>

1 for all j – let's say any orthonormal transform – is a solution to our optimization problem. Within these solutions we may find a privileged framework if we consider the eigenvectors of Σ_Y.

By definition (see Appendix D.1) eigenvectors are orthogonal and identify the n directions of residual maximal dispersion of the data in terms of the sum of quadratic distances from the origin along those directions, where the associate eigenvalues are proportional to the dispersions. Because of these properties the eigenvectors are said to identify the *principal components* of the data they refer to and their study is called *Principal Component Analysis* (PCA) [44]. It is clear from (7.7) that the only statistic of the data we are considering (apart from their normalization w.r.t. the mean vector) is the second order moment, say the distances of points in the feature space from their mean. Hence it is the l^2 metric which reflects our knowledge about data, substantially corresponding to the representation of their population through a (multivariate) Gaussian random variable after the Levin lemma (see (6.9)). Hence we frame the data item with feature vector $\boldsymbol{y}$ with a distribution of the form:

look for the most dispersed orthogonal components,

$$f_{\boldsymbol{Y}}(\boldsymbol{y}) \propto e^{-\boldsymbol{y}^T S \boldsymbol{y}}, \qquad (7.8)$$

with S being some non negative symmetric matrix, so that the contour lines of $f_{\boldsymbol{Y}}$ are ellipsoids as shown in Fig. 7.1. In this figure the ellipses represent the contour lines of the density function $f_{Y_1,Y_2}(y_1, y_2)$ of a two-dimensional Gaussian variable with covariance matrix $\Sigma_{\boldsymbol{Y}} = ((5,3),(2,2.5))$ (see Appendix B.2 for its general properties). From multivariate statistics theory [41], we know that their axes have the directions of the two

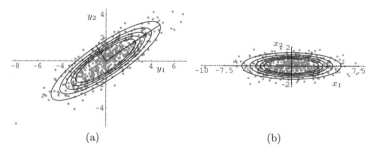

(a)　　　　　　　　　(b)

Fig. 7.1. (a) Contour lines and sample points of the density function $f_{Y_1,Y_2}(y_1, y_2)$ of a two-dimensional Gaussian variable with covariance matrix $\Sigma = ((5,3),(2,2.5))$ centered in the origin; (b) same as in (a) but rotated in the principal axes framework computed from the sample points.

Table 7.1. Confusion matrices of the experiments in Fig. 7.2 contrasted with conventional k-means algorithm (column All Components). The percentages cumulate both positive and negative errors.

	All Component	First Component	Second Component
Fig. 7.2 (I)	50 0 0 0 50 2 0 0 48 |1.3%	50 0 0 0 47 6 0 3 44|6%	0 0 0 3 32 2 47 18 48|46%
Fig. 7.2 (II)	21 0 20 25 50 4 4 0 26|35%	11 3 10 30 47 22 9 0 18|49%	23 8 13 14 42 0 13 0 37 |32%
Fig. 7.2 (III)	50 0 0 0 47 14 0 3 36 |11.3%	50 0 0 0 49 12 0 1 38 |8.6%	8 1 4 17 35 20 25 14 26 |54%

eigenvectors of Σ_Y, whereas the eigenvalues measure the variances of the coordinates along these axes. The identification of these axes meets both min/max goals of our clustering problems. Namely, in search of the saddle point maximizing between clusters distances and minimizing within clusters ones, we get the first target by considering at global level the coordinates with greatest dispersion w.r.t. a common center v, and the second target by considering at local level the coordinates with less dispersion w.r.t. the single centroids v_js.

i.e. the principal axes of the underlying Gaussian distribution.

Working with a sample $\{y_1, \ldots, y_m\}$, we compute their principal components on the basis of estimates of μ_Y and Σ_Y commonly given by

$$\widehat{\mu}_Y \equiv m_y = 1/m \sum_{i=1}^{m} y_i;\ \widehat{\Sigma}_Y \equiv S_{yy} = 1/m \sum_{i=1}^{m}(y_i - m_y)(y_i - m_y)^T. \tag{7.9}$$

Example 7.1. Let us look at the plots in Fig. 7.2. We get a good clustering by considering only the first principal component both with the former dataset made up of artificial data and with the latter referring to the Iris dataset. There is no way with the intermediate one, since the (artificial) data are too much confused. Confusion matrices in Table 7.1 give a quantitative feeling of these behaviors.

7.2 Independent Components

The l^2 metric and the corresponding Gaussian model did satisfy many of the clustering and regression goals in the past. The Central Limit Theorem (see [60]) provided a robust rationale to this

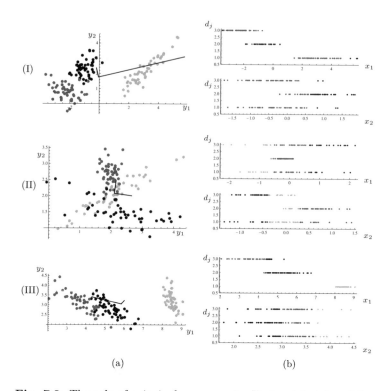

Fig. 7.2. The role of principal components. Horizontal strips of the figure refer to different datasets: (I) three well separated sets of Gaussian distributed data; (II) three poorly separated sets of Gaussian distributed data; (III) first two principal components of Iris dataset. Vertical strips: (a) data scatter plot and directions of the principal components, each denoted by a segment proportional to its standard deviation; (b) distribution of data along the single principal components, split in the three clusters d_j.

With different metrics: fact. In the world of symmetric distributions this metric covers symmetry and a general bias toward a central value of the data. For sure this is the symmetry achieved by the cumulation of a large number of random phenomena, possibly not singularly described by a symmetric distribution in its turn. What is crucial is that these many variables are reciprocally independent. At the current stage of data analysis we are not puzzled by the assumption on the large number – dealing with a random phenomenon, what against a decomposition into many atomic phenomena in turn? What it is often unbearable is the independence between the cumulating phenomena. Rather, we are often involved with structurally connected variables, whose sum preserves the symmetry in general, but moves away from the Gaussian distribution. This corresponds to realize that the relevant metric is not

l^2 but an l^q with q real number different from 2. A first consequence of this new situation is that we are dealing with random variables where linear independence, i.e. diagonal covariance matrix, does not imply that the variables are independent at all – so that for instance $f_{Y_1,Y_2} = f_{Y_1} f_{Y_2}$. This implication indeed holds only for multivariate Gaussian variables. Consequently, if we are looking for a set of independent variables $\{Y_1, \ldots, Y_\nu\}$, whose linear combination through a matrix W gives rise to the original variables $\{X_1, \ldots, X_\nu\}$, call them independent components, we must focus on other variables than the principal components discussed earlier. For short, you may separate the squares of the variables, say focusing on $y_1^2 + y_2^2$ since they are linearly independent, while you need to separate different powers, $y_1^r + y_2^r$, in order to have $f_{Y_1,Y_2}(y_1, y_2) = \exp y_1^r \exp y_2^r$, i.e. Y_1 and Y_2 to be independent random variable. In any case we may compute components of Y as a linear combination of X components as well. It remains to answer about why such a linear combination should prove useful to our clustering problem (7.1) and how to compute it.

<small>same target but with different dispersion measures.</small>

Let us come back to the splitting of the clustering problem like in (7.2). If we assume that the l^2 metric is not appropriate, we are correspondingly assuming that the data distribution law is of kinds like in the first and third rows of Fig. 7.3, where we distinguish between subgaussian distributions if they are rooted on l^q with $q < 2$ and supergaussian otherwise. Contrasting the graphs of the two-dimensional distribution arising from the product of two identical either Gaussian or supergaussian or subgaussian distributions, the feature that we will exploit is the maximal elongation of the contour lines from their centers. From the right column of Fig. 7.3 we see that, while Gaussian factors in the product give rise to circular contour lines, with subgaussians' you find the directions of maximal elongation in correspondence of the coordinate axes and those of minimal on the bisectors, *vice-versa* in the latter. Hence, willing either to maximize or minimize a cost function based on l^q metric we must select one of these directions. Coming back to our original data, we get close the above scenario after having: i) found the principal components of the data, ii) moved in the frame having these components as reference axes – hence having rotated the original frame – and, iii) rescaled these axes by dividing the components by the related eigenvalues – so as to have new components with equal variance. With these operations, that are known as the *whitening of the sample data*, we:

<small>Either more picked or more smooth w.r.t. a Gaussian bell</small>

<small>The standard three:</small>

- obtain a framework that is independent w.r.t. rotation. In the sense that any further rotation of the axes produces a

<small>whitening by first,</small>

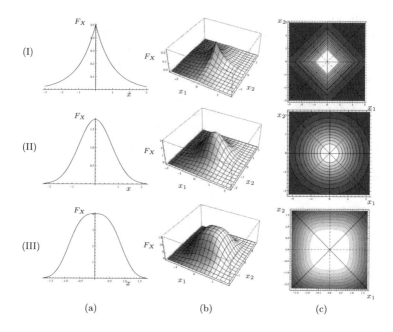

Fig. 7.3. Drifting from Gaussian shapes. Horizontal strips: (I) subgaussian; (II) Gaussian; (III) supergaussian distributions. Vertical strips: (a) one dimensional d.f.; (b) two dimensional d.f.; (c) contour plot of the latter. Crosses: dotted lines – minimal elongation axes; plain lines – maximal elongation axes.

new framework whose components are linearly independent. Indeed, starting from an $\boldsymbol{X}$ with unitary covariance matrix I, the rotated variable $\boldsymbol{Y} = W^T\boldsymbol{X}$, with W orthonormal matrix (see Appendix D.1), has still unitary covariance matrix, since:

$$\mathrm{E}[\boldsymbol{Y}\boldsymbol{Y}^T] = \mathrm{E}[W^T\boldsymbol{X}\boldsymbol{X}^TW] = W^TIW = I; \qquad (7.10)$$

- assume the linear independence of sample data as denoting the linear independence of the variables spanning the axes, i.e.

$$\mathrm{E}[Y_iY_j] = 0 \quad \forall i,j. \qquad (7.11)$$

Now, since independence implies linear independence, we expect to find the true Independent Components (IC for short) of the data lying over one of these orthogonal frameworks. So that our search is just for a suitable rotation of the framework identified through PCA. As a matter of fact the framework of our interest is either coinciding with those axes or with the bisectors of the orthant identified by them. The second question is more operational, i.e., we search an efficient way for identifying these directions. From a pure mechanical point of view we

Independent Components 241

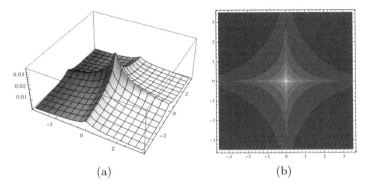

(a)   (b)

Fig. 7.4. A jig for identifying the independent framework: (a) 3D shape; (b) contour plot.

may think of a mask whose rotation will identify the direction. Namely, with reference to Fig. 7.3, we may figure that after whitening, you have a cross, like in the right strip of the figure, hidden under the density function f of data, but you do not know where, since it may highly differ from the principal components directions. Thus, you over-impose to it a radial symmetric function h periodic with the quadrant, as shown in Fig. 7.4, and rotate it until we get the maximum of the integral of the product of the two functions over the plane, i.e. their *angular cross-correlation* $f \star h$, as shown in Fig. 7.5. Of course, we have no way to produce meaningful results with a circular function, since the product is invariant versus rotations. Actually any non circular function is OK, the most proper one depending on the actual distribution of the sample. For instance you may obtain the mask in Fig. 7.4 just by considering the exponent of the subgaussian distribution in Fig. 7.3; an anologous mask may be obtained from the supergaussian distribution therein.

then look for a suitable mask,

and search for the highest cross-correlation direction.

In the literature people are used to work with masks made of the sum of nonlinear functions such as:

The most common masks

$$\chi_1 = \frac{1}{\alpha} \sum_{i=1}^{n} \log(\cosh \alpha a_i), \quad \chi_2 = \sum_{i=1}^{n} -\log \beta^2 + \log(\beta^2 + a_i^2),$$
(7.12)

where cosh denotes the hyperbolic cosine function, and α and β are suitable constants. Fig. 7.6(a) contrasts the single addends of these functions with the analogous of the masks considered in Figs. 7.4 and 7.5. Rather they refer to the absolute difference between these functions and a suitable constant c, namely:

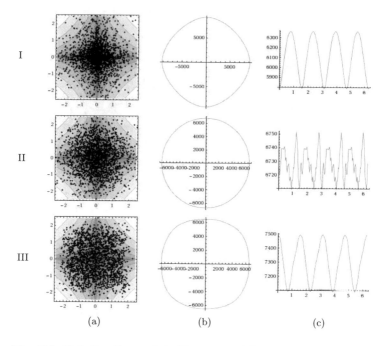

Fig. 7.5. Rotating the mask in Fig. 7.4 on different landscapes. Horizontal strips: same distributions as in Fig. 7.3. Vertical strips: (a) over-position of sample points on the mask; (b) polar graphs of the angular cross-correlation; (c) unfolding of the same quantities.

$$\chi_1' = \left| \frac{1}{\alpha} \sum_{i=1}^{n} \log(\cosh \alpha a_i) - c \right|, \quad \chi_2' = \left| \sum_{i=1}^{n} -\log \beta^2 + \log(\beta^2 + a_i^2) - c \right|,$$
(7.13)

thus having contour lines where dispersion minima of the previous functions flip to maxima. Actually, with these functions we have two families of equal relative maxima – the original maxima in correspondence of the independent axes and the flipped ones in correspondence of the bisectors. Moreover, a proper selection of the constant pairs (c, α) or (c, β) allows us to have the global maxima exactly pointing to the independent axes. We may realize it in the two dimensional frame looking at the χ_1' and χ_2' landscape in Figs. 7.6(b-c). We obtain the related angular cross correlation by summing the χ_i' on all sample points.

With these functions we are able to deal with both subgaussian and supergaussian distributions in a homogeneous manner, so that the rotation attaining its maximum always achieves,

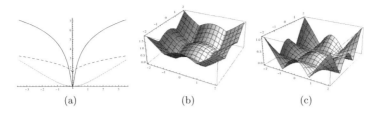

Fig. 7.6. (a) Contrasting mask cross sections. Plain curve: $l^{0.5}$ norm; dashed curve: χ_1 function; gray curve: χ_2 function; (b) plot of the sum of two χ_1' addends; (c) plot of two χ_2' addends'.

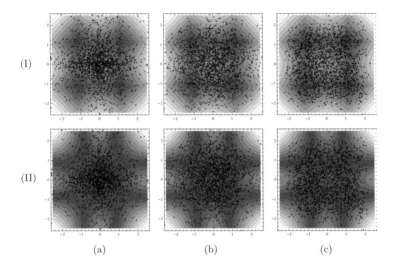

Fig. 7.7. Overimposing samples to the masks in Fig. 7.6(b-c). Horizontal strips: masks; vertical strips: distributions as in Fig. 7.5.

in the standard cases, the independent components frame, see Fig. 7.7. A suitable value for c is represented by the angular cross-correlation of χ_i' with a Gaussian d.f., representing the turn point of the behavior of the mask with sub and supergaussian distributions. The second coefficient in both distributions becomes critical only when the index q of l^q metric subtending the Y distribution law is close to 2. This is why we generally say that the above ICA functions are *robust* w.r.t. the coefficients selection. Moreover, as the maximum is searched through

Contrasting parameters with Gaussian touchstone,

plus a secondary parameter when the former is weak.

Optimization through a speedy iterative algorithm.

incremental algorithms, an additional benefit is represented by the simple form of the derivatives, namely:

$$\frac{\partial \chi_1}{\partial a_i} = \tanh a_i, \qquad \frac{\partial \chi_2}{\partial a_i} = \frac{2a_i}{\beta^2 + a_i^2}. \tag{7.14}$$

We obviously embed these derivatives into those of χ'_i with a further management of the sign of $(\chi_i - c)$. With non radial symmetric frameworks, generated for instance from the product of a symmetric distribution – say Gaussian – and a non symmetric one – say exponential – we obtain intermediate landscapes where the above cost functions are still capable in general of identifying the independent components directions. Things may get worse if the distributions have a more complex form, for instance in case they are multimodal.

Example 7.2. Continuing Example 7.1, Fig. 7.8 shows the companion role of ICA on both a synthetic case study and the Iris dataset. Namely, the case study samples have been drawn from distributions whose components d.f.s are proportional to $\mathrm{e}^{-|y_j - \mu_{cj}|^{0.5}}$, where $j \in \{1, 2\}$ spans the component indices and $c \in \{1, 2, 3\}$ the cluster indices, and the components are differently correlated with the different clusters. The mixture of samples seems to preserve the relevance of the $l^{0.5}$ metric, since if we: i) represent the points according to the projections on the independent directions computed from them using χ_1 (see (7.13)), and ii) group the new points through k-means algorithm where the assignment decision of $\boldsymbol{y}_i$ is based on the comparison between the quantities $\sum_{j=1}^{2} |y_{ij} - \mu_{cj}|^{0.5}$ computed for the different cs, then we improve the clustering as shown in Table 7.2. From this table we see that the same procedure works worse on the Iris dataset, though the clustering based on the sole first component allows the best clustering obtained in our experiments until now.

Finally, the cost function most preferred by statisticians comes in the form:

$$\chi_3 = \sum_{i=1}^{n} \frac{1}{12}\widehat{\mathrm{E}}[A_i^3]^2 + \frac{1}{48}\widehat{\mathrm{kurt}}[A]^2, \tag{7.15}$$

where kurt denotes the kurtosis of A (namely $\mathrm{E}[A^4] - 3\mathrm{E}[A^2]^2$), and $\widehat{h}$ denotes the sample estimate of the function h on A. The preference of this function is because of its interpretation within the Kolmogorov approach. Indeed, removing hats, the single addends in the sum are assumed to represent an approximation of the shift $H[Z] - H[A]$ of the actual entropy $H[A]$ of A with

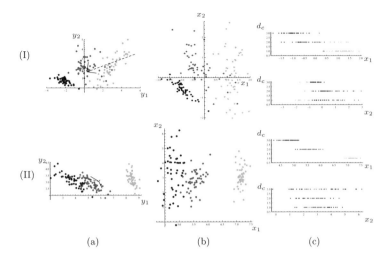

Fig. 7.8. The role of independent components. Horizontal strips: (I) refers to a synthetic dataset drawn from a subgaussian distribution; (II) refers to the first two components of Iris dataset. Vertical strips: (a) data scatter plot with directions of the principal (dashed line) and independent components (plain lines); (b) same scatter plots but with points represented in the independent components frame; (c) distribution of data along the single independent components, split in the three clusters d_c.

Table 7.2. Confusion matrices of the k-means algorithm on the datasets in Fig. 7.8 represented in the original, PCA and ICA frames. Same notations as in Table 7.1.

	Original data	PCA frame	ICA frame
Fig. 7.8 (I) both components	41 6 0 8 43 0 1 1 50 |10.7%	41 6 0 8 43 0 1 1 50 |10.7%	47 7 0 2 39 0 1 4 50 |9.3%
Fig. 7.8 (II) both components	50 1 0 0 45 13 0 4 37 |12%	50 0 0 0 47 14 0 3 36 |11.3%	50 5 0 0 31 18 0 14 32 |24.6%
Fig. 7.8 (II) first component			50 0 0 0 48 0 0 2 50 |1.3%

respect to a Normal random variable Z. The folklore is that, since Z has the maximum entropy within the family of variables with unitary variance, the more you move from this entropy and the more you find a nongaussian variable representing an independent component of Y.

<small>The nongaussianity folklore,</small>

and the entropic interpretation.

Specializing the entropic formulation (6.20) of the clustering problem to $\nu(\boldsymbol{y} - \boldsymbol{v}_c) = l^q(\boldsymbol{y})$, in case of non ambiguous clustering, so that each object is assigned to exactly one cluster, denoting with $I[\boldsymbol{Y}, C] = H[\boldsymbol{Y}] - H[\boldsymbol{Y}|C]$ the mutual information between patterns and clusters, we may prove the following

Lemma 7.1. *For any univocal mapping from $\boldsymbol{Y}$ to C,*

$$H[\boldsymbol{Y}|C] = H[\boldsymbol{Y}] - H[C], \tag{7.16}$$

$$I[\boldsymbol{Y}, C] = H[C]. \tag{7.17}$$

Claim (7.17) says that the more a clustering preserves information of patterns, the more the entropy of clusters is higher, i.e. the more they are specific. Claim (7.16) quantifies the gap between entropy of patterns and entropy of clusters that is managed by the clustering algorithm. As $H[\boldsymbol{Y}]$ is not up to us, the algorithm may decide to group the patterns so as to increase $H[C]$, in line with the claim (7.17) suggestion. In case the labels of the patterns are given and no other information about them is granted, we are harmless with respect to the $H[C]$ management. Rather, our problem is to find an algorithm that reproduces exactly the given labeling. If the task is hard, we may try to split it into two steps: i) improve the patterns representation so that their label may be more understandable, and ii) find a clustering algorithm within this new input. This corresponds to dividing the gap between $H[\boldsymbol{Y}]$ and $H[C]$ in two parts and, ultimately, looking for an encoding $\boldsymbol{Z}$ of $\boldsymbol{Y}$ minimizing $H[\boldsymbol{Z}|C]$, i.e. the residual gap $H[\boldsymbol{Z}] - H[C]$. We have two constraints to this minimization. One is to maintain the partitioning property of the final mapping. Hence, we have

Definition 7.1. *Consider a set $\mathfrak{Y}$ of $\boldsymbol{Y}$ patterns, each affected by a label b. We will say that an encoding $\boldsymbol{Z}$ of $\boldsymbol{Y}$ is correct if it never happens that two patterns of $\mathfrak{Y}$ with different labels receive the same codeword*[1].

The magic recipe: correct labels on independent components,

The second constraint is strategic: as we do not know the algorithm we will invent to cluster the patterns, we try to preserve almost all information that could be useful to a profitable running of the algorithm. This is a somehow fuzzy commitment, since *vice versa* our final goal is to reduce the mean information of the patterns exactly to its lower bound $H[C]$. A property that is operationally proof against the mentioned fuzziness is exactly the independence of the components of $\boldsymbol{Z}$. We may prove indeed the following lemma.

---

[1] Of course we will check this property on the available patterns, with no guarantee as to any future pattern we will meet.

Lemma 7.2. *Consider a set* $\mathfrak{Y}$ *of patterns and a probability distribution* π *over the patterns. Assume that for any mapping from* $\mathfrak{Y}$ *to* $\mathfrak{Z}$ *entailing an entropy* $H[\mathbf{Z}]$ *there exists a mapping from* $\mathfrak{Y}$ *to* $\mathfrak{Z}'$ *with* $H[\mathbf{Z}'] = H[\mathbf{Z}]$ *such that* $\mathbf{Z}'s$ *are independent. Then the function*

$$\widetilde{H}(\mathbf{Z}) = -\sum_{k=1}^{\nu}\sum_{j=1}^{r_k} p_{j,k} \ln p_{j,k} \qquad (7.18)$$

where ν *is the number of components, each assuming* r_k *values with marginal probability* $p_{j,k}$, *has minima over the above mappings in* $\mathbf{Z}s$ *having independent components.*

Independent components play the same role of principal components but with respect to a metric l^q for a q different from 2. It means that for an l^q suitable to the data at hand we may: i) split data into their independent components, and ii) use them to profitably cluster data with an algorithm, for instance k-means, computing distances according to the mentioned metric.

Finally, with the aim of lowering the vagueness of the entropic goal, we may decide to have Boolean $\mathbf{Z}$, as a way of forcing some reduction of data redundancy and information as well in the direction of taking a discrete decision, ultimately, the clustering. **if also Boolean, may be better.** This produces the side-benefit of a concise description of both patterns and clustering formulas, as a suitable premise for their semantic readability. Let us assume as cost function of the single pattern $\mathbf{y}_i$ in our coding problem the following function which we call *edge pulling* function:

$$\mathsf{C}_i = \log \left(\prod_{j=1}^{n} z_{ij}^{-z_{ij}} (1 - z_{ij})^{-(1-z_{ij})} \right), \qquad (7.19)$$

with z_{ij} being the j-th component of the encoding $\mathbf{z}_i$ of pattern $\mathbf{y}_i$ (see Fig. 7.9). Then:

$$\breve{H} \equiv \frac{1}{m}\sum_{i=1}^{m} \mathsf{C}_i \leq \sum_{j=1}^{n} (-\overline{z}_j \log \overline{z}_j - (1 - \overline{z}_j)\log(1 - \overline{z}_j)) \equiv \widehat{H}, \qquad (7.20)$$

where $\overline{z}_j = \sum_{i=1}^{m} \frac{z_{ij}}{m}$ and $\widehat{H}$ is an estimate of $H[\mathbf{Z}]$, i.e. the entropy of the $\mathbf{Z}$ patterns we want to cluster computed with the empirical distribution law, in the assumption that $\mathbf{Z}$ components are independent. This leads us to a representation of the patterns that is binary and with independent bits, which we call BICA representation, as reported in the following

Fact 7.1. *Any* $\mathbf{Z}$ *coding that is correct according to Definition 7.1 and meets assumptions in Lemma 7.2 while minimizing the edge*

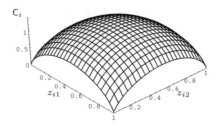

Fig. 7.9. Graph of the function C_i with $n = 2$.

pulling function (7.19) is expected to produce Boolean independent components minimizing (7.16) as well.

Note that an actual implementation of the whole binarization procedure requires the knowledge of the training data labels in order to check the correctness of this mapping. Thus BICA must be considered as a discriminant analysis, in place of clustering, procedure.

Example 7.3. Consider the Sonar benchmark. We divided the task of distinguishing metal targets from rocks in two phases: binary representation of the patterns and discrimination of the binary patterns. As for the former we map from the original 60 real variables to a 50 bit representation. Then we were able to separate the binary patterns with a classification rate of around 81%. Both mapping and discrimination have been realized with subsymbolic tools that will be discussed in the next chapter. Hence we will continue therein also the example.

7.3 Sparse Representations

From a good represenation a good clustering, and vice versa

A further move toward understanding the data is exactly to circularly using their clustering. Thus: you cluster the data, hence you assume that the morphology of a cluster identifies one privileged direction, for instance the direction from the origin to its center, that is exactly the coordinate characterizing the data that are gathered therein. Thus you have so many relevant coordinate axes as many clusters you find within the data. What you need to move to a higher dimensional space where these axes may be considered orthogonal is that the clustered points are extremely close to the related axis, so that their projections on the other axes are almost 0 and almost insensitive, consequently, to any rotation of the latter in order to become orthogonal each other. This is what characterizes the points as sparse. There could be numerous ways of identifying the axes directions in the original space. We essentially distinguish between geometrical and analytical families of methods.

7.3.1 Incremental Clustering

Assume you have clustered patterns $\{y_1, \ldots, y_m\}$ into k subgroups. Let $\{y_{c_1}, \ldots, y_{c_j}\}$ be the c-th group. Now, assume you are able to identify some privileged coordinates within each cluster, for instance those derived from local PCA or ICA. The divide is whether the new coordinates coincide with those used to determine the clusters or not. If they do, you have no innovation and consequently your possibility of further processing the data extinguishes. Otherwise, you may use the new coordinates for clustering, thus starting a new cycle of the procedure. In general, the way of using new coordinates is different for clustering purpose and for operating inside the clusters. However you have a great number of new coordinates, actually a reference framework for each cluster requiring a modification of the cluster algorithm in order to take it into account. In case the dual optimization problem (7.3) may be enunciated, the strategy remains the same: attribute the object to the cluster minimizing the cost function. The peculiarity is that we have an essentially different cost function for each cluster, possibly a function having the same shape but working with different coordinates of the points to be clustered. Now that we have clusters, we may find new coordinates that may show up to be suitable.

Use relevant axes in each cluster to build a higher dimensional framework,

A first option is to broadly identify the points with the centroids of the cluster, as in row (I) of Fig. 7.10 producing results as in the first row of Table 7.3. We are essentially assuming that the cluster mainly spreads along the direction joining its centroid with the origin, while shifts along other (possibly orthogonal as in our case, in any case relevant for some reason) directions represent a noise. Hence we attribute a point to the cluster where these shifts are minimal, with usual updating of the directions after renewed assignments of the points.

perhaps, the line joining origin with centroids, when points lie close to them,

Example 7.4. We get a noticeable improvement by applying this technique to the first two principal components of the Iris dataset, as shown in Fig. 7.10 (II) and reported in the second row of Table 7.3. These results are contrasted with those reported for a standard synthetic dataset in the first strip and line, respectively. A small variant of this procedure consists in normalizing the original dataset w.r.t. a suitable norm and then computing the sparse coordinate on the modified sample. Fig. 7.10 (III) and the third row of Table 7.3 report the results of the latter method applied to the Iris dataset. Naturally a similar performance can be observed

250   Suitably Representing Data

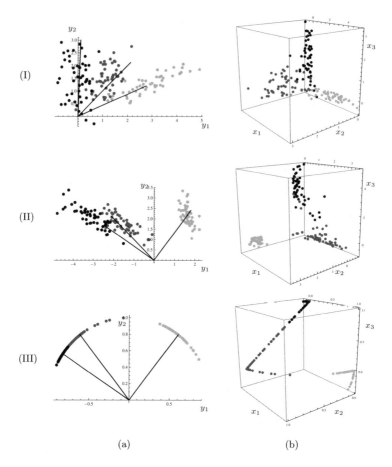

(a)   (b)

Fig. 7.10. Clustering with sparse data representation along the three axes hitting the centroids. Horizontal strip: (I) case study; (II) Iris dataset; (III) normalized Iris dataset. Vertical strips: (a) original representation with location of the new axes; (b) sparse representation. Grey levels distinguish the three cluster points.

if, instead of using l^2 metric, angular separation or correlation coefficient (see Table A.2) are used in the clustering algorithm. In particular, applying angular separation to the Iris dataset we obtain misclassification percentage equal to 3.3% corresponding to 5 misclassified points. This further improvement corresponds to measuring the distance from a centroid along the unitary circumference in place than perpendicularly to the various axes.

possibly the first principal components,   A second option w.r.t. a two dimensional framework is to identify the main directions with the first components of the data grouped in the single clusters and to measure the shifts from them along the second components. Extending the procedure

Table 7.3. Confusion matrices with clusters as in Fig. 7.10. Same notation as in Table 7.1.

	Original coordinates	Sparse coordinates
Fig. 7.10 (I)	32 0 0 18 49 17 0 1 33 |24%	47 1 0 3 49 17 0 3 33|14%
Fig. 7.10 (II)	50 0 0 0 47 14 0 3 36 |11.3%	50 0 0 0 47 5 0 3 45|5.3%
Fig. 7.10 (III)	50 0 0 0 43 0 0 7 50 |4.6%	50 0 0 0 43 0 0 7 50 |4.6%

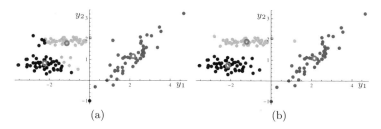

Fig. 7.11. Clustering with Mahalanobis distance: (a) preliminary output of the k-means algorithm; (b) same as in (a) but using Mahalanobis distance computed from the previous clusters.

to higher dimensional frameworks, this corresponds to a local whitening of the last principal components of the data, hence to a decision rule:

$$D = \arg\min_{c} \left\{ \sum_{j=1}^{n_c} \frac{(\boldsymbol{y} - \boldsymbol{v}_c)_j^2}{\lambda_{cj}} \right\}, \quad (7.21)$$

where the sum is commonly referred to as *Mahalanobis distance*, with $\boldsymbol{a}_j$ being the j-th among the n_c principal components considered for the c-th cluster, and λ_{cj} is the j-th eigenvalue computed within cluster c.

<small>using the last principal components for clustering.</small>

Example 7.5. With this rule we are able to correctly classify the synthetic dataset shown in Fig. 7.11 by exploiting the internal structure of the dataset as highlighted in the confusion matrices reported in Table 7.4. To appreciate the effect of this rule in Fig. 7.12, we whitened the components in order to compare point distances from the cluster centroids independently of the coordinate scales. Then, we rotate and rescale the data in Fig. 7.11(b): once with respect to the principal components of the first clus-

<small>Sometime we pay,</small>

Table 7.4. Confusion matrix with whitened data shown in Fig. 7.11.

	Classic k-means	Local Whitening
	48 1 0 2 39 17 0 10 44 |12.7%	50 0 0 0 50 0 0 0 50 |0%

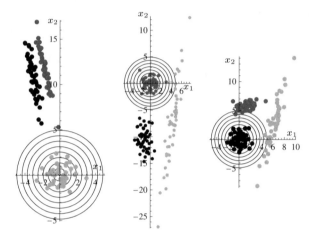

Fig. 7.12. Local whitening transformations w.r.t. the principal components of each one of the three clusters of Fig. 7.11.

ter, once with repect to the second cluster's, and finally to the third's. The above rule corresponds to compare the distances of each point in the three frames when deciding on the membership of the points.

Example 7.6. While we were not able to exploit the same benefits with the Iris dataset, where the local whitening procedure does not introduce an even minimal improvement, with MNIST things become more articulated. We move from an error rate of around 30% to 15% when we adopt the above local PCA algorithm in place of global PCA (see Fig. 7.13). Namely, in the 28×28 dimensional space representing the database records as vectors having one component per each pixel, we preliminary select the 30 principal components with highest eigenvalues. The k-means algorithm obtains on these data the mentioned error percentage. Then we move to a sparse representation in two ways. First of all we fixed k to 100 in order to get more homogeneous subgroups. Each cluster takes the label of the majority of its elements. Then we compute local PCA for each cluster and compare the distance from centroids of points representative of records,

Fig. 7.13. Subset of the 100 centroids obtained by applying local PCA k-means on the MNIST dataset.

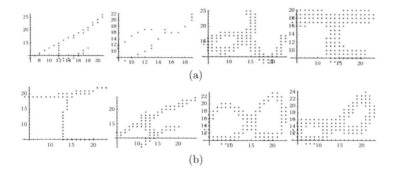

Fig. 7.14. Misclassified items (a) and subset of centroids (b) computed on the MNIST dataset when each digit is represented in the 2-dimensional plane. Each digit is rotated and reflected exactly as it is given in input to the k-means algorithm.

each computed in the framework constituted by the last principal components of each cluster, in a number sufficient to cover a given percentage of the cluster variance. With this procedure we got the most favorable percentage. We assume that there is no way in this thread to further significantly reduce it because the digit representations are not really sparse. We realize it when we manage the records as two dimensional pictures and deepened the matter by discriminating the digit 4 from the digit 8. In this case we decided for 20 clusters to allow sparsity, then we computed the bidimensional PCA frame for each cluster, using the lower eigenvalue component to decide the digit assignment. As a matter of fact, we arrived to the misclassification of only the 4 patterns in Fig. 7.14, with no hope however of improving the clustering. These records indeed constituted isolated mono-element clusters and the distance matrix showed that the closest records to them were representative of an opposite digit.

sometime we gain.

We note that the main directions associated to the clusters are not orthogonal but useful in any case. Non orthogonality is an

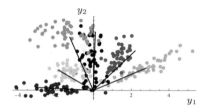

Fig. 7.15. The weak orthogonality of the sparse representation. Same notation as in Fig. 7.10.

<small>More axes need an enlarged space,</small>

unavoidable condition if the number of clusters is higher than the dimensionality of the points, as it happens in Fig. 7.10, and better highlighted in Fig. 7.15. We may have the following reading of the whole contrivance. Look at the k axes (3 in Fig. 7.10, 6 in Fig. 7.15 and so on) associated to the clusters as an overcomplete basis to describe the points. It means that you may use many subsets of its axes in order to identify a point y in the two dimensional plane. Looking for a sparse source representation requires you to select a subset with a minimal number of axes, hence a minimal number of non zero features x_j associated to the point. You will do it manually just by selecting a suitable pair of axes. If you are lucky y lies on one of these axes, so that you need only one non zero feature. You may also approximate this condition when y is extremely close to one axis so that you may assume the shift from it just as a disregardable noise. Otherwise you will pair the closest axis w_1 with another measuring the y shift from the former. A wise option is to select the axis mostly perpendicular to w_1 getting a twofold benefit: i) you get the smallest shift, possibly to be confused with noise, and ii) almost perpendicularity denotes almost linear independence between the involved features; moreover, since the shift is small, we expect that computations do not suffer exceedingly if we virtually stretch the angle between axes to $\pi/2$ to enforce orthogonality. Note that if you decide using the second option, thus identifying the main directions with the principal axes of the clusters, you get benefit from including in the enlarged framework also the other components of each clus-

<small>where to assess them orthogonally each other.</small>

ter. Indeed, these components are orthogonal, thus identifying the shifts with the last coordinate. It is exactly what we did in order to identify the new clusters. In any case, stretching the axes moves in the direction of further separating the points. With the new coordinates you re-cluster them. Possibly you iterate the procedure starting from the new groups, either in the original feature space or in the augmented one. Since we have no strong theorem about the convergence of the procedure you are free of deciding according to your personal feeling on the problem at hand.

When you are in a high-dimensional framework the minimization is achieved by solving a linear programming problem. By definition the search space is endowed with a metric l_0 (involving counting the number of non zero components of x), while the solution of the problem is easier when considering l_1 metric (taking into account also the value of the non zero components). A set of theoretical results ensures us that generally the two solutions coincide.

7.3.2 Analytical Solution

The problem of using a suitable vector $W^T y$ in (7.3) having a higher number of components than y itself is in general translated into the problem of identifying a matrix B with a number of rows greater than the number of columns and a x_i such that:

$$y_i = B^T x_i, \qquad (7.22)$$

for each i. The problem is *per se* undetermined. In order to reduce indeterminacy, on one hand we introduce the optimization target of having a solution in x_i with the least number of non zero components. On the other hand, we make some assumptions to facilitate the determination of the solution. The analytical details are rather hard. Here we limit ourselves to offer a simple guess of the solution strategies. The interested reader may refer to [35, 15, 64] for deeper insights. In reference to Fig. 7.15 we essentially make the assumption that the points are extremely close to the axes of the independent components, so that we need a very sharp mask to identify these axes, hence related to a norm l^q with $q < 1$ (entailing a hypothesis of supergaussianity).

Look for a wise subset of components,

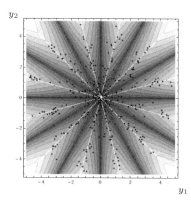

Fig. 7.16. A mask subtending the analytical formulation.

You may think of the analytical translation of a mask like in Fig. 7.16. Then the overcompleteness of basis B requires that we reduce the angle between the spokes from $\pi/2$ to a lower value in order to make room to further spokes corresponding to the additional coordinates. The way of computing the new angles is essentially iterative: from a set of angles, hence a given B, compute the optimal x_is. From them formed is a new B optimizing the cost function based on l^q, and so on.

from among a number of coordiantes augmented like above.

Obviously the sparseness is reached only if the base B is chosen appropriately. A computationally feasible method relates the search for a suitable base to that of k-means clustering, by chosing, as already shown in Fig. 7.15, as base vectors those corresponding to the normalized centroids. This makes sense in situations like those depicted in the picture, where the clusters move radially from the axes origin.

7.4 Nonlinear Mappings

You may think of reproducing many of the discussed mechanisms also with a nonlinear mapping allowing to identify polytopes with more complex edges. Consider for instance what happens in Fig. 7.17(a): here points are radially grouped but naive k-means shown in Fig. 7.17 (b) failed in finding the groups – giving an error of 44% – as the separate lines cannot describe surfaces more complex than a convex polytope. A smart way to recover the above results is to figure these mappings as nonlinear projections onto higher dimensional spaces where the above cluster methods are implemented. Let us come back to the above example and consider adding to the observed points a third coordinate whose value linearly depends on the distance of the points from the origin. With such a transform you obtain the picture in Fig. 7.17(c) where clusters can be identified by using the l^2 metric on this extended space, reducing the clustering mistakes to 0. In turn the separating polygons may be mapped back to the original plane in this way giving rise to the circular borders as shown in Fig. 7.17(d).

You may augment the space also to capture complex relations between data.

This procedure is quite general and such perspective is particularly useful when we work with distances in l^2 so that only second order moments are involved in the computations, and we may express them in terms of kernel functions, thanks to the *kernel trick* discussed in Appendix D.2. The choice of the kernel, free parameters included, is generally up to the user, who may exploit common wisdom rather than formal analysis. Rotation and threshold parameters are exactly inferred with the above methods referred to the higher dimensional space in principle, but implemented in the original space.

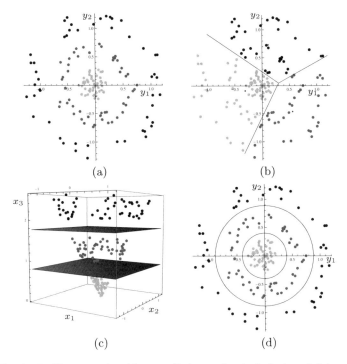

Fig. 7.17. K-means algorithm applied to a circularly featured dataset: (a) dataset scatter plot; (b) k-means clusters on original data; (c) scatter plot and separating surfaces when a third coordinate is added to the points; and (d) circular borders to be identified.

Kernel methods for clustering may be grouped in three families: i) *kernelization* of the metric [63], ii) implementation of k-means algorithm in the feature space [21], and iii) kernel methods based on support vector data description [5]. In the following we will give some hints on these procedures with no pretension of exhaustiveness, leaving the interested reader to the cited literature.

<small>Efficient methods when are based on kernels.</small>

Kernelization of metric. In case of distance functions induced by a dot product (see Appendix D) we can always define the former in terms of the latter, by noting that:

<small>k in place of l^2.</small>

$$d.(\boldsymbol{y}_1, \boldsymbol{y}_2) = \boldsymbol{y}_1 \cdot \boldsymbol{y}_1 + \boldsymbol{y}_2 \cdot \boldsymbol{y}_2 - 2(\boldsymbol{y}_1 \cdot \boldsymbol{y}_2). \quad (7.23)$$

This means that we obtain new metrics by substituting the dot product expression with proper kernels k, i.e.,

$$d_k(\boldsymbol{y}_1, \boldsymbol{y}_2) = k(\boldsymbol{y}_1, \boldsymbol{y}_1) + k(\boldsymbol{y}_2, \boldsymbol{y}_2) - 2k(\boldsymbol{y}_1, \boldsymbol{y}_2). \quad (7.24)$$

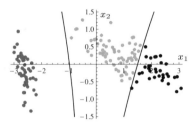

Fig. 7.18. K-means algorithm applied to the first two principal components of the Iris dataset when using a kernelized metric $d_k(\boldsymbol{x}_1, \boldsymbol{x}_2) = (x_{11} - x_{21})^2 + (x_{12} - x_{22})^2 + (x_{12}^2 - x_{22}^2)^2$.

Of course kernelization of metrics may be implemented in both global and local scenarios, as the ones based on local PCA methods.

Example 7.7. Fig. 7.18 shows the application of k-means algorithm to the first two principal components of the Iris dataset when defining the metric in terms of the same polynomial kernel as the one we used to obtain Fig. 7.17(c). While in the former instance we notably improve the clustering procedure, in this case on the contrary we attain an error of 0.12 which, compared with naive k-means algorithm (providing a misclassification percentage of 11%) performs relatively worse.

Projection on feature space. Nonlinear mapping may involve projections toward both high dimensional and low-dimensional space. Typical examples are provided by the huge family of generalizations of linear PCA to the nonlinear case. These include unsupervised neural network algorithms [43], autoassociative multilayer perceptrons [16], Kernel PCA [50] and principal curve [25].

For instance, principal curve approach [25] estimates a curve (or a surface) capturing the structure of the data, by projecting data onto the curve, in such a way that each point on the curve is the average of all data points projecting onto it. More generally, Principal Surfaces method is a latent variable model defining a nonlinear mapping from a low-dimensional latent space to the data space. Low dimensionality of the latent space guarantees (under continuity and differentiability hypothesis of the mapping function) that it corresponds to a Q-dimensional manifold nonlinearly embedded in the data space, where Q is the dimension of the latent space. In case of Probabilistic Principal Surfaces (PPS) [11] the problem becomes that of estimating the parameters of a family of maps (usually mixture of Gaussians). By starting from a 3-dimensional latent space we obtain a spherical manifold: the projection of the data items onto it may provide a useful visualization in terms of a first understanding of data aggregation [53].

Nonlinear mapping is a favorite functionality of neural networks, a computational paradigm that we will discuss in Chapter 8. Actually, many of the mentioned clustering methods have been featured in terms of neural networks *unsupervised learning* procedures in order to better organize the involved computations. The archetypal implementation of them is represented by Self Organizing Maps (SOM) [33]. However, as will be discussed later on, the preprocessing of data is generally forced with a more strength in *supervised learning* procedures, hence in correspondence with discriminant analysis problems.

Support Vector Clustering. Such family of methods exploits synergies from the structure hidden in the data and the special information content of some of them playing the role of support vectors (see Section 3.2.4 for an introduction to SVM).

Support Vector Clustering (SVC) algorithm [5] was first developed as an extension of single-class SVM [49], an algorithm inferring the support of the distribution underlying the observed sample, intuitively speaking the smallest sets having a fixed measure μ, say 90%, containing these points. The idea is the following: consider a kernel implicitly mapping points in a small and well defined region of the feature space and find the optimal margin hyperplane separating these points from the origin. Maximizing this margin corresponds to compute a function which is supposed to capture regions in input space where the probability density is large, i.e. a function that is nonzero in a region where most of the data are located. In turn, the possibly different contours identifying the support region may be regarded as cluster boundaries, where an increase in μ translates in both a reduction of the clusters width, and a possibly increase in their number.

A kernel having the above property is the Gaussian kernel k_G (see Appendix D.2), whose analytical form corresponds to a function ϕ mapping all points $\boldsymbol{y}$ in the positive orthant. In fact, $0 \leq k_G(\boldsymbol{y}_i, \boldsymbol{y}_j) \leq 1$ for each i, j, being equal to 1 only when $i = j$. This property, which comes from the dependence of k_G on $\boldsymbol{y}_i - \boldsymbol{y}_j$, guarantees that all points will lie in a hypersphere in feature space, so that maximizing the margin of separation from the origin corresponds in finding the smallest sphere enclosing the data [49].

Formally we aim at solving the following constrained minimization problem:

$$\min_{R,c} R^2 + \grave{\alpha} \sum_{i=1}^{m} \xi_i, \qquad (7.25)$$

$$l^2(\phi(\boldsymbol{y}_i) - \boldsymbol{c}) \leq R^2 + \xi_i, \quad \forall i, \qquad (7.26)$$

<small>possibly inducing regular shapes in the feature space.</small>

where $\boldsymbol{c}$ is the center of the sphere having radius R, ξ_is are slack variables softening the constraints (see Section 12.1.1 for the application of slack variables to standard SVM algorithm) and $\grave{\alpha}$ is a constant determining the trade-off between large regularization (minimization of the sphere radius) and small training error (fraction of points lying outside the sphere). The power of this approach is that this problem can be completely reformulated in terms of dot products in the feature space, or kernel in the data space. Moreover, once found the optimal radius R_{opt} of the sphere, the contours that enclose the points in the data space can be obtained by the contour lines of level R_{opt} of the function:

$$r(\boldsymbol{y}) = l^2(\phi(\boldsymbol{y}) - \boldsymbol{c}) =$$
$$= k_G(\boldsymbol{y},\boldsymbol{y}) - 2\sum_{i=1}^{m}\alpha_i k_G(\boldsymbol{y}_i,\boldsymbol{y}) + \sum_{i,j=1}^{m}\alpha_i\alpha_j k_G(\boldsymbol{y}_i,\boldsymbol{y}_j),$$
$$(7.27)$$

describing the sphere in the feature space and projecting it back onto the data space. The shape of the clusters depends on two parameters: i) the width of the Gaussian kernel (h in Appendix D.2), and ii) the soft margin constant $\grave{\alpha}$, where $1/(m\grave{\alpha})$ asymptotically approaches the fraction of outliers. Intuitively, a decrease of h goes in the same direction of the μ effect of increasing the number of disconnected contours in data space, hence the number of clusters. Setting the true values of the parameters is a hard task. Usually they are adjusted by looking at the number of support vectors, as the smaller is their number and the higher is the guarantee of smooth boundaries. Alternatively one may set their value according to the stability of cluster assignments.

Example 7.8. Fig. 7.19 traces the cluster boundaries of the first two principal components of the Iris dataset for two different values of shape and soft margin parameters. By setting $h = 1$ and $\grave{\alpha} = 1$ from Fig. 7.19(a) we realize that while

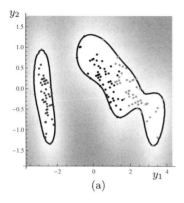 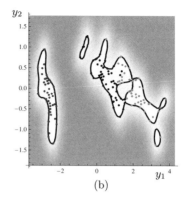

Fig. 7.19. Support Vector Clustering applied to the first two principal components of the Iris dataset for different values of parameters h and $\grave{\alpha}$: (a) $h = \grave{\alpha} = 1$; and (b) optimal assignments $h = 6$ and $\grave{\alpha} = 0.6$.

Iris Setosa class, linearly separable from the remaining two classes, is correctly identified, the latter are not able to split. Note that a value $\grave{\alpha} = 1$ means that not even one point (at least asymptotically) would be placed outside of a connected region. Chosing $h = 6$ and $\grave{\alpha} = 0.6$ the two overlapping clusters were separated yet they split into two groups each (see Fig. 7.19(b)). In this case we observe 11 misclassifications w.r.t. correctly joined subclusters. Other choices of parameters are possible. Note that a value of $\grave{\alpha} = 1$ stands for zero asymptotical error frequency.

7.5 Conclusions

The clustering methods discussed in the previous chapter are functionally quadratic, coming from the pairing of points a distance measure is fed on. We aim at preserving this simplicity by allowing only linear transforms of the data, a facility that allows to identify their independent components.

If a set of variables are independent, then they are linearly independent as well. This allows us to narrow our search for a reference framework of our dataset with independent coordinates within the frameworks with linear independent coordinates. Thus, in the approximation of first and second order population moments with the analogous sample moments, you obtain solution with a sequence of linear transforms made up of a framework rotation, then scale change, then another rotation and finally another scale change. With the former you identify principal components of the patterns, with the first scale change you whiten them so that each component has unitary variance. For the whitened dataset each orthogonal framework is made up of principal components, hence linearly independent coordinates. The second rotation identifies the independent coordinates and a further scale change eventually de-whitens the patterns restituting them the original scale. First rotation is identified by the eigenvectors of the patterns, by definition. The second rotation is questionable. We identify them in a *mechanical way* with the directions of either maximal or minimal elongation of the contour curves of the patterns joint d.f. The method is devised for the case that the single independent component d.f.s are symmetric; but we expect it working enough well also in the presence of some asymmetry. In this perspective, the functions in (7.13) normally employed with standard methods such as FastICA are aimed at incrementally discovering these directions. As a matter of fact minima and maxima are located on directions with clear reciprocal positions, and the preferability of the former w.r.t. the latter depends on the use we will do of the patterns. Independent components have a great entropic value; however for clustering purpose we may be more interested to the dispersion of the components than to their independence.

With sparse datasets these suitable components may be identified in an augmented reference frame as a follow-out of the clustering of the patterns. In this case one or more axes relevant to each single cluster may be gathered in a new space where they are *stretched* in order to be orthogonal each other, so as to result useful both for entropic tasks and for possible improvement of the clustering itself.

We consider in this chapter also further either augmentation or reduction of the reference framework obtained through nonlinear transforms of the data. The general strategy is to employ the methods discussed till now to cluster yet applied to nonlinearly stretched patterns. You recover this stretching through linear transforms again but in a higher or lower dimensional space.

7.6 Exercises

1. Use the formulas of the marginal distribution of X_2 and the density $f_{X_1/x_2}(x)$ of conditional random variable X_1 given $X_2 = x_2$ (see Appendix B.2), when X_1, X_2 are components of a two-dimensional Gaussian variable $\boldsymbol{X}$, to generate a sample of 100 specifications (x_{1i}, x_{2i}) through the sampling mechanisms of both X_2 and the distribution of X_1 conditioned by X_2 specifications. For the joint distribution use

$\boldsymbol{\mu_X} = (10, 20)$ and $\boldsymbol{\Sigma_X} = ((1, 0.7), (0.7, 1.5))$. Draw the scatterplot of these points.

2. From the points generated in Exercise 1 compute: i) $\boldsymbol{m_x}$ and $S_{\boldsymbol{xx}}$ as in (7.9), and ii) the eigenvalues and eigenvectors of $S_{\boldsymbol{xx}}$. Draw eigenvectors with origin $\boldsymbol{m_X}$ in the scatterplot drawn in Exercise 1. Generate new points as specifications of the random variable $\boldsymbol{X'}$ having coordinates: i) x'_{1i} specifications of a Gaussian variable with $\mu_{X'_1} = \mu_{X_1}$ and $\sigma^2_{X'_1}$ equal to the first eigenvalue of $S_{\boldsymbol{xx}}$, and ii) x'_{2i} specifications of a Gaussian variable with $\mu_{X'_2} = \mu_{X_2}$ and $\sigma^2_{X'_2}$ equal to the second eigenvalue of $S_{\boldsymbol{xx}}$. Compare the scatterplots of the first and second set of points.

3. Compute the three principal components of the Iris dataset and use them in order to produce a three-dimensional scatterplot.

4. Denote with $\boldsymbol{y}_i$ the i-th record of the SMSA dataset, having both excluded the first variable (city), and whitened the data w.r.t. the single features (which corresponds to standardize each column of the dataset so as to have null mean and unitary variance). Compute the 16 eigenvalues from $S_{\boldsymbol{yy}}$ and their sum. Sort the obtained value in decreasing order and take the first r components that cumulate 90% of the sum. Construct the matrix W having columns constituted by the eigenvectors of $S_{\boldsymbol{yy}}$ corresponding to these eigenvalues. For each i compute $\boldsymbol{x}_i = W^T \boldsymbol{y}_i$, then get $\boldsymbol{y}'_i = \widetilde{W}^{-1T} \boldsymbol{x}_i$, where $\widetilde{W}^{-1}$ is a pseudoinverse of W that you may compute using any utility on the WEB. Analyze the differences between $\boldsymbol{y}_i$ and $\boldsymbol{y}'_i$.

5. Repeat the experiments of Exercise 1, but with $f_{X_1/x_2}(x) = 1/2 \mathrm{e}^{-|x - x_2|}$. *Withen* the data as in Section 7.2 and apply them the masks (7.12) and (7.15). Identify in this way the independent components of the data and draw their directions in the scatterplot of the whitened data.

6. Apply ICA to the pattern drawn from SMSA dataset in Exercise 4. You can find latest release of FastICA algorithm together with both the way of encoding $\boldsymbol{y}_i$s into its independent component and the way of inverting the transform in [20]. Compute the empirical entropy for each component on a suitable discretization of its domain, i.e. having partitioned the x_i domain in intervals $\{[I_i, I_{i+1}], i = 1, \ldots q\}$, compute $H[X_i] = -\sum_{i=1}^{q} \hat{p}_i \log \hat{p}_i$, with $\hat{p}_i$ being the frequency of items in the SMSA dataset whose i-th components lies in $[I_i, I_{i+1}]$. Retain more relevant those features having low values of entropy and select, after a normalization procedure w.r.t. their sum, those components whose cumulative

relevance sum up to 10%, starting from lower relevance values. Finally use the same procedure as in Example 4 to come back to the original coordinates using only these components, and compare the obtained values with the originals ones.

7. On the same data obtained in Exercise 4, i.e. after computation of SMSA dataset principal components and selection of the greatest variance ones, apply ICA to them and compare the indipendent components so obtained with the ones computed in the previous exercise.

8. Consider the records referring to three provinces in the Swiss dataset. Cluster the patterns with a k-means algorithm with $k = 5$ and for each cluster consider the principal component with maximum eigenvalue. Reapply the k-means algorithm encoding each record with the projection of the patterns along the above directions.

9. Discuss about the role played by h in the Support Vector Clustering algorithm described in Section 7.4 when applied to the same records of Exercise 8.

Further Reading

As a search tool of supersets containing granules of information, clustering has constituted a philosopher's sophisticated sieve to drill the universe [14]. From an operational perspective it has been a main decision support tool of Artificial Intelligence since Von Neumann time [57]. A reference book of the modern way of conceiving data processing led by cluster analysis dates however 1973 [17] as a sign of the real availability of computing machines for managing masses of data in a comfortable manner. With the common granularity feature of the information at the basis of many operational problems we are currently called to solve, clustering routines are present in most numerical procedures running today on computers. They are so pervasive that, like for an algebraic operation, it proves difficult distinguishing their intervention as utility from the global functionality of the procedure. In turn, it is difficult as well to distinguish clustering core of the utility from all contrivances which are set up in order to make it efficient. In this part of the book, we identify the core with the conceptual tools, extracting information from data finalized to classify. We did it in a very essential way, thus the reader has a vast variety of options to deepening the matter and enlarging the scope. However, the clustering lesson is exactly that we must hit the core if we want to understand data and use them profitably. Discarding technicalities, we may suggest three veins to be followed on the border of the core.

A former concerns *data preprocessing* that precedes the representation problem. The points of the Euclidean space we are dealt with in this part actually are the images of a suitable transform of input data that depends on the problem at hand. It reckons the non granular part of the information. For instance, with time signals we know that there is a time structure underlying data that we may highlight using Fourier transform [7], so that clustering will apply directly to the coefficients of this transform.

Also in this case an efficiency key is represented by the compression rate we achieve with transforms. High values are desirable both to soften the computational load and to help understanding. A last deal in this direction is represented by wavelet transform [45] that proves particularly suitable with image processing [10] and produces wavelet coefficients as features of our clustering patterns. Wavelet represents an intermediate layer between analogical features, like the sinusoids in Fourier transform, and the logical features which were used, with a scarce success, in syntactical pattern recognition [65], for instance to recognize a chair on the basis of geometrical relationship between subpatterns representing legs, back, etc. This kind of recognition has represented a clever challenge for computer scientists but is exceedingly time consuming.

A second vein concerns *stochastic minimization* of the clustering cost function. Optimization is another big issue *per se* that we carefully avoid in this book. However we underline the suitable pairing of granular information and stochastic optimization procedures. By definition, Granules hide many details of the cost function landscape. Randomness grains injected in the procedure represent a proper antidote to relieve the knowledge lack linked to the missing details in the realistic goal of reaching a reasonable solution. In this sense Simulated Annealing [32] did constitute a foundation of family of algorithms so randomly drawn as to escape local traps and so problem-driven as to track good solutions. A relatively recent member of this family is represented by Superparamegnetic clustering [6] specially suited to deal with groups affecting homogeneously the cost function in analogy with what we will see in Chapter 8 with genetic algorithms [40]. A variant inspired by quantum mechanics is represented by Quantum Annealing. Discovered in the eighties [1], this method has today a relatively large diffusion [13].

Finally *data representation* is a timely research topic of high practical relevance. The fact is that, with the overtaking of technologies w.r.t. methodologies, we are submerged of data often without efficient tools to process them. Then a big aim is to quickly reduce their amount preserving a reasonable part of their information. Said in other words, the paradigm is as follows: they are too many because, like with intronic segments in genome strings, most of them are useless; hence the first commitment is to select the sole useful sparse features, then we will see how to use the latter. This gave rise to *universal rules* such as the search for *non Gaussian* components [27] which might generate some misunderstanding. Confirming the extreme importance of finding a proper representation and the great value of the results obtained in this field (try to query ICA on any search engine and you will obtain thousands of references), in this monograph we link its solution to both the general knowledge we have about the data we are processing and the use we plan to do of them. With the customary style, we constrained ourselves to fundamentals, thus omitting sophisticated representation methods and feature selection techniques, such those connected to Spectral Clustering [42] as a graph theoretic version of sparse representation.

References

1. Apolloni, B., Caravalho, N., De Falco, D.: Quantum stochastic optimization. Stochastic Processes and their Applications 33, 233–244 (1989)
2. Arthur, D., Vassilvitskii, S.: How slow is the k-means method? In: Proceedings of SCG 2006, Sedona, Arizona, USA. ACM, New York (2006)
3. Ball, G., Hall, D.: Isodata, an iterative method of multivariate analysis and pattern classification. In: IFIPS Congress (1965)
4. Bang-Jensen, J., Gutin, G., Yeo, A.: When the greedy algorithm fails. Discrete Optimization 1, 121–127 (2004)
5. Ben-Hur, A., Horn, D., Siegelmann, H.T., Vapnik, V.: Support vector clustering. Journal of Machine Learning Research 2, 125–137 (2001)
6. Blatt, M., Wiseman, S., Domanym, E.: Super-paramagnetic clustering of data. Physical Review Letters 76, 3251 (1996)
7. Bracewell, R.N.: The Fourier Transform and Its Applications, 3rd edn. McGraw Hill, Boston (2000)
8. Brin, S.: Near neighbor search in large metric spaces. In: Proceedings of the Twenty First International Conference on Very Large Databases (VLDB 1995), pp. 574–584 (2005)
9. Calinski, R.B., Harabasz, J.: A dendrite method for cluster analysis. Communications in Statistics 3, 1–27 (1974)
10. Chan, T.F., Shen, J.J.: Image Processing and Analysis - Variational, PDE, Wavelet, and Stochastic Methods. Paperback, Society of Applied Mathematics (2005)
11. Chang, K., Ghosh, J.: A unified model for probabilistic principal surfaces. IEEE Transactions on Pattern Analysis and Machine Intelligence 23(1) (2001)
12. Cover, T., Hart, P.: Nearest neighbor pattern classification. IEEE Transactions on Information Theory 13, 21–27 (1967)
13. Das, A., Chakrabarti, B.K.: Quantum Annealing and Related Optimization Methods. Lecture Note in Physics, vol. 679. Springer, Heidelberg (2005)
14. Davis, M.: Weighing the universe. Nature 410, 153–154 (2001)

15. Delgado, K.K., Murray, J.F., Rao, B.D., Engan, K., Lee, T.W., Sejnowski, T.J.: Dictionary learning algorithms for sparse representation. Neural Computation 15, 349–396 (2003)
16. Diamantaras, K.I., Kung, S.Y.: Principal component neural networks. Wiley, New York (1996)
17. Duda, R.O., Hart, P.E.: Pattern classification and scene analysis. John Wiley & Sons, New York (1973)
18. Duda, R.O., Hart, P.E., Stork, D.G.: Pattern classification. Wiley Interscience Publication, Chichester (2000)
19. Garey, M.R., Johnson, D.S.: Computer and Intractability: a Guide to the Theory of NP-Completeness. W. H. Freeman, San Francisco (1978)
20. Gavert, H., Hurri, J., Sarela, J., Hyvarinen, A.: The fastica package for matlab (2005)
21. Girolami, M.: Mercer kernel based clustering in feature space. IEEE Transactions on Neural Networks 13(3), 780–784 (2002)
22. Glover, F., Laguna, M.: Tabu Search. Kluwer, Norwell (1997)
23. Har-Peled, S., Sadri, B.: How fast is the k-means method? Algorithmica 41(3), 185–202 (2005)
24. Hartigan, J.: Clustering Algorithms. Wiley, New York (1975)
25. Hastie, T.J., Stuetzle, W.: Principal curves. Journal of the American Statistical Associations 84(406), 502–516 (1989)
26. Heyer, L.J., Kruglyak, S., Yooseph, S.: Exploring expression data: Identification and analysis of coexpressed genes. Genome Research 9, 1106–1115 (1999)
27. Hyvärinen, A., Kahunen, J., Oja, E.: Independent Component Analysis. John Wiley & Sons, Chichester (2001)
28. Jessop, A.: Informed assessments: an introduction to information, entropy and statistics. Ellis Horwood, New York (1995)
29. Johnson, S.C.: Hierarchical clustering schemes. Psychometrika 2, 241–254 (1967)
30. Jolliffe, I.T.: Principal Component Analysis. Springer, Heidelberg (1986)
31. Kaufman, L., Rousseeuw, P.: Finding groups in data: an introduction to cluster analysis. Wiley, New York (1990)
32. Kirkpatrick, S., Gelatt, C.D., Vecchi, M.P.: Optimization by simulated annealing. Science 220, 671–680 (1983)
33. Kohonen, T.: Self-Organizing Maps, 3rd edn. Springer Series in Information Sciences. Springer, Berlin (2001)
34. Krzanowski, W.J., Lai, Y.T.: A criterion for determining the number of groups in a data set using sum of squares clustering. Biometrika 44, 23–34 (1985)
35. Li, Y., Cichocki, A., Amari, S.: Analysis of sparse representation and blind source separation. Neural Computation 16, 1193–1234 (2004)
36. Lloyd, S.P.: Least squares quantization in pcm. IEEE Transactions on Information Theory 28, 129–137 (1982)
37. MacQueen, J.: Some methods for classification and analysis of multivariate observations. In: Proceedings of the Fifth Berkeley Symposium on Mathematical Statistics and Probability, Berkeley, CA, vol. 1, pp. 281–296 (1967)

38. Milligan, G.W., Cooper, M.C.: An examination of procedures for determining the number of clusters in a data set. Psychometrika 50, 159–179 (1985)
39. Minaei-Bidgoli, B., Topchy, A., Punch, W.F.: Ensembles of partitions via data resampling (to be discovered, 2005)
40. Mitchell, M.: An Introduction to Genetic Algorithms. MIT Press, Cambridge (1996)
41. Morrison, D.F.: Multivariate Statistical Methods. McGraw-Hill, New York (1967)
42. Ng, A., Jordan, M., Weiss, Y.: On spectral clustering: analysis and an algorithm. In: Becker, S., Dietterich, T., Ghahramani, Z. (eds.) Advances in Neural Information Processing Systems, vol. 14. MIT Press, Cambridge (2002)
43. Oja, E.: A simplified neuron model as a principal component analyzer. Journal of Mathematical Biology 15, 267–273 (1982)
44. Pearson, K.: Principal components analysis. The London, Edinburgh, and Dublin Philosophical Magazine and Journal of Science 6(2), 559 (1901)
45. Resnikoff, H.L., Wells, R.O.: Wavelet analysis: the scalable structure of information. Springer, Berlin (1998)
46. Ristad, E.S., Yianilos, P.N.: Learning string edit distance. IEEE Transactions on Pattern Analysis and Machine Intelligence 20(5), 522–532 (1998)
47. Rohatgi, V.K.: An Introduction to Probablity Theory and Mathematical Statistics. Wiley Series in Probability and Mathematical Statistics. John Wiley & Sons, New York (1976)
48. Rubner, Y., Tomasi, C.: Texture-based image retrieval without segmentation. In: The Proceedings of the Seventh IEEE International Conference on Computer Vision 1999, Kerkyra, Greece, vol. 2, pp. 1018–1024 (1999)
49. Scholkopf, B., Platt, J., Shawe-Taylor, J., Smola, A.J., Williamson, R.C.: Estimating the support of a high-dimensional distribution. Neural Computation 13(7) (2001)
50. Scholkopf, B., Smola, A., Muller, K.R., Scholz, M., Ratsch, G.: Kernel pca and de-noising in feature space. In: Advances in Neural Information Processing Systems, vol. 11, pp. 536–542 (1999)
51. Smyth, P.: Clustering sequences with hidden markov models. In: Advances in Neural Information Processing Systems, vol. 9, pp. 648–654 (1997)
52. Spielman, D.A., Teng, S.H.: Smoothed analysis of algorithms: Why the simplex algorithm usually takes polynomial time. Journal ACM 51(3), 385–463 (2004)
53. Staiano, A., De Vinco, L., Ciaramella, A., Raiconi, G., Tagliaferri, R.: Probabilistic principal surfaces for yeast gene microarray data mining. In: Perner, P. (ed.) ICDM 2004. LNCS (LNAI), vol. 3275. Springer, Heidelberg (2004)
54. Tibshirani, R., Walther, G., Hastie, T.: Estimating the number of clusters in a data set via the gap statistic. Journal of the Royal Statistical Society: Series B (Statistical Methodology) 63(2), 411–423 (2001)

55. Turner, J.R.: Introduction to analysis of variance: Design, analysis, and interpretation. Sage Publications, Thousand Oaks (2001)
56. Von Neumann, J.: Mathematical Foundations of Quantum Mechanics. Princeton University Press, Princeton (1955)
57. Von Neumann, J.: The Computer and the Brain. Yale University Press, New Haven (1958)
58. Voronoi, G.: Nouvelles applications del paramètres continus à la théorie des formes quadratiques. Journal für die Reine und Angewandte Mathematik 133, 97–179 (1908)
59. Ward, J.H.: Hierarchical grouping to optimize an objective function. Journal of the American Statistical Association 58, 236–244 (1963)
60. Wilks, S.S.: Mathematical Statistics. Wiley Publications in Statistics. John Wiley, New York (1962)
61. Wishart, D.: ClustanGraphics Primer: A Guide to Clustaer Analysis, 2nd edn. Clustan Ltd., Edinburgh (2003)
62. Wojna, A.: Center-based indexing in vector and metric spaces. Fundamenta Informaticae 56(3), 285–310 (2003)
63. Yu, K., Ji, L., Zhang, X.: Kernel nearest-neighbor algorithm. Neural Processing Letters 15(2), 147–156 (2002)
64. Zibulevsky, M., Pearlmutter, B.A.: Blind source separation by sparse decomposition in a signal dictionary. Neural Computation 13, 863–882 (2001)
65. Zucker, S.W.: Local structure, consistency, and continuous relaxation. In: Haralick, R., Simon, J.C. (eds.) Digital Image Processing Noordhoff International, Leyden (1980)

Part IV
Directing Populations

Introduction

Generally speaking, our strategy to face information granules has been the following. If you have enough information consider scenarios that are compatible with the observed data and give them a compatibility measure as a basis for future decisions (see for instance Chapter 3); otherwise make directly on a basis of data decisions that are most convenient to us (most part of Chapter 6). A third way arising in the last years is a merge of the former: mould populations that are compatible with our convenience. Thus, on the one hand we use either implicit or explicit sources of randomness for seeding populations, on the other we invent sampling mechanisms that both comply with the acquired knowledge about the data and are in agreement with the ultimate target of our computational effort.

Design sampling mechanisms complying both with data and with our own convenience

Drawing from the genetic metaphor we may say that we move from a genotypes' population, that we partly know, to a phenotypes' population, that we may observe but in general cannot be able to put in an exact reverse relation with the former. Thus we modify the former, observe the latter and look for trends suggesting us direct modifications on genotypes so that what we observe is more and more performing in respect to a fixed goal. The ways of implementing these steps are definitely numerous. The common thread, however, looks to be the idea that optimization from granular information is a somehow *social* phenomenon: it may prove a velleity to hit at an optimal solution by selecting it from a huge set of candidate solutions. Rather, we may look at these circumstances for its emerging from the set after a dynamic made of so complex as unpredictable interactions between the elements in the set. It seems to be a mandatory way of profiting of granular information that we may only loosely control with general purpose strategies. This a typical phenomenon you meet with your attic or your writing desk after an intense interaction period with them. If you want to pick up a specific object

Commit this task to social phenomena in a population of computing elements,

they are so autonomous as genotypes, and so obscure to be controllable only indirectly at a phenotypical level.

or a given document, either it emerges quickly from the mess (the most common behavior) as a consequence of dynamics that you are unable to explain yourself, or you become lost in systematically sieving all items in the heap without any guarantee of finishing this job in a reasonable time.

In this part of the book we will consider a few paradigms of population moulding, just in their most elementary enunciation with the aim of offering templates of a rich taxonomy. In this way we plan to cover variations in: **1)** the genotypical elements, such as: i) input of an unknown function ruled or not by a distribution law, ii) set of candidate solutions of an optimization problem randomly generated or wisely collected, iii) components of so complex as unpredictable functions, and so on; and **2)** the phenotypical elements, such as: i) error function specifications denoting the accuracy with which a given goal has been approached, ii) fitness values as a reward of the computation performed on the input, iii) distance between distribution laws denoting how far the phenotypes' population is from the target toward which it is pushed, and so on. We will devote one chapter to some basic aspects and the two principal paradigms of population directions, namely the genetic algorithms and the neural networks. In a second chapter we will excerpt some new paradigms in the same vein that are not equally widespread though offer cues for new perspectives jointly with some useful application examples. As usual in the facts of science, they are ridiculous in general the efforts of reducing all techniques and methodologies around a given problem to a unique scheme like the *true* key of understanding the universe. Rather, with the anticipation of not being exhaustive, we will outline of these templates the aspects concerning the information management and check whether and how the above ingredients, random seeds and sampling mechanism, are a part of this management.

A hard task leaving many options open.

8 The Main Paradigms of Social Computation

In its primary form of connectionist paradigms, social computation raised in the eighties of the last century with the role of bringing an answer to the failure of the parallel computation as a general way of overcoming the speed limits of sequential computers. In that time these limits where mainly theoretical, encountered when people did try to solve exactly combinatorial optimization problems such as crew scheduling or bin packing. The general solution was:

If we are not able to find a quick solution ask directly the machine for this.

> if we are not able to define a procedure solving the problem in a feasible time, let the computer find by itself a good solution in a way bearable by its computing resources.

This is the principle of modern machine learning, that in a realistic acceptance looks for sophisticated regression procedures in very high dimensional parameter spaces and basing on a highly moldable function family (hence equipped with derivatives of any order, composed of relatively simple modular functions, and so on). In this frame you initialize your learning machine, give it the fitting of a specific function of sample data as optimization goal and an incremental optimization procedure, and wait for the *emergence* of the regression curve. You may regress the solution function from sample data, in which case you speak of learning algorithms. On the contrary, in the optimization problems the sample data are generated by the procedure itself.

We supply examples, machine fit them optimally.

With the time these kind of problems have been recovered also thanks to the computer power growing with accordance to the Moore's law (roughly by a factor of 2 every 18 months). Nevertheless, a further need for social computation is connected with interfaces. Indeed, while on the one hand the Moore's law for the computational power starts to be barred by the quantum phenomena posing problems that are far to be solved, on the other one the utilities expected by the computers, understood as executors of explicit commands, are almost saturated both in industrial and domestic environments. Rather there is wide room to new computer functionalities mainly related to the communications/interactions between computers and, over all, direct interaction between computers and human beings. With direct interaction it is understood a non symbolic interaction that may paradigmatically denote the absence of keyboard and any of its surrogates (such as speech-to-text devices). Again the job of interpreting non symbolic signals collected by environmental sensors is demanded by the computer itself. In more abstract terms, to fully benefit from the potential of information granules as well as all constructs emerging there, there is a genuine need for effective mechanisms of global optimization. The most visible feature of most – if not all – algorithms devoted to this task, is that they pursue an optimization goal by relying on a collection of individuals which interact between themselves in the synchronization of joint activities to find solutions. They communicate between themselves by exchanging their local findings. They are also influenced by each other. In this chapter we consider the main theoretical ingredients of these algorithms in terms of stochastic processes, problem representation and the incremental algorithm that animates the ensemble of computing individuals.

Even worse if solution concerns man-machine interactions:

a solution may come from a lot of machines collaborating for the success.

8.1 Extending Sampling Mechanisms to Generate Tracks

With our data generation paradigm – random seed plus explaining function – so far we have maintained a *static* reading of data. This does not prevent us from considering dynamic phenomena where the explaining function depends on some quantitative variables such as space or time. Say, the Hansel habit of trowing pebbles is different in early morning or at noon. What renders *dynamic* an explaining function is its dependence on a set of ordered seeds, each stroking the time of the evolution of a data track as a specification of a *random process*.

When seeds are ordered a sample is a track of a process.

Being more descriptive, let us consider the following experiment. You have n urns each containing a certain number of balls, each one labeled by the index of one of the urns (the one containing the ball included). Let N_{ij} be the number of balls labeled by j in the urn i. The experiment consists in a never-ending sequence of picking up the balls: at time t on a given urn, you draw a ball and move it at time $t+1$ into the urn whose index is marked on the ball, and repeat the process. Considering a grid with repeated columns sequentially associated to a timestamp, where each cell in the column is a urn, you may draw an infinite number of trajectories starting at time 0 from any of the n urns according to a probability distribution $\boldsymbol{\psi}(0)$. You may decide the starting urn with a sampling mechanism like the one given by Fact 1.4. Namely, denoting with $\psi_i(0)$ the probability of starting with urn i, you toss a specification u of a $[0,1]$ uniform random variable U and start with the k-th urn if $\sum_{i=1}^{k-1} \psi_i(0) \leq u < \sum_{i=1}^{k} \psi_i(0)$. The subsequent moves may be explained by an analogous selection mechanism. Actually you may virtually determine the trajectories using U specifications and the same urn assignment using the probability $p_{ij} = \frac{N_{ij}}{\sum_{k=1}^{n} N_{ik}}$ in place of $\psi_j(0)$, where i is the urn from where the trajectory moves and j the urn which it reaches. In this way the asymptotic frequency (over the infinitely many trajectories) of passing from urn i to urn j converges to p_{ij} defined earlier. Related to the experiment you have a family of asymptotic frequencies of crossing the urns with time, that denote probabilities $\boldsymbol{\psi}(t)$ whose sampling mechanism depends on the log of the seeds drawn fron time 0 to time $t-1$.

Main ingredient: transition probability from one state to another.

We consider now a very elementary process, known as Markov chain, that will constitute the theoretical substrate of many considerations in the next sections.

Definition 8.1. *Given a finite set of sites* $\mathfrak{X} = \{x_1, \ldots, x_n\}$, *a discrete time Markov chain is a pair of infinite sequences:*

- $\boldsymbol{\psi} = (\boldsymbol{\psi}(0), \ldots, \boldsymbol{\psi}(t), \ldots)$ *of state vectors* $\boldsymbol{\psi}(t)$, *where each* $\boldsymbol{\psi}(t)$ *is a probability distribution* $\{p_{x_1}(t), \ldots, p_{x_n}(t)\}$ *over* $\mathfrak{X}$, *and*
- $\boldsymbol{P} = (P(0), \ldots, P(t), \ldots)$ *of transition matrices* $P(t)$, *where* $P(t)$ *is a matrix whose* (i,j)-*th element* $P_{ij}(t)$ *measures the conditional probability of observing* x_j *at time* $t+1$ *once* x_i *has been observed at time* t.

Henceforth we will focus on Markov chains, that we denote *homogeneous*, with $P(t) = P$ for all t.

Fact 8.1. *In a Markov chain:*

- *the state vector is a stochastic vector; i.e. all components are* ≥ 0 *and their sum equals 1.*
- *the transition matrix is a stochastic matrix; i.e. its rows are stochastic vectors.*

Fact 8.2. *Given the meaning of the transition matrix, the dynamics of the state vector is the following:*

$$\boldsymbol{\psi}^T(t+1) = \boldsymbol{\psi}^T(t) P. \qquad (8.1)$$

A simple dynamic on probability evolution,

Hence:

$$\psi_i(t+1) = \sum_{j=1}^{n} \psi_j(t) P_{ji} \tag{8.2}$$

$$\psi(t+l) = \psi^T(t) P^l = \psi^T(0) P^{t+l} \tag{8.3}$$

$$P_{ij}^{t+l} = \sum_{k=1}^{n} P_{ik}^t P_{kj}^l \quad \textit{(Chapman-Kolmogorov equation)}. \tag{8.4}$$

Fact 8.3. *The flow of transitions across site i at a certain time t is governed by the following equation:*

$$\psi_i(t+1) - \psi_i(t) = \sum_{j=1}^{n} (\psi_j(t) P_{ji} - \psi_i(t) P_{ij}) \quad \textit{(master equation)}. \tag{8.5}$$

It derives from the fact that:

$$\psi_i(t+1) - \psi_i(t) = \sum_{j=1}^{n} \psi_j(t) P_{ji} - \psi_i(t) \left(\sum_{j=1}^{n} P_{ij} \right), \tag{8.6}$$

where the sum in the brackets equals 1.

Remark 8.1. The master equation states that a probability increment at site i is due to the difference between the probabilities of incoming and outcoming this state. A stationarity condition occurs when this difference is null in each site. This means that $\psi(t)$ does not change at time t, i.e. from time t on, given the dynamics (8.1) of the process. A sufficient condition for having this difference vanishing is what we call the *detailed balance*, which occurs when:

$$\psi_j(t) P_{ji} - \psi_i(t) P_{ij} = 0 \quad \forall i, j. \tag{8.7}$$

Still given the dynamics (8.1), the evolution of the state vector of a Markov chain is fully determined by its initial value $\psi(0)$ and the transition matrix P. In particular, the conditions for having an asymptotic stationary state affect only P. Namely:

Theorem 8.1. *If P is an $n \times n$ stochastic matrix such that:*

- *for at least one $i \in \{1, \ldots, n\}$, $P_{ii} > 0$*
- *for each $h, k \in \{1, \ldots, n\}$ there exists an m such that $P_{hk}^m > 0$*

then any Markov chain having transition matrix P has one and only one stationary state ψ_∞ whose components are all greater than 0.

deserving asymptotic properties.

We will not prove the theorem. The interested reader can find the complete proof in [40]. Instead, we make the following remarks.

Remark 8.2. For an operational counterpart of Theorem 8.1, we note that the stationary state ψ_∞ does not depend on the initial state $\psi(0)$; hence it depends only on the transition matrix. The conditions of Theorem 8.1 ensure that each site of $\mathfrak{X}$ will be visited by the chain with probability greater than 0 (absence of prejudice).

8.2 Knowledge Representation: From Phenotype to Genotype Space

As mentioned in Chapter 7, suitable problem representation is a key issue to direct populations that predetermines success of the leading optimization process and implies the quality of the produced solution. From an operational perspective this corresponds to an inverse transform from the phenotype space $\mathfrak{P}$ – where results are appreciated – to genotype space $\mathfrak{G}$ – where the optimization is carried out – that is realized with the use of some encoding and decoding procedures, as shown in Fig. 8.1. In a more descriptive way, we could view these encodings as a matter of a suitable representation of our knowledge available about the nature of the goal of the procedure. Knowledge representation is a truly multifaceted problem and as such one has to proceed with prudence realizing that the effectiveness of this scheme implies the quality of solution. <small>From phenotypes to genotypes: a hard representation task,</small>

<small>using as codewords:</small>

In what follows, several examples of encoding and decoding serve as an illustration of the diversity of possible ways of knowledge representation.

1. Binary encoding and decoding. Any parameter assuming real values can be represented in the form of a corresponding binary approximation. This binary coding is used quite commonly in digital systems. The strings of bits are then subject to evolutionary operations. The result is decoded into the <small>simply bits – getting a rather long codes,</small>

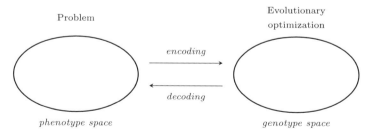

Fig. 8.1. From phenotype space to genotype space: links between optimization problem and its representation in evolutionary optimization.

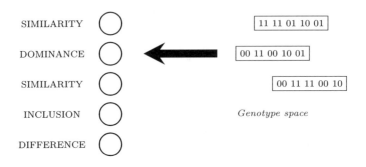

Fig. 8.2. Binary encoding of the fuzzy logic network.

corresponding decimal equivalent. More formally, the genotype space is the hypercube $\mathfrak{G} = \{0,1\}^n$, where n stands for the dimensionality of the space and depends on the number of parameters encoded in this way and a resolution (number of bits) used to complete the encoding.

2. Floating point (real) encoding and decoding. Here we represent values of parameters of the system under optimization using real numbers. Typically, to avoid occurrence of numbers positioned in different ranges, all of them are scaled (e.g., linearly) to the unit interval so in effect the genotype space is the unit hypercube $\mathfrak{G} = [0,1]^q$, with q denoting the number of parameters. The resulting string of real numbers is re-transformed into the original ranges of the parameters.

reals – coding normalized values,

3. Structure representation of fuzzy logic hierarchy. Fuzzy logic hierarchies exhibit a diversity of topologies. In particular, this variety becomes visible when they map into graphs with referential nodes. Given four types of referential nodes, that is similarity, difference, inclusion, and dominance, we can consider several ways of representation of the structure in the genotype space: (a) one can view a binary encoding where we use two bits with the following assignment: 00 - similarity, 01 - difference, 10 - inclusion, and 11 - dominance. Alternatively, we can consider a real coding and in this case we can accept the decoding that takes into consideration ranges of values in the unit interval, say, $[0.00, 0.25)$ - similarity, $[0.25, 0.50)$ - difference, $[0.50, 0.75)$ - inclusion, and $[0.75, 1.00]$ - dominance. The dimensionality of the genotype space depends on the number of the referential nodes used in the graph. An example of the binary encoding for the fuzzy logic hierarchy with 5 referential nodes is illustrated in Fig. 8.2.

vertical structures – to denote relationships,

4. Structure representation of subsets of variables. In many cases, in order to reduce problem dimensionality, we might consider a problem of selecting a subset of input variables.

horizontal structures – to denote subsets,

Knowledge Representation 283

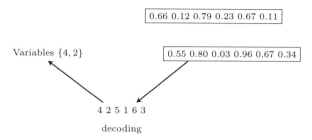

Fig. 8.3. Variable selection through ranking the entries of the vectors of the genotype space $\mathfrak{G}$; here the total number of variables under consideration is 5 and we are concerned about choosing 2 variables.

For instance, when dealing with hundreds of variables, practically we can envision the use of a handful subset of them; say 10 or so, in the development of the fuzzy system (say, a rule-based system). Given these 10 variables, as the genotype components, we develop a network and assess its performance. This performance to index could be regarded as a suitable fitness function to be used in moulding the genotypes population. Let us also note that the practicality of a plain enumeration of combinations of such variables is out of question; say, choosing 10 variables out of 200 variables leads to $\binom{200}{10}$ possible combinations. Here the representation of the structure can be realized by forming 200-dimensional strings of real numbers, that is $\mathfrak{G} = [0,1]^{200}$. To decode the result, we rank the entries of the vector and pick its first 10 entries. For 100 variables and 10 variables to be selected, we end up with around $1.731 \cdot 10^{13}$ possible alternatives. An example of this representation of the genotype space is illustrated in Fig. 8.3. Note that the plain selection of the entries decoded by using intervals of the unit interval (say, $[0, 1/200)$ for variable x_1, $[1/200, 2/200)$ for variable x_2, ...) will not work as we could quite easily en-

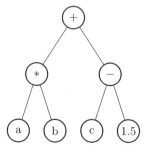

Fig. 8.4. Tree representation of the genotype space.

counter duplicates of the same variable. This could be particularly visible in the case of a large number of variables.

trees – to denote logical dependences.

5. Tree representation of the genotype space: This form of knowledge representation is commonly encountered in genetic programming. Trees such as the one shown in Fig. 8.4 are used to encode algebraic expressions. For instance, Fig. 8.4 encodes the expression $(a*b) + (c - 1.5)$.

8.3 Starting from an Artificial Population

An illuminating case study to introduce the mentioned third way to granularity management is constituted by the knapsack problem: given a set of objects, each characterized by a *weight* and a *profit*, find within subsets of them whose total weight does not exceed the *capacity* of a knapsack the one maximizing cumulative profit. Let us fix notation:

Imposing a threshold to a sum of terms

Definition 8.2. *An instance for the* 0-1 *knapsack problem is a four-tuple* $s = (n, W, Z, b)$, *where*

- $n \in \mathbb{N}$;
- $W = \{w_1, \ldots, w_n\} \in \mathbb{N}^n$;
- $Z = \{z_1, \ldots, z_n\} \in \mathbb{N}^n$;
- $b \in \mathbb{N}$ *is such that* $w_i \leq b$ *for each* $i \in \{1, \ldots, n\}$ *but* $\sum_{i=1}^n w_i > b$.

Given a knapsack instance s, *an ordered* n*-ple* $\boldsymbol{x} = (x_1, \ldots, x_n) \in \{0,1\}^n$ *is called a* feasible solution *for* s *if* $\sum_{i=1}^n w_i x_i \leq b$, *and* $y = h(\boldsymbol{x}) = \sum_{i=1}^n x_i z_i$ *is called its* value. *Denoted with* $\mathrm{Sol}(s)$ *the set of all the feasible solutions for* s, *an optimal solution of the knapsack problem on* s *maximizes* h *over* $\mathrm{Sol}(s)$, *i.e. is an* n*-ple:*

$$\boldsymbol{x}^* = \arg\max_{\boldsymbol{x} \in \mathrm{Sol}(s)} h(\boldsymbol{x}).$$

Approximate solutions produce values close to the maximum. Denoted by y^* *and* $\widehat{y}$ *the values of an optimal and an approximate solution respectively,* $\eta = \frac{y^* - \widehat{y}}{y^*}$ *is the approximation degree provided by the solution.*

The main characteristic of this problem is to be so common to be met in the everyday life as hard to be solved. The fact is that, because of the bound on the bag capacity, the natural order of the objects w.r.t. either the weight or the profit or elementary functions of them is not useful for searching the optimal solution. Thus blocks or blobs like the ones shown in Fig. 1.3 prove misleading if our interest focuses on the best values which may be assumed by the solutions. Said in other words, we are interested in a variable y that is a function of $\boldsymbol{x}$. Since we have no tool for determining the

upsets any order relationship within them,

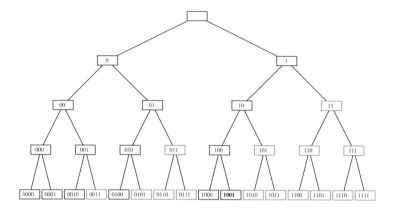

Fig. 8.5. A search tree for a knapsack problem. Each layer adds one bit to the candidate solution strings in the leaves. Gray nodes refer to unfeasible solutions; the optimal configuration is typeset in bold.

best value of y, we come to consider a random variable Y. Since the mapping h from $\boldsymbol{x}$ to y has no analytical expression, and not even a numerical procedure for inverting it, i.e. for computing h^{-1}, is feasible in terms of computational resources, we cannot deduce the distribution law of Y from $\boldsymbol{x}$'s as in Section 2.1.1. At this point we are free to give $\boldsymbol{X}$ any distribution law, in principle a uniform distribution law for the sake of simplicity, and will use $\boldsymbol{X}$ specifications as seeds of Y specifications.

In order to perceive the degree of unawareness about h^{-1}, we mention the knapsack problem belongs to the NP-hard complexity class [57]. Hence, if we were capable of solving it in a reasonable time then we would be able of solving almost all search problems we may meet in our computations in feasible time. A simple explanation of this hardness comes from the analysis of Fig. 8.5. Each leaf of the decision tree is a candidate solution, but only a few ones are feasible and optimal, and generally there is no way of deciding at any node how to continue the visit of the tree. Hence you just pick up a set of these leaves to form an exploring population, remove those whose cumulative weight exceeds the sack, and wisely modify the remaining ones.

hence any priority criterion for processing them.

8.3.1 A Very Schematic Genetic Algorithm

The crux of the evolutionary optimization process lies in the use of a finite population of n representatives of the search space $\mathfrak{G}$, picked as mentioned before, whose evolution leads to an optimal solution. The population-based optimization is an outstanding feature of the evolutionary optimization and is practically present in all its variants we can encounter today. The value the

Promote the excellence, without neglecting the rest,

function to be optimized takes on a population element constitutes the fitness of the latter. Populations evolve by wise modifications of their elements and selection of a subset of them. The whole construct boils up to the way of connecting fitness to the survival probability of an element. Like with other local optimization procedures, we agree with privileging elements with higher fitness, according to a loose continuity assumption of the fitness landscape. We do not exclude a situation, however, that a relatively worse element may be at the origin of a successful modification, since this is exactly the nature of hard problems like knapsack. The equilibrium between the two criteria is variously pursued by the different methods present in the literature and used in practice. In a very schematic and abstract way, a computing skeleton of evolutionary optimization can be described as in the Algorithm 8.1. Accomplishing

is the recipe of a fruitful evolution.

Algorithm 8.1. Template of evolutionary optimization algorithms.
1. *initialize* the genotypes population
2. until termination triggers do
    a) *generate* local populations
    b) *synthesize* overall population
    c) *improve* the new generation

---

the initialization step involves two facilities: problem representation and seeding the population. As for the former, we define a genotype as a string of symbols (genes), each representing a feature whose specification we want to involve in the computation. We have the same considerations presented in Chapter 5, following the line: the better you understand data, the better you can solve the problem. In this case however it is convenient to maintain a tight connection of representation with the tools we will use in the generation of the subsequent populations. In the knapsack case study it becomes evident on how to represent candidate solutions, in the form of genotypes, through binary strings where each bit is associated to an object with the following semantics: $1 \to$ the object is placed in the sack, $0 \to$ the object is left out. Seeding looks obvious as well. As we have no preference in advance for any of these solutions, in the sense that those we know to fill the sack too much (thus exceeding its volume) or leaving too much vacant space (thus reducing potential profit) may be discarded from the beginning, we fill up the initial population uniformly with the other strings (hence each string may belong to the population with the same probability).

Meaningful encoding and fair selection of the first generation, as for starting;

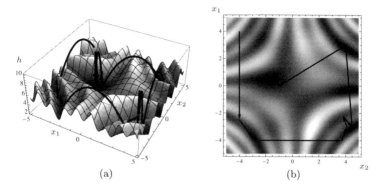

Fig. 8.6. Different views of exploration (big jumps) and exploitation (neighboring visits) alternating on a typical fitness landscape.

Then we proceed with an incremental improvement of the population until we are satisfied according to some predefined criterion. The criterion defines functions of the genotype through which we can appreciate it. The collection of their evaluations constitute the phenotypes corresponding to the genotypes at hand, where any appreciation represents the capacity of the genotype of fitting a given goal, i.e. its fitness for short. The general goal is to move from an initial distribution reflecting features of the genotype space reverberating on unbiased distribution of the phenotypes – for instance a Gaussian-like distribution of the knapsack value in our case study – to distributions peaked in the maximum value achievable by the phenotypes, hence the optimal solution value in our optimization problem. Since we are blind to the corresponding genotypes population, we change the latter adopting typical strategies we used to regulate a phenomenon, for instance the sound of your radio through analogical knobs: a slow transition through neighboring settings that do not perturb drastically the current status of the apparatus. This passes through the following clues.

<small>then, gently move next generation toward highest fitting elements'.</small>

- *Neighboring notion.* Following a single process, such as the regulation of your knobs, you typically connect neighborhood to nearness in the feature space (say, the knob rotation angles). Facing a population of processes you can dare to do more, with the twofold objective of exploring close settings in far locations in the feature space, eventually jumping into far new settings as well in order to explore new regulations (see Fig. 8.6). What typically happens with genetic algorithms is that from a set of genotypes (candidate solution of the knapsack) you generate new items either by local changes of a genotype (flips of bit from 0 to 1 and *vice*

<small>Normally explore neighborhood, seldom jump far away,</small>

288   The Main Paradigms of Social Computation

> *to have local populations' gemmation,*

versa) or by exchanges of segments of genes between pairs of genotypes (interchanges of substrings between candidate solutions). The former operation, known as *mutation*, accounts for some corruption during the chemical replication of a genomic sequence, while the latter is denoted *crossover* and generally contemplates the twisting of one or two segments between genotypes. Segment motion is rooted in the assumption that the sequence with which genes are assembled obeys a sequential coding rule (an assumption not highly convincing – neither disrupting, in any case – in our case study). Depending on the representation of the genotype space, the evolutionary operators come with different realizations. As an example, consider a mutation operator. In the case of binary encoding and decoding, mutation is realized by simply flipping the value of the specific entry of the bit vector. In the real number encoding and decoding, we may use the complement operator, viz. replacing a certain value in $[0, 1]$ by its complement, say $1 - x$, with x being the original component of the vector.

- *Absence of prejudice.* It is customary to ask that the neighboring rules and their implementation guarantee the principled possibility of tossing any genotype allowed by the problem definition (hence any candidate solution of our knapsack).

> *always without preclusions.*

In technical terms, we say that the transition matrix of the Markov process underlying the routine deciding which neighboring string to toss has a stationary state in the uniform distribution on the allowed genotypes. Hence, in the dummy procedure consisting of moving from one genotype to a neighboring one according to this process you would pass through each genotype asymptotically with the same frequency. From an operational perspective you uniformly pick items from the neighborhood of the genotype at hand, according to a fair notion of the neighborhood.

- *Overall appreciation.* Collecting the local populations gem-

> *Synthesizing populations is not simply forming a heap.*

mated from the individuals of past generation is not only a matter of merging them. You must state an overall ordering of the new individuals based on a *relative*, hence properly normalized, appreciation of their fitness. This require inserting fitness within a model that takes into account both current and prospective features of the populations describing generations.

- *Gentle biases.* We must decide an intermediate scenario between the conservative one: don't trow anything off (maintain a continuously enlarging population), and the greedy strategy: maintain only the most performing strings, whose number complies with your computational power. The general idea is that, since you do not know the phenotypical value landscape, so that none of orderings on the genotypes is meaningful, you cannot exclude that a neighbor of a less performing item contains, on the contrary, a highly performing, possibly the best, one. However, under a loose continuity hypothesis you expect that these jumps are rare – especially because many discontinuity sources have been recovered in the neighboring rules, while it is more probable to have smooth slopes in the value landscape. Thus, you evolve your population over time in the direction of increasing the probabilities of most performing genotypes at the expenses of the others.

- Last, but not least, the system must stop. This is a general problem of all algorithms considered in this chapter. *Per se* we do not have any universal program saying whether or not any other computer program will halt or not [55]. This certifies that the halting of a computer program is undecidable. We bypass completely this problem with our algorithms: no any automatism is embedded into the algorithm decreeing its stop. Rather the halting of the procedure is completely up to us. We may ground our halting decision on some more or less sophisticated statistic suggesting that no result improvement is expectable from further running of the evolution. Alternatively, we may realize that we have no more computational resources available. In any case you need a *stopping criterion* in order to complete your algorithm.

Improving generations with a global shift toward best fitting elements,

having an end term.

We may schematize the whole discussion as follows. At a given generation, you use its elements as centers of clusters of points generated from the former according to the above neighborhood and picking rules. You really do not know the local distribution of the value within each cluster, so that you miss the overall distribution of the population coming from the union of the clusters. Thus you generally decide to reshape this distribution by wisely picking items from the population. Knowing the value of each element in the population, in principle you could manage for having a final histogram of the values according to a fixed distribution, for instance truncated Gaussian-like as in Fig. 8.7(a). Common rules, like *roulette* or *tournament selection* [48] fill the mentioned equilibrium with the rule of thumb: privilege genotypes with high phenotype, but not too much. Hence they accept a new genotype after tossing its preferability either on the whole population of the new generated genotypes or on current subset of them just after their generation. For instance,

You may toss preferability

either on single item d.f.,

290   The Main Paradigms of Social Computation

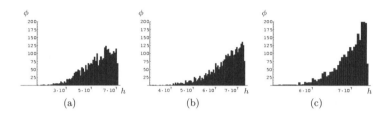

Fig. 8.7. Evolution snapshots in solving a knapsack problem on $n = 20$ objects. The graphs report the histograms of the fitness $h(x)$ after (a): 1, (b9: 3, and (c): 5 offsprings have been generated.

- with the former rule the acceptance probability may be computed as:
$$p(x) = h(x)/\sum_{\xi \in \mathfrak{X}} h(\xi); \tag{8.8}$$
- with the latter rule we may have the same formula but for subsets of $\mathfrak{X}$.

We suggest local rules based on statistics that loosely guarantee the shape of the next generation. This brings us to compute an on-line x acceptation probabilities of the form:

$$p(x) = g(F_{X_{\breve{\theta}}}(x)), \tag{8.9}$$

where g is a suitable function, possibly ranging from $\min[F_{X_{\breve{\theta}}}(x), 1-F_{X_{\breve{\theta}}}(x)]$ to $\max[F_{X_{\breve{\theta}}}(x), 1-F_{X_{\breve{\theta}}}(x)]$. Its rationale comes from the fact that in case of a perfect replica of the previous generation, you obtain its empirical cumulative distribution law from a decimated population just using for g the mean between F_X and $1 - F_X$, which brings to a constant acceptation probability (1/2 or anything else if you multiply g by a constant). On the contrary you may generally recover the shrinking induced *[or on overall c.d.f.,]* by the new local populations with a function comprised between mean and max, just because you recover the original symmetry of F_X around 0.5. The parameter $\breve{\theta}$ is a suitable manipulation of the items of the population of compatible θs with the statistics drawn in the previous generation. For instance, with the knapsack optimization problem we have introduced to exemplify the procedure, in Fig. 8.7 we have that the Gaussian fitness distribution is truncated at a certain value because of the constraint on the sack capacity. Our goal is to maintain the population *[with the reward of remaining in the same family of distribution laws.]* spread enough in order to explore a rich variety of solutions. Hence the general philosophy is to consider all populations compatible with the observed set of values y. Our strategy consists in evolving the population of sum of values y disregarding the satisfaction of the constraint on the bound on $\sum w_i x_i$. Since Y is a sum of independent replicas of Z (see Definition 8.2), we may assume its distribution to belong to the family of Gaussian distributions. Hence we may pivot master equations around the statistics $\sum y_i$ and $\sum y_i^2$ that prove sufficient for the parameters μ and σ respectively, in order to compute compatible populations of these parameters that we denote μ^* and σ^* respectively (see Section 3.1.3). We evolve this distribution through mutation and crossover operators and select the new xs according to the probability:

$$p(\boldsymbol{x}) = \min\left[F_{\boldsymbol{X}_{\breve{\mu},\breve{\sigma}}}(\boldsymbol{x}), (1 - F_{\boldsymbol{X}_{\breve{\mu},\breve{\sigma}}}(\boldsymbol{x}))\right], \quad (8.10)$$

where parameters with ˘ equal the starred one, apart from possible positive increments in order to privilege values closer to the unknown optimum.

8.4 The Seeds of Unawareness

What we have exposed in the previous section are strategies to manage singularly the population items. Boltzmann machines, on the contrary, represent a typical way of directing populations in parallel through their parameters. Moreover, the goal of the process is an entire population rather than a single optimal point, so that the cost function is obviously represented by the distance between actual and target population, that we measure through the Kullback function [39]. Namely, denoting by $\pi(\boldsymbol{x})$ and $\phi(\boldsymbol{x})$ the probability functions of the two populations, respectively, and by $d_{\pi\|\phi}$ the above distance, we have by definition:

Moving populations synchronously

$$d_{\pi\|\phi} = \sum_{\boldsymbol{x} \in \mathcal{X}} \phi(\boldsymbol{x}) \log \frac{\phi(\boldsymbol{x})}{\pi(\boldsymbol{x})}, \quad (8.11)$$

that we appreciate through its sample version:

$$\widehat{d}_{\pi\|\phi} = \sum_{i=1}^{m} \phi(\boldsymbol{x}_i) \log \frac{\phi(\boldsymbol{x}_i)}{\pi(\boldsymbol{x}_i)}. \quad (8.12)$$

Actually we may assume $d_{\pi\|\phi}$ to be a dissimilarity measure in place of a distance since it lacks symmetry on the two arguments. The keen point of this paradigm is that we represent both target and actual distribution through a second order function in the Boolean representation of $\boldsymbol{x}$ such that sufficient statistics of its parameters are represented by first and second order sample moments:

through a pseudodistance between populations and sufficient statistics,

$$s_{1_j} = \sum_i x_{ij}; \quad s_{2_{jk}} = \sum_i x_{ij} x_{ik} \quad (8.13)$$

where x_{ij} denotes the j-th component of the vector $\boldsymbol{x}_i$. The analogous of the SVM kernel trick here is constituted by the fact that we split $\boldsymbol{x}$ into two parts: a visible one, say $\boldsymbol{x}^v$, actually representing what you observed of $\boldsymbol{X}$ in the population, and a hidden part $\boldsymbol{x}^h$ constituted by dummy variables that we need in order to represent the generic $\phi(\boldsymbol{x}^v)$ through a quadratic function of $\boldsymbol{x}$. In greater detail you have:

$$\phi(\boldsymbol{x}^v) = \frac{1}{c_{\widetilde{\beta}}} \sum_{\boldsymbol{x}^h \in \mathcal{X}^h} e^{-\boldsymbol{x}^T \widetilde{W} \boldsymbol{x} - \boldsymbol{x}^T \widetilde{\gamma}} = \frac{1}{c_{\widetilde{\beta}}} \sum_{\boldsymbol{x}^h \in \mathcal{X}^h} e^{-\sum_{i,j} x_i \widetilde{w}_{ij} x_j - \sum_i x_i \widetilde{\gamma}_i} \tag{8.14}$$

$$\pi(\boldsymbol{x}^v) = \frac{1}{c_{\beta}} \sum_{\boldsymbol{x}^h \in \mathcal{X}^h} e^{-\boldsymbol{x}^T W \boldsymbol{x} - \boldsymbol{x}^T \gamma} = \frac{1}{c_{\beta}} \sum_{\boldsymbol{x}^h \in \mathcal{X}^h} e^{-\sum_{i,j} x_i w_{ij} x_j - \sum_i x_i \gamma_i}, \tag{8.15}$$

for suitable normalization constants c_β and $c_{\widetilde{\beta}}$.

Remark 8.3. From theory we know that for every probability distribution there is a pair $(\widetilde{W}, \widetilde{\gamma})$ so that ϕ approximates the distribution with an accuracy that you may have fixed [60].

<small>in order to reach an observed population</small>

Differently from genetic algorithms (GA) framework, the target population is not artificial. Rather we observe it through a sample $\{\boldsymbol{x}_1^v, \ldots, \boldsymbol{x}_m^v\}$, so that our goal is an inference task aimed at estimating the parameters $\widetilde{W}$ and $\widetilde{\gamma}$ specifying ϕ. The inner quadratic dependence of ϕ on $\boldsymbol{x}$ allows us to define a computing framework, the neural networks, that is much efficient and flexible to be employed in many variants of the inference problem. In the next section we will give a very streamlined description of the framework, before solving our optimization problem.

8.4.1 The Neural Networks Paradigm

<small>What better than a melting pot of either symbolic or subsymbolic processing elements?</small>

We refer to a neural network as a set of processing elements (PEs) variously interconnected like shown in Fig. 8.8. Denote by a square those that stably compute a function, call them *symbolic* PEs (usual processors, like a PC we may buy at a store) and by a circle the other units. We will say that the latter perform a subsymbolic computation, just because we cannot describe in an analytic (i.e., symbolic) way the function they compute. Call them *subsymbolic* PEs. The life of this tricky computing machine is the following: start with an initial set of messages flowing through the connections. Then on the basis of current incoming signals, each processor individually computes its output, which it sends to the connected companions. To give some order to this chaotic processing mode, with reference to Fig. 8.8 we fix a terminology and some specifications without diminishing in principle its generality.

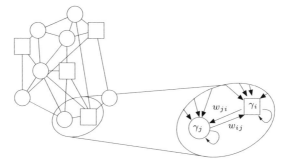

Fig. 8.8. The general architecture of a neural network. Squares: symbolic PEs; circles: subsymbolic PEs. γ_i denotes the threshold of PE i. Analogously, w_{ij} and w_{ji} denote the weight of the connection from PE j to PE i and *vice versa*.

- Neural network: a graph constituted of ν PE's (either symbolic or subsymbolic) connected by oriented arcs – thus the arc connecting PE i to PE j is different from the one connecting PE j to PE i. The graph is not necessarily fully connected, in the sense that it may lack a connection between some pairs of PEs. The PEs may be either symbolic or subsymbolic. In case both types of processors appear in the network, we generally speak of *hybrid systems*. **A grid of PEs**

- State $\boldsymbol{\tau} = (\tau_1, \ldots, \tau_\nu)$: the current ordered set of messages in output of each processor, where each message τ_i takes values in a set $\mathfrak{T} \subseteq \mathbb{R}$. We introduce an order to identify the messages of the single PEs, thus we refer to a *state vector*. If we want to specify a time dependence we will denote the state with $\boldsymbol{\tau}(t)$, and a sub-state related to PEs' location L (capital letter) with $\boldsymbol{\tau}^L$. **computing a state vector**

- Free parameters $(\mathbf{w}, \boldsymbol{\gamma})$: the weight vector $\mathbf{w} = (w_{11}, \ldots, w_{ij}, \ldots, w_{\nu\nu}) \in \mathbb{R}^{\nu\nu}$ where w_{ij} is associated in the matrix W to the connection from processor j to processor i ($w_{ij} = 0$ if no connection exists from j to i), and the inner parameters vector $\boldsymbol{\gamma} = (\gamma_1, \ldots, \gamma_\nu)$, where γ_i is associated to processor i. Depending on its use, typically according to (8.16) later on, when no ambiguity occurs we will refer to the sole vector $\mathbf{w}$ by adding a dummy PE piping a signal constantly equal to 1 to the i-th PE through a connection with weight set to γ_i. **depending on free parameters,**

- Activation function vector: the ordered set of functions $\mathbf{h}_i : \mathfrak{T}^\nu \mapsto \mathfrak{T}$ computing a new state $\tau_i = \mathbf{h}_i(\boldsymbol{\tau})$ of the i-th PE in function of the current state vector (as a component of the hypothesis h computed by the whole network). **affecting the function computed by the PE's**

294   The Main Paradigms of Social Computation

with different timings
- Activation mode vector: the synchronization order of the single processors. We may distinguish, for instance, between the following activation modes:
  1. parallel: the PE updates its state at the same time as the other PEs synchronized with it.
  2. asynchronous: the PE updates its state according to an inner clock.
  3. random: the i-th PE tosses a (non necessarily fair) coin. It renews its state on the head.
  4. delayed: the PE updates its state a given time after the updating of the afferent PEs. In particular, instantaneous mode means a delay equal to 0, hence coinciding with the parallel mode.

An algorithm for molding the free parameters,
- Training algorithm: an algorithm modifying the free parameters according to some utility function. In a broad sense we may consider also the architecture of the network as a free parameter, which may be modified by some training algorithm.

on the basis of given examples,
- Training set: the set of examples in input to the training algorithm.

checked on new examples
- Test set: a set of examples not considered for training the network, but used to verify how the trained network behaves. Some training algorithms use this set for tuning the parameter modifications and a third set, called *validation set*, for the above checking.

in terms of performance of the molded network.
- Generalization: the network's attitude of behaving well on the test or validation set.

In the case of subsymbolic processors the notation specifies as follows:

- PE → neuron;
- arc → connection;

A set of parameters plus
- free parameters: w_{ij} → connection weight, γ_i → threshold;

an activation function,
- activation function: $h_i(\tau) = \sigma(a_i(\tau))$, where

$$a_i(\tau) = \sum_{j=1}^{\nu} w_{ij}\tau_j + \gamma_i \qquad (8.16)$$

is the *net input* to neuron i. Omitting the neuron index, the most common expressions of σ are the following:

possibly, a linear function,
1. the simplest one is a *linear* function (see Fig. 8.9(a)):

$$\sigma(a) = \beta a, \qquad (8.17)$$

with $\beta \in \mathbb{R}$;

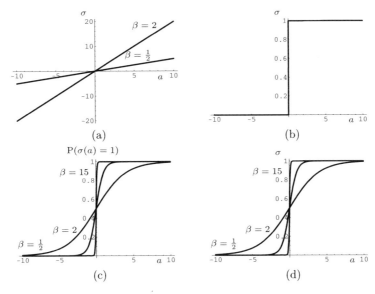

Fig. 8.9. Typical shapes of the activation function σ: (a) linear as in (8.17); (b) nonlinear, non continuous as in (8.18); (c) probabilistic, as in (8.19); and (d) nonlinear, continuous, as in (8.20). β constitutes a curve parameter.

2. the primary nonlinear one is the *Heaviside* function (see Fig. 8.9(b)):

   $$\sigma(a) = \begin{cases} 1 & \text{if } a \geq 0 \\ 0 & \text{otherwise,} \end{cases} \qquad (8.18)$$

   which smooths in two directions described in the following two points;

   possibly, a step function,

3. the primary probabilistic one (see Fig. 8.9(c)) is described as follows:

   $$P(\sigma(a) = 1) = \frac{1}{1 + e^{-\beta a}}, \qquad P(\sigma(a) = 0) = \frac{1}{1 + e^{\beta a}}, \qquad (8.19)$$

   with $\beta \in \mathbb{R}^+$, which smooths function (8.18) in terms of random events, coinciding with the original function for $\beta \to +\infty$. Hence the meaning of β is the inverse of a temperature θ of a thermodynamic process determining the value of τ [3];

   preferably, probabilistic smooth, or

4. the primary continuous one (see Fig. 8.9(d)) is the so-called *sigmoid* function:

   $$\sigma(a) = \frac{1}{1 + e^{-\beta a}}, \qquad (8.20)$$

   with an analogous smoothing effect, $\sigma(a)$ in (8.20) being the expected value of $\sigma(a)$ in (8.19).

   continuously smooth.

We generally introduce a neural network through its architecture, node activation function features, and a learning algorithm to determine parameters. In this chapter we deal with supervised algorithms where a *teacher* shows the network the signals it should produce on deputed nodes as a response of specific external solicitations. A very schematic snapshot of these learning procedures is described in Algorithm 8.2.

Algorithm 8.2. Template of supervised learning algorithms.
1. *initialize* the vector ψ of the free parameters
2. until termination triggers do
    a) *run* the network with current ψ on the training examples
    b) *update* a cost C as a function of shifts from expected to produced signals
    c) *change* ψ into $\psi' = \mathsf{h}(\psi, \mathsf{C})$ with the goal of lowering C

8.4.2 Training a Boltzmann Machine

A random activation

In order to produce a distribution on the visible nodes we adopt the activation function (8.19) which, in terms of a stochastic process on the Boolean vector $\boldsymbol{T}(t)$, reads as follows:

$$P(T_i(t+1) = 1 | \boldsymbol{T}(t) = \boldsymbol{\tau}) = \frac{1}{1 + e^{-\beta a_i(\boldsymbol{\tau})}}. \quad (8.21)$$

determining a transition matrix of a Markov chain

Thus the configuration $\boldsymbol{T}(t)$ of the network performs a random walk with t on $\mathfrak{X} = \{0, 1\}^\nu$. We describe this process through a Markov chain with transition matrix whose generic element detailing the transition from $\boldsymbol{\tau}$ to $\boldsymbol{\tau}'$ is expressed as:

$$p_{\boldsymbol{\tau},\boldsymbol{\tau}'} = \mathrm{P}\left(\boldsymbol{T}(t+1) = \boldsymbol{\tau}' | \boldsymbol{T}(t) = \boldsymbol{\tau}\right) = \prod_{i=1}^{\nu} \mathrm{P}\left(T_i(t+1) = \tau_i' | \boldsymbol{T}(t) = \boldsymbol{\tau}\right)$$

$$= \prod_{i=1}^{\nu} \frac{1}{1 + e^{\beta(1 - 2\tau_i')a_i(\boldsymbol{\tau})}}. \quad (8.22)$$

When the connection matrix is symmetric, i.e. $w_{ij} = w_{ji}$ for each i and j, and the diagonal elements are null, i.e. $w_{ii} = 0$ for all i, the transition matrix ensures the convergence to the stationary distribution law:

with certified equilibrium distribution.

$$\pi_\beta(\boldsymbol{\tau}) = \frac{\sum_{\sigma \in \mathfrak{X}} e^{-\beta \mathscr{L}(\sigma, \boldsymbol{\tau})}}{\sum_{(\rho, \sigma) \in \mathfrak{X} \times \mathfrak{X}} e^{-\beta \mathscr{L}(\sigma, \rho)}}, \quad (8.23)$$

where:

$$\mathscr{L}(\boldsymbol{\tau}', \boldsymbol{\tau}) = -\boldsymbol{\tau}'^T W \boldsymbol{\tau} - \boldsymbol{\gamma}^T(\boldsymbol{\tau} + \boldsymbol{\tau}'). \quad (8.24)$$

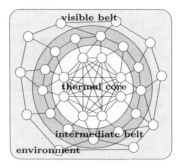

Fig. 8.10. Thermodynamical view of a Boltzmann Machine.

Setting $C = d_{\pi\|\phi}$ in Algorithm 8.2, our goal is to directing the above π_β asymptotically produced by the neural network toward the target ϕ. We obtain it simply by moving (W, γ) toward $(\widetilde{W}, \widetilde{\gamma})$ as in Remark 8.3. Since s_1 and s_2 are monotone with these parameters (they are sufficient statistics indeed) this corresponds to make the statistics taken during the running of the network, hence according to π_β, close to those coming from ϕ. We can sketch the matter as in Fig. 8.10, where the three distinguished clusters of PEs are so characterized:

- **thermal core:** provides dynamics to the machine state;
- **intermediate belt:** gives structure to the above dynamics;
- **visible belt:** simulates the environmental distribution law.

<small>The three belts metaphor</small>

Like in any sampling mechanism, the thermal core provides a seed U that, for a proper value of $\beta > 0$, a sufficient number of neurons and non biased free parameters, spans with almost equal probability the vertices of a ν_b-dimensional hypercube, where ν_b is the number of neurons on the borderline with the intermediate belt. The latter represents the explaining function of the random binary vector $\boldsymbol{T}^v$ produced on the *visible belt*. We expect a training algorithm to mold the connection structure of the intermediate belt enabling it to produce a $\boldsymbol{T}^v$ population almost undistinguishable from a sample $\{\boldsymbol{\tau}_1^v, \ldots, \boldsymbol{\tau}_m^v\}$ supplied to the algorithm as a training set. When we expect results from the asymptotic distribution (8.23) we must also collect statistic from it. This is generally obtained through a long run of the machine in order to observe, after each change of parameters, this distribution. In order to realize this task, we distinguish between

<small>Again an inverse transform to produce random variables</small>

A dream phase insensitive to external stimuli, and an awake phase affected by them,

two evolution modes. In the *unclamped* mode the whole state is left free to evolve starting from a random distribution law; after quite a while its distribution law assumes a form close to (8.23). In the *clamped* mode the situation is the same, except that the states of the visible nodes are maintained fixed to a vector of values supplied by the environmental distribution law ϕ. This is why folklore calls the former the *dream* phase, and the latter the *awake* phase. At the end of each phase we collect statistics s_1 and s_2 as in (8.13). It is easy to prove that the gradient of $d_{\pi\|\phi}$ is a function of the differences between the values of the statistics s_1 and s_2 collected at the end of the two phases. Hence these differences are used both to modify W and γ and to decide when the learning may be concluded. Omitting considering the *batch size problem*, i.e the timing with which interleaving the processing of the examples in the clumped stage with the updating of W and γ, the general scheme of Algorithm 8.2 specializes with this learning method as in Algorithm 8.3.

Algorithm 8.3. Template of learning algorithm with Boltzmann machine

1. *initialize* the free parameters W and γ
2. until termination triggers do
   a) run the network with current W and γ in the *unclamped* mode,
      collect statistics s_1 for each node and s_2 for each pair of connected nodes
   b) run the network with current W and γ in the *clamped* mode,
      collect statistics s'_1 for each node and s'_2 for each pair of connected nodes
   c) *change* (W, γ) into $(W', \gamma') = \mathsf{h}(W, \gamma, \{s_1 - s'_1, s_2 - s'_2\})$

---

generally requiring a long time to converge.

Boltzmann machines are a powerful theoretical case study but a worst implementation tool. Its drawback comes with the need of winding off their evolutions through asymptotical states. This requires long simulation times to allow a network to adsorb any perturbation the learning process induces on the stationary state. Indeed, apart from rare exceptions, its effectiveness has been really tossed only on toy examples like the one shown below.

Example 8.1. In order to learn the XOR function described in Table 8.1, consider the 2-2-1 architecture in Fig. 8.11(a) (see next section for multilayer structures). In essence, it has the thermal core coinciding with an intermediate belt made up of two neurons and the visible belt consisting of the two input and one

Table 8.1. The XOR truth table.

v_1	v_2	$\text{XOR}(v_1, v_2)$
0	0	0
0	1	1
1	0	1
1	1	0

output neuron. Then proceed according to the following steps: i) require symmetric connection weights, i.e. $w_{ij} = w_{ji}$; ii) activate the neurons sequentially with the probabilistic rule (8.19) so that the overall algorithm constitutes a sequential Boltzmann machine; and iii) train the network through sample means as described previously (see [4] for the analytical expression of the training procedure). For fixed $\beta = 5$ and $\eta = 0.001$, Fig. 8.11(b) shows the course of the percentage of success with the training cycles. A success is recorded when a correct pair $((\tau_1, \tau_2), \tau_0)$ is exhibited by the machine in the unclamped phase. A training cycle corresponds to the exhibition of a randomly selected correct pair and the corresponding computations; we compute the percentages every ten cycles after having increased β to 100 in order to reduce the variance of the output. Frequency 0.5 denotes a machine that essentially tosses a coin for associating τ_0 to (τ_1, τ_2); with 0.7 the machine loses one of the four examples. Due to the randomness of the network, even in case of a

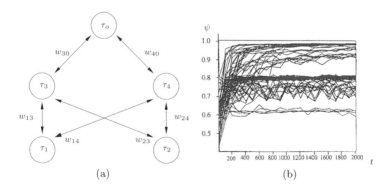

Fig. 8.11. (a) A 2-2-1 stochastic multilayer perceptron: $\tau_i \in \{0, 1\}$; the double arrows indicate the symmetry of the coupling constants $w_{ij} = w_{ji}$; and (b) course of the frequency of successes ψ vs. number of steps t when the network in (a) is trained to learn the XOR function through sequential learning rule and with initial w_{ij} drawn uniformly from $(-9, 9)$.

complete success, the percentage is a somewhat lower than 1. Better results may be obtained with a different initialization of the weights; for instance this occurs when no *prejudice* is involved in this operation, i.e. when all weights are initially set to 0.

Starting with tabula rasa condition generally works better.

8.4.3 Multilayer Perceptron

To avoid drawbacks of the above machine we usually refer to the deterministic version of the network which essentially works with the mean of the above target distribution. Actually we substitute (8.19) by the activation function (8.20) producing on each input a the expected value of the bit distributed according to the former. Simple application of Jensen inequality [61] says that we cannot expect to get in such a way the expected value of ϕ as well. In return, for the visible nodes we obtain a function that is highly sensitive to the neural network parameters, since the sigmoid function has derivatives of any order w.r.t. its argument. Thus, we substitute $d_{\pi\|\psi}$ by more suitable cost functions in Algorithm 8.2, such as the quadratic error:

A deterministic activation to bypass the convergence time.

$$\mathsf{C} = \sum_{i=1}^{m} l^2(x_i^v - \widetilde{x}_i^v) \quad (8.25)$$

with respect to a set of values that we expect to have on the visible nodes. Moreover, since we have no more random seeds to make the network running, we commit some nodes to receive input from outside the network. At the end of the day, we may think of the neural network as a computing device g equipped with input and output devices – each constituted by a set of specific neurons where to upload input x and read output y – implementing a function $y = g(x)$. Hence the problem of learning g comes as a regression problem without a suitable neither probabilistic nor fuzzy model. In principle you are free of choosing any strategy to minimize a cost function like C. The general understatement is that in the twisting argument:

$$T_\pi > t \Rightarrow \Pi \geq \pi \Rightarrow T_\pi > t + \mu, \quad (8.26)$$

we expect that by minimizing cost function T on the training set we minimize goal function Π as well (left implication). In absence of robust probabilistic companion results, we check the minimization of Π by checking the minimization of T (right implication) on the test set. Actually, the methods we generally use apparently lose the *social* features we highlighted in GA and we will find with other paradigms later on. As a matter of fact the functionality emergence is demanded here to the interactions

Collaboration comes from PE interactions,

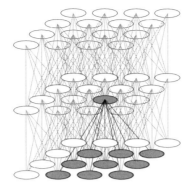

Fig. 8.12. A 3-layer feed-forward neural network. Each neuron in the upward layer receives signals from randomly connected neighboring neurons in the downward layer, as shown by the gray neurons linked by black wires.

between PEs. *Per se* they are deterministic. The complexity of their composition, however, gives rise to an unforecastable behavior that proves random to the user. In other words, within the neurons non deputed to input-output functionality we have again a thermal core, even though the seed of their production is exogenic, being supplied by the external inputs to the network. At their coming into the network generally they are not random, but after a few shaking inside it they result random to the rest of the computation.

<small>randomness from its complexity</small>

The most popular learning scheme in this framework is *Back-Propagation* [56] that one may find in many mathematical software libraries and can easily download from the web as well. In its basic version, the method refers to a neural network with a multilayer layout (usually referred to multilayer perceptron, MLP, for short) like the one shown in Fig. 8.12. At the bottom you have the input layer whose neurons have states set to external values coding x_i (i-th input signal as a part of the visible state x^v). These values are piped up through the various layers up to the last one (say L) having the function of output layer (see left picture in Fig. 8.13(a)). At each cross of a neuron the signal is updated according to the activation rules so that the output neurons code a signal that we compare with the y_i associated to x_i in the training set (as the remaining part of x^v) and the quadratic difference between them is added in (8.25). The learning method consists in the steepest descent along the C landscape in the parameter space. The basic ingredient is the computation of the derivatives of C w.r.t. the single parameters w_{ij}s and γ_is. This is obtained through the derivative chain rule

<small>An ordered flow of data in one direction, and</small>

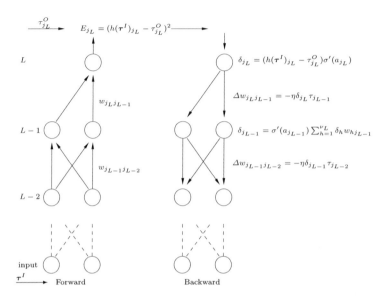

Fig. 8.13. Forward and backward phase of back-propagation algorithm.

<small>an ordered flow of corrections in the opposite direction.</small>

rising a *back-propagation* of the error terms from output to input layer through the intermediate layers, denoted as *hidden* layers, that gives the name to the corresponding algorithms' family (see right picture in Fig. 8.13). The *very popular* expressions of the derivatives are the following:

$$\frac{\partial \mathsf{C}(\boldsymbol{w})}{\partial w_{kj}} = \delta_k \tau_j, \text{ with} \qquad (8.27)$$

$$\delta_k = \begin{cases} \frac{\partial \mathsf{C}}{\partial \mathsf{h}(\tau^I)_k} \sigma'(a_k) & \text{if } k \text{ denotes an output neuron} \\ \sigma'(a_k) \sum_{h=1}^{\nu} \delta_h w_{hj} & \text{if } k \text{ denotes a hidden neuron,} \end{cases}$$

where the expressions for γ_is are deduced from the mentioned equivalent network with these parameters replaced by weights of connections from a dummy node with fixed state set to 1. The derivatives are variously used in the learning algorithms in conjunction with tuning coefficients in the province of the implementer. A typical update rule of the connection weights is:

$$w_{ij} = w_{ij} - \eta \frac{\partial \mathsf{C}}{\partial w_{ij}}. \qquad (8.28)$$

It is a very elementary way of descending the C landscape. We take the direction denoted by the derivatives under the assumption of a local independence of the single weight influences – which calls

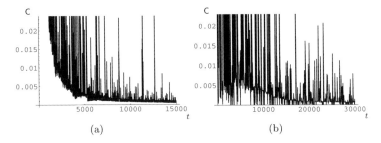

Fig. 8.14. Error trajectories during the training of a multilayer perceptron on the variants of two digits in the MNIST dataset: (a) initial training phase; (b) subsequent retraining after insertion in the training set of the misclassified test examples.

for an analogous separability of the representation produced by the hidden layer, rather than by the linear transforms as those used in Section 7.2. Then we modulate the length of the move along this direction through the factor η and correct the direction with various artifices of ingenuity, such as a momentum term taking into account previous parameter variations. Actually, the descent along the C landscape may be led by employing more sophisticated methods, such as the conjugate gradient [36] exploiting C second derivatives and other techniques, such as Levenberg-Marquardt (LM) [32] taking into account higher order derivatives. We may also introduce randomness during the descent via methods such as Simulated Annealing [2]. As a matter of fact, these techniques prove very efficient with specially featured landscape, while they introduce numerical complications elsewhere. Practitioners are mainly drawn by methods adapting learning coefficients such as η (called the learning rate) or the analogous coefficient α weighting the momentum term to the current run of the algorithm. Thus they plan to either jump a local minimum by increasing the value of η, or reduce waving of the trajectories by increasing α, or even explore with greater accuracy a given valley by reducing η, and so on. Their ingenuity is mostly attracted by automatic rules for quantifying these changes. Fig. 8.14 reports the typical control instruments during the training of a neural network represented by the descent of the error function (8.25) with the number of weights updating interations according to (8.28).

In line with previous algorithms sketched in this chapter, still omitting considering the batch-size problem, we may synthesize Back-Propagation as in Algorithm 8.4.

Example 8.2. The neural network entailing Fig. 8.14 was trained to distinguish the digit 4 from the digit 8 within the MNIST

> It makes a well organized computation

> susceptible of many practitioners' interventions,

Algorithm 8.4. Template of Back-Propagation algorithm on multilayer perceptron.

1. *initialize* the free parameters W and γ
2. until termination triggers do
   a) pipe forward external signal from Input layer to Output layer, compute cost function C as a function of the error in computing output
   b) propagate backward the error signal from Output layer to Input layer, compute derivatives of $\partial C(\boldsymbol{w})/\partial \theta_{kj}$ versus the layer parameters as shown in (8.28)
   c) *change* (W, γ) into $(W', \gamma') = \mathsf{h}(W, \gamma, \{\partial C(\boldsymbol{w})/\partial \theta_{kj}\})$

---

dataset; a task solved with a poor accuracy (7% of mistakes at the best) with the clustering procedures. The actual implementation of the back-propagation algorithm required to fix a certain number of details, among which we mention the following:

1. the network architecture. We devoted one neuron of the input layer to each of the 28 × 28 pixels coding the digits with gray levels in the database. We also decided to commit one neuron per digit (hence 2 in total) to output 1 when the digit code 4 is put in input to the network, 0 otherwise (which provides redundant but reliable coding of the output). As for the hidden layers we did rely on the rule of thumb: "many hidden neurons to favor generalization, but not too many to avoid overfitting", and decided for 10 hidden neurons;
2. training strategy. We trained the network in two phases, according to a principled *boosting* procedure [20]. In the former we randomly drawn a training set of 192 digits from the 1956 records representing in an equal number either 4 or 8 in MNIST. On this set we run the algorithm for many learning cycles, where each cycle consists of: i) presenting the whole set in input to the network, ii) computing the error function, and iii) updating the network weights accordingly. In the second phase we changed the training set by substituting randomly picked elements with those records (in number of 54) out of the training set that were bad classified by the network at the end of the first phase. After a long retraining of the network with this set we obtained a better classification of the whole 1956 long dataset with an error percentage $= 0.85\%$;
3. technical details. A momentum term has been added to the second member of (8.28) which recursively reads: $\Delta w_j =$

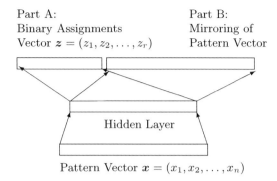

Fig. 8.15. Layout of the hourglass neural network mapping features to symbols.

$-\eta \frac{\partial C}{\partial w_{ij}} + \alpha \Delta w_j$. The total number of training cycle approached 45000.

Even though the implementation we did was very immediate, without involving the many tricks in the experience of the practitioners, the error percentage shows the great benefit coming from the knowledge of the label w.r.t. the clustering methods.

Example 8.3. Continuing Example 7.3 we get the binary mapping of the Sonar signals through a neural network with the hourglass architecture shown in Fig. 8.15 with $n = 60$ and $r = 50$. It shares the same input and hidden layer with the two output segments A and B, computing the Boolean assignments z and a copy of the input x, respectively.

Mirroring is a usual functional requirement for an MLP [53]. Hiding for a moment part A, the remaining network has a three-layer architecture with the same number of units in both input and output layers and a smaller number of units in the hidden layer. Therefore the hidden layer constitutes a bottleneck which collects in the state of its nodes a compressed representation of the input. This part of the network is trained according to the usual quadratic error function (8.25). Things are different for the units of part A of the output. In this case we require that the network minimizes an error represented by the edge pulling function (7.19). The error which is backpropagated from the units of part A is:

at architectural level, or

$$\delta_{ik} = \sigma'(a_i)\gamma_{ik}, \qquad (8.29)$$

with:

$$\gamma_{ik} = \pi_A \left(\theta_{ik} - \frac{\partial C_i}{\partial z_{ik}} \right), \qquad (8.30)$$

where $\frac{\partial C_i}{\partial z_{ik}} = -\log\left(\frac{z_{ik}}{1-z_{ik}}\right)$ and θ_{ik} forces the coding z_i of the pattern x_i to be correct according to Definition 7.1. The general idea is to insert into the γ expression an extra term which has the form of directed noise added to the initial value of γ when we are not satisfied with the correctness of the result. The effect is to shake the network in order to search for a new equilibrium.

in concern of specific training rule expedients.

We activate this punishment each time we find a pair of patterns with different labels to be coded with binary vectors having the Hamming distance below a given threshold (say 1 or 2). Namely, a positive random term ρ_{ik} is generated in correspondence to the incorrectly coded pattern x_i, specifically in those nodes (possibly all nodes) that do not increase the Hamming distance from the mate. Its value contributes to γ_{ik} with the following function:

$$\theta_{ik} = (1 - 2\Gamma(z_{ik}))\rho_{ik}, \qquad (8.31)$$

where Γ is a threshold function. The first term in the brackets specifies the sign of θ_{ik} so that the contribution to the network parameters goes in the opposite direction from the one the unit is moving in. Finally, π_A is a tuning parameter to balance the mutual relevance of corrections coming from parts B and A.

Special computational power is achieved by introducing recursion,

A more prominent variant concerns the introduction of recursion in the computation of the neuron states. This is done by introducing some backward connections piping back some states in order to influence the value of next states in the same neuron (see Fig. 8.16). The companion recursion in the derivatives' chain rules has the same form of the static case, provided we unfold

but training traps as well;

the network as in Fig. 8.16(b). Its loop, however, often primes instability in the learning algorithm whose remedy requires sophisticated techniques. In these cases hybrid architectures prove more reliable both as for convergence of the learning cycles, and for a wider exploitation of the available knowledge. Thus, the most general version of a neural network figures as a mix of traditional artificial neurons interacting with a set of symbolic PEs wiring the known part of the dynamic system under question. Though the complexity of these interactions vehiculated by the

hybrid architectures may relieve the latter.

connection web between PEs does not allow to fix in advance the entire function computed by the network [8], we have no problem for computing derivatives if the transfer functions from input to output of the symbolic PEs are (*quasi*) continuous in the parameters of our interest. Hence, we may mould these parameters so that we may check the suitability of the emerging function, at least on a given test set. For instance, Fig. 8.17 reports the 0 set point tracking of the yaw, pitch and roll attitude angles of a geostationary satellite (see Fig. 8.18) controlled by a recursive

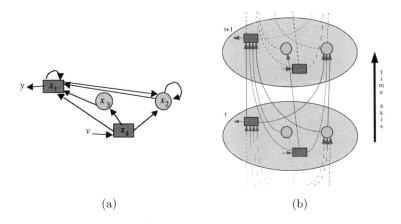

Fig. 8.16. Two representations of a hybrid system: (a) recurrent network and (b) its unfolded companion. Rectangles: symbolic PE's; cirlces: subsymbolic PE's; black arrows: running connections; plain gray arrows: current unfolded connections; dashed gray arrows: past or future unfolded connections.

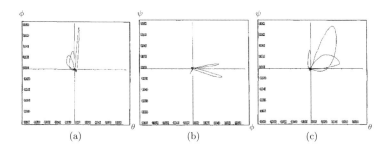

Fig. 8.17. Polar plots of the roll ϕ, pitch θ and yaw ψ attitude angle of satellite in Fig. 8.18.

hybrid architectures [7] trained by a general purpose software available at [31].

8.4.4 From a Population of Genes to a Population of Experts

Even though we generally abandoned the paradigm of a randomly activated neural network, we did not deny the role of randomness in computing functions. Actually we are more satisfied with relying on it both for seeding differences in PE behaviors and as a glue of the PE computations rather than on a complex weft of connections having the high drawback of hiding the role

> Better a random merge of computations than a complicate one.

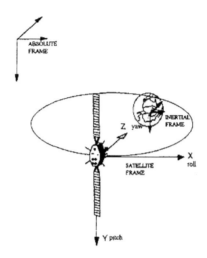

Fig. 8.18. Reference frames of an earth centered satellite.

of the former. Thus in recent years the *ensemble learning* became a new deal of connectionism.

The concept is the following. You train many replicas of the same PE but with random initializations in the expectation that they will correspond to different and equally prominent sensitivities to the training data. Then you adopt statistical criteria to get rid of the randomness of their output in respect to the same input. The most common rule in case of Boolean function is the majority voting: "output 1 if the majority of PEs does it, 0 otherwise". What could sound puzzling is that the functions computed by the single PEs are linear functions or a linear composition of functions of single variables in the most complex cases. Thanks to their elementary nature, indeed, these functions are perfectly controllable.

<small>an extended sampling mechanism:</small>

<small>pick so simple as reliable PEs,</small>

<small>for instance perceptrons or SVM,</small>

- We have the perceptron theorem [51] saying that if two sets of points on the vertices of any Boolean hypercube $\mathfrak{X} \subseteq \{0,1\}^\nu$ are linearly separable, i.e. there exists a function $\widetilde{y} = A\boldsymbol{x} + b$ such that $\widetilde{y} \geq 0$ for all items in the first set and $\widetilde{y} < 0$ in the other, or vice versa, then you may quickly train a neuron with activation function $y = \text{sign}(W\boldsymbol{x} + \gamma)$. Namely you obtain a neuron outputting the same sign of $\widetilde{y}$ after a series of updating:

$$W = \begin{cases} W + \boldsymbol{x} & \text{if } y = 0 \text{ and } \widetilde{y} = 1 \\ W - \boldsymbol{x} & \text{if } y = 1 \text{ and } \widetilde{y} = 0, \end{cases} \quad (8.32)$$

that are in a number polynomial in the optimal margin (see Section 3.2.4) you may obtain with a hyperplane.
- You solve the same problem in a time quadratic in the number of items to be separated using the SVM technique, as explained in Section 3.2.4.
- You linearly fit points with compatible hyperplanes with a confidence that you may compute as in Section 3.2.1.
- You extend the above results to nonlinear curves via the kernel trick explained in Section 7.4.

Also the effects of combining rules such as majority voting are controllable under loose hypotheses. For instance:

combine their results in a so elementary as fair way, for instance, through majority voting,

- For ys Boolean and independent with parameter $p = P(Y = 1)$, when p is greater than 0.5 you have an enhancement of its gap with 0.5, hence an increase of probability P that the whole ensemble produces 1, which goes as [14]:

$$P = \sum_{i=\lceil \frac{n}{2}+1 \rceil}^{n} \binom{n}{i} p^i (1-p)^{n-i}, \qquad (8.33)$$

with n number of experts, as shown in Fig. 8.19.
- If we give a different weight to the voters, because of their reliability, hence the system outputs 1 if $\sum_{i=1}^{n} w_i y_i > \theta$ where w_is are the weights and θ a suitable threshold, but we ignore which voter in the pool performs well on which instance, then by adaptively reducing the weights of those making a mistake could boost the performance of the overall algorithm near the mistake bound of the best voter in the pool [43].
- As for the threshold θ, its value may by determined from the balance of false positive and false negatives on a training set. Namely denoting with Z_0 the above sum (possibly with all weights set to 1) when the expected answer of the ensemble is 0, and Z_1 analogously, we may estimate their c.d.f.s and suitably balance $P(Z_0 \leq \theta)$ with $P(Z_1 > \theta)$. In the hypothesis that both variables follows a Gaussian distribution law, the θ equalizing the two probabilities is given by:

$$\theta = \frac{\widehat{\mu}_- \widehat{\sigma}_+ + \widehat{\mu}_+ \widehat{\sigma}_-}{\widehat{\sigma}_- + \widehat{\sigma}_+}, \qquad (8.34)$$

where $\widehat{\mu}_-$ and $\widehat{\sigma}_-$ are, respectively, the sample estimates of parameters μ_- and σ_- of the negative distribution; idem for the positive distribution.

Example 8.4. Concluding Example 8.3 we use an ensemble of SVMs as discriminating function. Namely, a single SVM draws

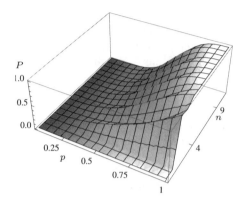

Fig. 8.19. Course of the probability P that the majority over n voters produces 1 for different values of n and probability p that a single voter says 1.

and you obtain often outperforming results, a hyperplane to linearly separate positive from negative vertices of the Boolean hypercube leaving the highest margin between points and plane. This operation is carried out as described in Section 3.2.4 on kernelized patterns. BICA vectors, however, are not a univocal representation of the original data. Rather, we have different representations if we start from differently initialized hourglass networks. Moreover, given the randomness of the network parameters we use for initializations, we expect that different BICA vectors may enhance different features of the data to be clustered, so that some features are more appropriate for discriminating a certain subset of them, while other features serve the same purpose for other subsets. Hence on each BICA representation we train an SVM, considered as a *base learner*. Then we combine the classification proposed by each SVM through a majority rule with threshold computed as in (8.34). We submit any record to each base learner of the ensemble and count the frequency with which it receives label 1. We classify the record as positive if this frequency overcomes the threshold, negative otherwise. As visualized in Fig. 8.20 we appreciated the performance of the whole discriminating procedure through a multiple hold-out scheme. We randomly split 50 times the Sonar dataset in two equally-sized training and test sets. Then, on each one of the 50 instances we train 50 hourglass networks, obtaining 50 BICA encodings. On each encoding of a training set we train an SVM getting an ensemble of 50 SVMs decreeing the cluster of each pattern in the training set. Data in Table 8.2 are the synthesis of discrimination performance on the 50 instances. We compare the parameters outputted by the proposed method with the

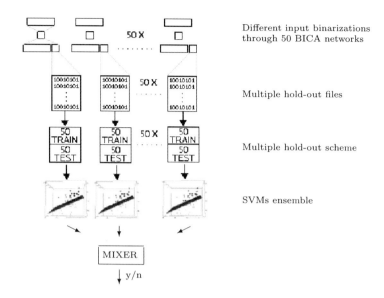

Fig. 8.20. A pictorial image of the classification through a SVM ensemble on binarized data.

results obtained in the literature with a three-layer perceptron [29] expressly trained on Sonar dataset with the back-propagation method. In a conservative way, we conventionally attribute a length of 4 bits to the original real signals, in order to account for their lengths in comparison with those compressed by BICA. We assume indeed that these bits are sufficient to the above methods to discriminate the data w.r.t. the classification problem they are questioned on. Moreover, with the former method we indicate the number of BICA components per single coding. The true length would emerge after a feature selection step that we do not consider however in this experiment. The accuracy is denoted through the percentage of correctly classified test set patterns. Because of the adopted multiple hold-out scheme, this percentage is expressed in terms of average μ and standard deviation σ. From this table we observe that compression offers a good preprocessing allowing to obtain accuracy percentage comparable with the results obtained through neural networks, where the absence of symbolic constraints represents an obvious benefit of the latter.

Ensemble learning algorithms are usually used in conjunction with resampling techniques in order to improve accuracy percentage of the base learner algorithms. The simplest method to achieve such enhancement is represented by *bootstrap aggregating* **even better if in conjunction with resampling techniques.**

Table 8.2. BICA experimental profile in comparison with three-layer neural network.

	BICA			Neural Network		
	Length	Correct %		Length	Correct %	
		μ	σ		μ	σ
Sonar	50	81.2	3.70	240	84.7	5.7

technique [11] (for short *bagging*). The main idea is to train each learner on a different training set obtained from the original using resampling techniques – such as the Efron nonparametric bootstrap procedure [24] – and combining the output by averaging (in case of regression) or voting (in case of classification). According to the law of large numbers bagging reduces the variance of predictions without changing the bias and, moreover, in case of unstable algorithms (i.e. where a small change in the training set yields large variations in the classification) improves the estimate accuracy.

A smarter choice of both resampling method and output combination leads to more interesting meta-algorithms, as those defining the *boosting* family [27]. The idea is to boost *weak* learning algorithms, performing slightly better than random guessing, into a *strong* algorithm, having an arbitrarily high accuracy over the training set. To achieve this, on the one hand flat resampling is modified so that harder to classify points are choosen more often than easier points, in such a way that the algorithm concentrates on the most difficult part of the sample space. On the other, the final output is built by weighted majority, where weights are proportional to the accuracy of each algorithm. Even if the error rate on the training instances approaches zero exponentially quickly as the number of experts increases, although boosting generally increases overall generalization accuracy, it may leads to a deterioration on some datasets, usually for highly noised and small-sized ones [28].

Finally, resampling methods can be applied directly to the feature space. A typical example is represented by gene expression analysis, where the major problem is the high dimensionality and low cardinality of the data, from which the *curse of dimensionality* problem may arise. *Random Subspace Methods* [33] work by generating different subspaces from the original sample space, where only a small number of randomly drawn features is considered. Base learners are then trained on such different subspaces and output are aggregated by majority voting.

8.5 Conclusions

The main question we address in this chapter is how to convince an individuals' population to serve our profit while conducting their own existence. This passes through three items:

1. they must be alive. This means that each individual must receive stimuli and react to them according to its own rules;
2. we must formulate our profit in terms that may be beneficial of their actions, and
3. they must feel satisfying an own need while adapting themselves to better serve our profit.

These ingredients are not quite different from those flavoring in the past treatises on the art of ruling people such as *The Principe* [45]. With our artificial populations, they are variously seasoned in an integrated way depending on the social computation paradigm we adopt.

With genetic algorithms individuals compute their contribution to a fitness function coinciding with our benefit by default (like slaves); they are so proud of this contribution that autonomously adapt themselves with the help of feedback from the rest of the population, to improve it (a very distributed algorithm). The global fitness in turn coincides with our profit (a true emergent functionality).

With neural networks individuals work under external stimuli. They may come either from contiguous individuals or from environment. Their own goals are different, in principle, from ours, so that we must directly intervene to indirectly modify their goals (of each one in relation to the others) by changing some parameters of their work – an activity that we call training. With Boltzmann machines in particular we are in an intermediate position where the population goal must be adapted to ours, but the training may be operated in a distributed way by the individuals themselves.

With ensemble learning we renounce to mutual feedbacks between individuals. Rather, like for genetic algorithms, each of them compute its contribution to our benefit and adapt himself to improve it. *A posteriori* we merge these contributions on our convenience.

Having in mind the ingredients, we may specify and arrange them in a lot of other ways suggested by our experience and environment conditions. Still more, we may realize them within other computational paradigms arising in our application fields.

8.6 Exercises

1. Starting from the initial state $\psi(0) = (0.2, 0.5, 0.2, 0.1)$ compute $\psi(10)$ through the evolution determined by the constant transition matrix:

$$\begin{pmatrix} 0.2 & 0.3 & 0.1 & 0.4 \\ 0.3 & 0.2 & 0.2 & 0.3 \\ 0.1 & 0.2 & 0.6 & 0.1 \\ 0.4 & 0.3 & 0.1 & 0.2 \end{pmatrix}$$

   Repeat the same exercise starting from $\psi'(0) = (0.1, 0.6, 0.3, 0.5)$. Elaborate on the consequences of Theorem 8.1 in your experiment.

2. Given a graph made of n, say 30, nodes and a set of one-way connections, encode paths that, starting from a node, link the greatest number of other nodes through the above connections. The goal is to feed a genetic algorithm with a limited set of these codes as genotypes in order to get the minimum spanning tree, i.e. a subgraph which is a tree and connects all the vertices together.

3. Define mutation, cross-over, and selection rules for the genotypes in Exercise 2, possibly with variants of your invention w.r.t. those enunciated in Section 8.3.1.

4. Run the algorithm defined in the previous two exercises and draw the course of the fitness with number of generations. Motivate a stopping rule as well.

5. Consider a 4 nodes network with the following matrix of connection weights:

$$W = \begin{pmatrix} 5 & 3 & 6 & -2 \\ 3 & 0 & 0 & 1 \\ 6 & 0 & -1 & -2 \\ -2 & 1 & -2 & 4 \end{pmatrix}$$

   and threshold $\gamma = (-2, 2, -2, 2)$. Run 20 evolution steps of a Boltzmann machine with $\beta = 10$, parallel activation, and initial configuration $\tau = (1, 0, 0, 1)$.

6. Evaluate on the basis of a sample $\{5, 12, 7, 15, 4\}$ the relative entropy between two Gaussian distribution of parameters $\mu_1 = 8, \sigma_1 = 3$ and $\mu_2 = 6, \sigma_1 = 5$, respectively, according to (8.12). Check that in the case that $\mu_1 = \mu_2 = 8, \sigma_1 = \sigma_2 = 3$ both the relative entropy and its appreciation through the sample are 0.

7. Design a 5-layer perceptron for mirroring 16 bits signals, according to the description of this task given in Example 8.2. What is the minimal number of nodes in the third layer?

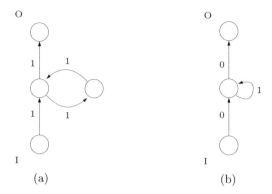

Fig. 8.21. Equivalence between network dynamics: (a) standard network with delay 1 on each edge; (b) equivalent network with non homogeneous delays.

Initialize randomly in $[-0.01, +0.01]$ weights and thresholds and run one forward step and one backward step of your machine.

8. With reference to the Guinea Pig dataset, divide the 80 pairs glucoside concentration-uptake in a training set, test set and validation set to be used for training a multilayer perceptron via back-propagation algorithm. Elaborate on the relative sizes of the three sets and on the batch size to process before each updating of the network parameters.

9. Consider the two elementary networks in Fig. 8.21. The left network is operated in the standard way consuming one timestep to cross one connection (the flag 1 on the connections). The right one assumes that the signal flows with no time consumption (delay 0) along the axis from input I to output neuron O, while covering the self connection takes one timestep. Elaborate on the fact that the two networks make the same computations. Moreover, compute the derivative of the activation function of the intermediate node w.r.t. the selfconnection weight.

10. Design an ensemble learning scheme to repeat the experiment in Example 8.2, taking into account the difficulties met to learn some special digit images. Elaborate, in particular, on the whole ensemble learning strategy (bagging, boosting, etc.), the number of base learner, the tools for training them, and the rule for assigning label to the digits.

9 If also Ants Are Able...

Learning, as a single value inference facility is unavoidably connected with an optimization problem. You fix a cost function compliant with your probability model, if any, and look for a solution in the parameter space that minimizes the function up to a given approximation. You search in the parameter space since you want to assess a function to be profitably used in the variable space. Otherwise you may be interested in optimizing a given function directly in the variable space. A typical example where the two optimization targets are in symbiosis is represented by the Boltzmann machines. You move along the increasing direction of the function (8.24) with the machine updating its state according to (8.22). When you are close to the maximum of this function you are close to the equilibrium, hence you may collect statistics feeding the learning rule in the parameter space. As a matter of fact, many optimization techniques do not change when you move from one space to the other, and we often speak of learning facility even when we are optimizing in the variable space. Actually with the same GA techniques you may learn the best fitting function by working on its code, like with the chromosome sequence determining our proteins. Thus learning is coming to achieve the widest acceptance of emergence of functionality, possibly specific – as the ability of optimizing a given function – possibly broad-band – as the ability of assessing an optimal function.

An optimization task at the basis of learning, either in the parameter or in the variable space.

Biologically inspired optimization offers a wealth of optimization mechanisms which tend to fulfill these essential needs. The underlying principles of these algorithms relate to the biologically motivated schemes of system emergence, survival, and refinement. The idea is that if some elementary algorithm works so well in Nature, it should bring benefits also in artificial problems. Thus, the inspiring role of various mechanisms encountered

To fulfill the task: "there is strength in numbers"

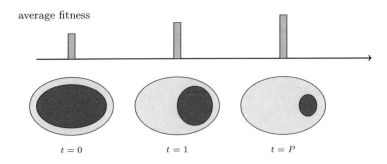

Fig. 9.1. A schematic view at evolutionary optimization; note a more focused populations of individuals (dark gray sets) over the course of evolution and the increase in average fitness of the entire population.

in the Nature are considered as pillars of the methodology and algorithms. The most visible feature of most, if not all, such algorithms, is that in their optimization pursuits they rely on a collection of individuals which interact between themselves in the synchronization of joint activities in order to find solutions. They communicate between themselves by exchanging their local findings. In this way, they are also influenced by each other in a sort of *anti-entropic* behavior. Indeed the energy exchange in a wealth of gas molecules generates an increase of entropy. *Vice versa*, the information exchange in the social life of the individuals in our populations reduces the above growth within a vital cycle where an entropy increase favors the exploration of possible solutions and the subsequent reduction highlights the most promising paths. A general visualization of the evolutionary nature of these algorithms is shown in Fig. 9.1.

9.1 Swarm Intelligence

Recent biologically inspired optimization techniques are examples of the modern search heuristics which belong to the category of so-called Swarm Intelligence methods [22, 37, 52]. The underlying principle deals with a population-based search in which individuals representing possible solutions carry out collective search by exchanging their individual findings while taking into consideration their own experience and evaluating their own performance. In this sense, we encounter four fundamental aspects of the search strategy.

An ensemble of elementary individuals,

1. The one deals with a social facet of the search; according to this, individuals adjust their behavior according to the successful beliefs of individuals occurring in their neighborhood.

2. The cognition aspect of the search underlines the importance of the individual experience where the element of population is focused on its own history of performance and makes adjustments accordingly.

    with a multifacet behavior.

3. The self organization rooting the collective behavior is more *volatile* than with connectionist or GA paradigms, in the sense that it emerges from the state of the system, as a collection of the states of the single agents, rather than from a structured memory owned by the latter. This introduces the last element that is

4. the age of the system and the course of its behavior with it. The state of each agent depends on the state of the system, contributing to the maturity of the computed function.

9.1.1 Ant Colony

A widespread representative of these investigations is represented by the algorithms that are assumed to be used by ants to find an optimal path from the nest to the food – a target not far from the Hansel aim of getting back home. You may think of the system raised up by ants as a neural network with changing connections. The update mechanisms are very simple, so that the ideal optimization problem solved by this system is the traveling salesman problem (TSP) [10, 9] that could look as a multi-heap multi-nest version of the original metaphor. Namely, given a set of cities connected by a network of roads, the problem is to find a shortest path bringing the traveler visiting all the cities. Since it is the search version of a decision problem (asking whether is there a path shorter than given kilometers with the mentioned property) enjoying a great generality (it belongs to the class of NP-complete problems), a great deal of optimization problems may be translated into the TSP, hence finding solution from ant colony system as well. However the quality of the solution may prove not so good with the other problems. Look at the basic problem of shortly connecting nest with food heap. We may imagine ants wandering around the nest at the beginning. If they find food they bring back it to the nest nest to store it. Like Hansel, during its movement an ant releases an information pebble, namely pheromone trails, so that the event "food found" translates into a thickening of the successful track, since it is used to go back to the nest. Under the assumption that ants preferably follow tracks with a great load of pheromone laid on, the food discovery primes a preferential marking of tracks that are more efficient for finding food, say the shorter paths. This effect is enhanced by the pheromone volatility which causes a proportional decrease of the pheromone load with the time of

Individual successful tracks,

stored on a volatile support,

staying on the ground. There are many variants of this scheme when it is implemented on a computer [19]. They are introduced in function of the specific optimization problem with the usual aims of local search algorithms, namely the avoidance of getting stuck in local minima and the speed of the search. The basic rules concern, in any case, the choice of next ant move and the consequent updating of the pheromone load. In the case of the TSP it could be [18]:

<small>make the solution to emerge,</small>

- Selection of next road in the network, say the arc e_{ij} within all arcs e_{il} moving from node i to neighboring node l. It occurs according to a probability p_{ij} defined as follows:

$$p_{ij} = \frac{\tau_{ij}^\alpha \eta_{ij}^\beta}{\sum_l \tau_{il}^\alpha \eta_{il}^\beta}, \qquad (9.1)$$

where τ_{ij} is the pheromone load on arc e_{ij}, η_{ij} is an *a priori* probability (say desirability) of selecting the arc as an inverse function of the cost d_{ij} of the arc, while α and β are ad hoc control parameters.

- Updating of the pheromone load is ruled by:

$$\tau_{ij} = \rho \tau_{ij} + \epsilon_k I_{ij}(k) \qquad (9.2)$$

$$I_{ij}(k) = \begin{cases} r_{ij} & \text{if ant } k \text{ crosses arc } e_{ij} \\ 0 & \text{otherwise,} \end{cases}$$

where ρ is the evaporation factor, ϵ_k is the quantity of pheromone laid by ant k, and r_{ij} is some profit constant of this arc.

<small>possibly with some external intervention.</small>

- Daemon action. This is a centralized intervention that, in contrast with the general distributed philosophy of the algorithm, could be triggered by the user, for instance because a chasm did interrupt a road. An example of daemon action is a local optimization procedure applied to an individual ants solution, or the collection of global information to update the pheromone mapping, such as identification of the best solution in order to give this solutions components an extra amount of pheromone.
- Final evaluation s_k of the current k-th solution is linked to the global pheromone charge of a solution, hence:

$$s_k = \sum_{i,j} \tau_{ij} Q_{ij}(k) \qquad (9.3)$$

$$Q_{ij}(k) = \begin{cases} 1 & \text{if arc } e_{ij} \text{ belongs to solution } k \\ 0 & \text{otherwise.} \end{cases}$$

- Final solution σ is provided by the most fitting one in terms of global pheromone charge, i.e.,

$$\sigma = \arg \max_{k=1,\ldots,m} \{s_k\}. \tag{9.4}$$

- Cost C of the solution σ is given by the sum of the costs of its arcs:

$$C = \sum_{i,j} d_{ij} Q_{ij}(\sigma). \tag{9.5}$$

Generalizing the procedure to build the solution of a general optimization instance, we obtain the pseudo code of ant colony optimization (ACO for short) in Algorithm 9.1.

Algorithm 9.1. Ant colony optimization

Given a set of components $\mathfrak{C} = \{c_1,\ldots,c_n\}$ and the completion function h on set $2^{\mathfrak{C}}$ of subsets of $\mathfrak{C}$

Initialize the desirability matrix with component η_{ij} associated to the move from component c_i to component c_j and set 0 the pheromone load τ_{ij} of this move, $i, j = 1,\ldots,n$.
Generate randomly initial moves associated to m ants.
repeat
  for $k = 1,\ldots,m$ **do**
    repeat select next move according to (9.1) **until** solution is completed.
    for $(i, j = 1,\ldots,n)$ update pheromone load according to (9.2)
    Daemon action
  end for
until termination criterion is met.
Compute $\sigma =$ best fitting substet of $\mathfrak{C}$ according to (9.3-9.4).
return σ

Example 9.1. Consider the three instances of the TSP problems shown in Fig. 9.1 drawn from the TSPLIB database of optimization problems [54]. Namely, the instance kroA100 is based on a case study set of 100 cities, att532 on 532 cities of the United States, and rl5934 on 5934 German cities. To implement Algorithm 9.1 we used the ACOTSP facility [58], with d_{ij} represented by the Euclidean distance between city i and city j, and choosing a number of ants $m = 25$. To obtain the statistics reported in Table 9.1, we iterated the whole procedure for a total of 10 trials, each with a stopping time of the ant moves set to 10 seconds of the server machine. Note that the best solutions obtained with the ACO equate to the best known results obtained so far in the first two datasets, while the lowest cost of 556045 is far to be reached in rl5934 problem - a drawback that could be overcome

Traveling salesperson is the template problem.

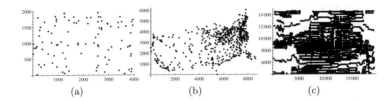

Fig. 9.2. Three instances of symmetric TSP composed of: (a) 100, (b) 532, and (c) 5934 cities. The cost function coincides with Euclidean distance between pair of cities, rescaled by a factor 10 in case (b).

Table 9.1. Performance of Algorithm 9.1 on the three TSP datasets shown in Fig. 9.1. First two columns refer respectively to the cost of the global best and worst solution, while last two columns refer to the mean and standard deviation of the best solutions found at each trial.

	Best try	Worst try	Average-Best	Std-Best
kroA100	21282	21282	21282.0	0
att532	27686	27756	27721.4	22.75
rl5934	590992	598283	594593.20	2429.79

with a better implemetation of the algorithm both in terms of tuning parameters and involved computational resources (hence, number of ants and the stopping time).

9.1.2 Swarm Optimization

A deterministic version of the above scheme is represented by Particle Swarm Optimization (PSO) [23]. In this case particles of a swarm move deterministically, at some velocity, according to an attraction force toward both local and global optimal points. Namely, denote the m vectors of the variables (particles) positioned in the n-dimensional search space by $x_1, x_2, \ldots, x_m$. There are m particles involved in the search, leading to the concept of a swarm. The performance $h(x_i)$ of each particle is described by an objective function referred to as a fitness (or objective) function.

A set of individuals, The PSO is conceptually simple, easy to implement, and computationally efficient. Unlike the other heuristic techniques, PSO has a flexible and well-balanced mechanism to enhance the global and local exploration abilities. As in the case of evolutionary optimization, the generic elements of the PSO technique involve:

each one pursuing an own objective,
1. performance. Each particle is characterized by some value of the underlying performance index or fitness. This is a tangible indicator stating how well the particle is doing in the search process. The fitness is reflective of the nature of the

problem for which an optimal solution is being looked for. Depending upon the nature of the problem at hand, the fitness function can be either minimized or maximized;

2. best particles. As a particle wanders through the search space, we compare its fitness at the current position with the best fitness value it has ever attained so far. This is done for each element in the swarm. The location of the i-th particle at which it has attained the best fitness is denoted by x_{best_i}. Similarly, by x_{best} we denote the best location attained among all x_{best_i};

just sharing the best result,

3. velocity. The particle is moving in the search space with some velocity which plays a pivotal role in the search process. Denoting the velocity of the i-th particle by v_i, at each updating we have:

and partially trusting it,

$$x_i = x_i + v_i. \qquad (9.6)$$

From iteration to iteration, the velocity of the particle is governed by the following expression:

$$v_i = \alpha v_i + c_1 r_1 (x_{\text{best}_i} - x_i) + c_2 r_2 (x_{\text{best}} - x_i), \qquad (9.7)$$

where r_1 and r_2 are two random values located in $[0, 1]$, and c_1 and c_2 are positive constants, called the acceleration constants. They are referred to as cognitive and social parameters, respectively. As the above expression shows, c_1 and c_2 reflect on the weighting of the stochastic acceleration terms that pull the i-th particle toward x_{best_i} and x_{best} positions, respectively. Low values allow particles to roam far from the target regions before being tugged back. High values of c_1 and c_2 result in abrupt movement toward, or past, target regions. Typically, the values of these constants are kept close to 2.0. The inertia factor α is a control parameter that is used to establish the impact of the previous velocity on the current velocity. Hence, it influences the trade-off between the global and local exploration abilities of the particles. For the initial phase of the search process, large values enhancing the global exploration of the space are recommended. As the search progresses, the values of α are gradually reduced to achieve better exploration at the local level. Denoting by t the generic iteration and $t_{\max}$ the planned number of iterations, we may schedule a linear decay from an initial $\alpha_{\max}$ to a final $\alpha_{\min}$ through:

are pushed toward a common success.

$$\alpha = \alpha_{\max} - \frac{\alpha_{\max} - \alpha_{\min}}{t_{\max}} t. \qquad (9.8)$$

We may also require to clip the x_i and v_i to be positioned within a required region.

As the PSO is an iterative search strategy, we proceed with it until either there is no substantial improvement of the fitness or we have exhausted the number of iterations allowed in this search.

Overall, the algorithm can be outlined as the following sequence of steps encoded in Algorithm 9.2.

Algorithm 9.2. Particle swarm optimization

Given the objective function h on $\mathfrak{X}$

Initialize to 0 the set $\{x_{\text{best}_i}, i = 1, \ldots, m\}$ and x_{best}.
Generate randomly m particles x_i and their velocities v_i.
repeat
  for $i = 1, \ldots, m$ **do**
    Set $x_{\text{best}_i} = \arg\max\{h(x_{\text{best}_i}), h(x_i)\}$.
    Update x_i, v_i and α according to respectively (9.6), (9.7), and (9.8).
    Eventually clip values.
  end for
  Set $x_{\text{best}} = \arg\max\{h(x_{\text{best}}), h(x_i)\}$.
until termination criterion is met.
return x_{best}.

Example 9.2. Continuing Example 9.1, Table 9.2 reports the performance of Algorithm 9.2 implemented through MAOS_TSP facility [62], when applied to the three TSP instances shown in Fig. 9.1. For consistency of comparison we choose to work with

Table 9.2. Performance of Algorithm 9.2 on the three TSP datasets shown in Fig. 9.1.

	Best try	Worst try	Average-Best	Std-Best
kroA100	21282	21305	21284.3	7.27
att532	27690	27709	27699.5	7.15
rl5934	559824	562458	561247.0	1051.83

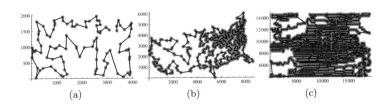

(a)   (b)   (c)

Fig. 9.3. Path computed by PSO algorithm on the three TSP datasets shown in Fig. 9.1.

$m = 25$ particles and a total of 10 trials. From this table it appears that no algorithm outperforms the other on all datasets, not even a monotonic relation exists between standard deviation of the best solution obtained in each trial. The same argument we made before holds, i.e. smarter implementations of PSO together with a better choice of the free parameters may provide better results. Finally Fig. 9.1.2 shows the path connecting all cities as computed by PSO algorithm.

9.2 Immune Systems

Immune systems share a *social* behavior of particles and the moulding features of evolutionary systems in order to produce emerging functionalities with prominent optimization attitudes. You may think of them as subsymbolic systems with some fixed points through which the identification of the goal function is realized. From a cognitive perspective, the management of the information passes through a phase of identification of the target, to which a phase of achievement of the goal follows. In the immune systems metaphor the former corresponds to the antigen recognition, while the latter consists of raising up a more or less complex procedure to grow and activate the antibodies. Thus, differently from ants who innately follow pheromone tracks, here B- and T-cells must identify the target cell to be destroyed [13]. Related artificial systems will need special pattern recognition algorithms to achieve a similar result, provided that a suitable encoding of the objects playing the role of antigens has been realized. As antigens are represented by strings of characters, pattern recognition generally comes to measuring a suitable distance between them and analogous strings on the part of the deputed cells. Once antigens have been identified, a population of items able to cope with them, the antibodies, is produced. Its improvement proceeds according to the intuitively appealing fitness function denoting the affinity of the single items with antigens, i.e. their capability of destroying the latter. Antibodies are also strings of characters, so that the mentioned distances are at the basis of the fitness functions. This binds and simplifies, at the same time, the definition of fitness, while leaves open all evolutionary tools for improving the population in the subsequent generations. This is what is called the *maturation* phase of the Immune System algorithm, where traditionally no mutation operator is involved, while an aging factor *reinforces* most fitting antibodies guaranteeing them a greater survival probability with time, so that they may represent the memory of the artificial system. The last aspect is explained by Immune Network theory through the concept of idiotypic network – i.e. an immune

A more articulated social goal:

first, identify the fitness function,

then maximize it,

with a special attention to maintain the optimum with time.

network of interactions between receptors – linking receptors of an antigen. Similarly to Boltzmann machines, in the absence of foreign antigens, the immune system must display a behaviour or activity resulting from interactions with itself, and from these interactions immunological behaviour such as tolerance and memory emerge [16]. Finally, Danger Theory [46], a relatively new addition to the field of immunology, removes the need to define what is self, but adds the necessity to define danger, safe and pathogen signals.

A biological immune system is extremely complex and even not yet fully understood, so it should not be surprising that a huge number of variants have appeared at the computational end of the concept. Here we quote two of them which are widely implemented.

Many variants to fill the task

1. The simplest form of artificial immune system is represented by Negative Selection algorithms [26], inspired by the main mechanism in the thymus that produces a set of mature T-cells capable of binding only non-self antigens. The main idea is to define a set of self strings $\mathfrak{S}$ and a set of detectors $\mathfrak{D}$ able to recognize the complement of $\mathfrak{S}$. Anti-spam filtering or intrusion detection algorithms are all typical scenarios where these algorithms find a successful application.

2. Clonal Selection is perhaps the best known algorithm in the field, performing both optimization and pattern recognition tasks. The main principles come from the antigen driven affinity maturation process of B-cells with its associated hypermutation mechanism, and the presence of memory cells to retain good solutions to the problem being solved [15]. The skeleton of this algorithm, which in any case proves more detailed than the analogous for Genetic Algorithms, is sketched as Algorithm 9.3. Note how: i) the proliferation of B-cells is proportional to the affinity of the antigen that binds it, and ii) the mutations suffered by the antibody of a B-cell are inversely proportional to the affinity of the antigen it binds. When applied to pattern matching, a set of patterns to be matched are considered to be antigens.

Example 9.3. Continuing our leading example on TSP instances we tried to apply Clonal Selection algorithm (implemented through OAT software package [12]) to the datasets in Fig. 9.1, being aware of the fact that immune systems usually behaves well on both pattern matching and multi-modal function optimization, and TSP does not belong to these categories. In fact, the best tours for the first two datasets (shown in Fig. 9.4) with a cost of 46949 and 1220968, respectively, approach not even remotely the optimal solutions. Note that we obtain these results by running the algorithm for 15 minutes, compared with

Algorithm 9.3. Clonal selection

Given the set of antigenic patterns $\mathfrak{G}$ and an affinity function h

---

Initialize the population of antibodies $\mathfrak{B}$
Split the $\mathfrak{B}$ population into the subpopulations $\mathfrak{B}_m$ of memory antibodies and $\mathfrak{B}_r$ of reservoir ones
while stop criterion is false **do**
  for all $g_i \in \mathfrak{G}$ **do**
    – Select from $\mathfrak{B}$ the n highest affinity antibodies forming the set $\mathfrak{B}_{\{n\}}$ (*selection*)
    – Generate from $\mathfrak{B}_{\{n\}}$ a clone population $\mathfrak{C}$ privileging the higher affinity antibodies   (*cloning*)
    – Get population $\mathfrak{C}^*$ from $\mathfrak{C}$ by mutating its antibodies to a degree inversely proportional to their affinity   (*affinity maturation*)
    – Select antibodies within $\mathfrak{C}^*$ from among those with higher affinity as candidates memory abtibodies $\mathfrak{B}_m$ (*re-selection – candidature*)
    – Replace any antibody already in $\mathfrak{B}_m$ with a candidate memory cell whose affinity is higher   (*replacement*)
    – Generate the reservoir antibodies population $\mathfrak{B}_r$ by replacing within the remaining antibodies those with lower affinity with new generated ones   (*reservoir generation*)
  end for
end while
return $\mathfrak{B}_m$

---

less than 2 minutes for both ACO and PSO (remember each of the 10 trials was run for 10 seconds).

9.3 Aging the Organisms

We may review also the original neural network paradigm in terms or more elementary and more autonomous PEs. The main support of this paradigm is claimed to be the fact that this machinery works like the human central nervous system. Effectively, the firing mechanism of the neurons may be resumed in many cases without great oversimplification by the Heaviside function characterizing the activation of a binary threshold neuron. What highly diverges in the artificial networks from the biological ones is the learning mechanism. Apart from Boltzmann machines, the learning algorithms we have examined for neural networks in Chapter 8 all require both a global synchronization of the activities of the single neurons in order to collect statistics, and a set point tracking in order to increase or lower the weights of the single connections. Both tasks are too complex to be realized

Learning without set points:

by the biological system, and to be implemented in a large scale (million of neurons) neural network as well. We may enhance the analogy with biological systems by augmenting the role played by the threshold, a parameter that is local to the single neuron. We model it as an *age* parameter moving from small values toward high values. The former denote youth of the neuron, an age when it is very sensible to external stimuli. This means it fires often under their solicitation, with a side effect of propagating its sensitivity to surrounding neurons (i.e. awakening these neurons). As its age increases the neuron becomes resistant against these stimuli, hence being less and less prone to fire and to propagate as well, till becoming actually dead w.r.t. the current flow of signals. The set of neurons that may be awakened by the current one identifies the layer in a multilayer structure, where each layer cooperates in transmitting signals from a previous layer to a subsequent one. Awakening a neuron means reducing its threshold from a high value, which puts it in a quiescent status, to a low value, allowing it to react to the external stimuli. Hence synapses functionalities are different if they connect neurons on a same layer (triggering functionality), or neurons on other layers (communication piping functionality).

<small>a monotone change with time, hence with the neuron age,</small>

The aging mechanism determines a long-term homeostasys of the system. We assume that the single neuron increases its threshold by discrete quantities as a reaction to the difference between its age and the mean age of the other live neurons in the layer, i.e. of the difference between its firing rate w.r.t. the mean firing rate. At a higher level the neurons' aging seems to correspond to the overlapping of thought cycles: a set of low threshold neurons primes the cycle by reacting to the external stimuli and pushing surrounding neurons to do the same. We may figure a pre-climax phase where this mechanism is self-excitatory, followed by a post-climax phase where neurons tend to be quiescent and do not awake other ones. Hence the thought cycle ends, with some feeble fallout, and related groups of neurons become quiescent. They remain so until they are awakened by neurons from other zones of the brain involving them in another thought cycle.

<small>regulated by contrast with the environment.</small>

9.3.1 The Neuron's Life

Further taking inspiration from nature, we may frame the new neurons within a predator-prey scenario constituted by two layers of a multilayer feed-forward neural network architecture embedded in liquid medium able to uniformly vehiculate elementary signals between neurons. The upward layer attacks the downward one with signals that may cause its neurons two forms of deaths: a true *physiological death* caused by an incremental stress due to

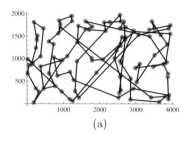
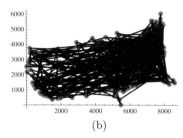

Fig. 9.4. Best tour found by Clonal Selection algorithm on the first two datasets in Fig. 9.1, respectively, with population and selection size equal to 50, clone factor of 0.1 and mutate percentage of 2.5%, randomly replacing 5 antibodies in the population.

the neuron firings; and *functional death* denoted by a complete non reactivity of the neuron to any signal coming from the upward layer. Downward neurons (*d-neurons* for short) play the role of prey. Their role is to avoid the former type of death with the side effect of promoting the latter. In particular, we consider the d-neuron threshold as a lever for regulating the partition of its energy between defense and reproductive activity. Defense reads insensitivity to stimuli from upward layer neurons (*u-neurons*). Reproduction has to be intended in functional terms. Each neuron may prime the reactivity of other neurons which it is connected to (hence forming the layer) each time it produces a spike. So from this time onward other neurons bear the task of elaborating the upward signal, thus increasing the bulk of the thought. At each solicitation the live neuron is pushed to increase its defense attitude, unless it is refrained each time its threshold is high enough in comparison with a ground state. By identifying the individual defense attitude of neuron B with its threshold γ_B (call it *strength* as well), we assume the ground state γ_g to coincide with the mean strength $\overline{\gamma}_B$ of a set of surrounding[1] neurons representing the *colony* of neurons B is embedded in. This comparison is made possible in the brain by the volume transmission operating therein [1].

A balance between reproduction and defense,

From an ecological perspective, B is only pushed by its own need to survive, being prone to increment its defense effort in diminution of the attitude to propagate the thought. The sole "social" mindset consists in *deciding* to stop increasing its strength when it realizes it is behaving relatively better than the rest of the colony. The rest is pure mechanics: the number of effective transmitting units decreases as much as they are made useless by the B

having death and social conveniences as ingredients.

---

[1] Here surrounding is a late notion due to the possible large range of synapses in the brain reflected in the sparseness of the connection matrix.

neuron threshold; each B primes another neuron with a probability decreasing with the strength, etc.

Hence we model the dynamics of the population through a procedure that iteratively updates 2 patterns, the number n of spikes produced by the d-neuron layer in a temporal window and the local distribution of neurons producing the spikes, as follows:

1. set of d-neurons. We work with a set of neurons in the downward layer connected through an incomplete connection matrix. As a former approximation we fix connections randomly, by assuming each neuron connected with the others of the layer with a given probability ρ, the latter being a function of the distance of the neurons on the grid. We distinguish live neurons from dead ones, where the latter may be either physiologically or functionally dead. In any case, a neuron never producing a spike in a given temporal window is considered dead. A functionally dead neuron turns live if primed by a spiking neuron along an existing connection;

2. set of u-neurons. They behave as transmitting units, each possibly sending a unitary signal to the downward neurons depending on the evolution of the whole brain state, provided that a connection exists between the unit and the receiver neuron. As a prelimiary approximation, the topological map describing the interconnections between couples of neurons in two different layers is fixed *a priori* and depends on the distance of the paired neurons. Moreover, to maintain the model as simple as possible, at least in a first approximation, a signal is sent from a unit to a receiver neuron with a given probability b. Analogously, the connection weight has a random value w in a given interval centered around 0;

3. fate of the single d-neuron. It depends on its basic functionality of computing the Heaviside function of its potential v. Namely, at each instant a d-neuron B tries to communicate with a u-neuron, say j, receiving either a non null signal $X_j(\text{B}) = 1$ with a probability b or a null signal $X_j(\text{B}) = 0$ (with probability $1 - b$). Thus signals are random variables. The linked unit is randomly selected according to a uniform distribution on the u-neurons connected to B. The transmitted signal $w_j X_j$ is null with probability $1 - b$, otherwise coincides with the weight w_j of the link. If $v + w$ trespasses the threshold γ_B the neuron fires ($S(\text{B}) = 1$) and sets its potential to the resting value v_0. Otherwise $v = v + w$, or a smoothed form of this sum to take into account the decay of the signal over time, for instance:

$$v = v + w - \beta(v - v_0), \qquad (9.9)$$

with β playing the role of decay rate. If the neuron fires, then there is a probability $\rho\alpha^t$ that it awakes another d-neuron B′ connected to it according to the connection matrix as in point 2, getting $L(B') = 1$, where L is the living indicator of the neurons. The first term ρ takes into account the connectivity of the neurons (as explained in point 1), while $\alpha < 1$ is a dumping term having the effect of reducing the awaking probability of a neuron B′ that has already been risen t times. Alternatively, we have $L(B') = 0$ in the case of either physiological or functional death. Namely,

$$P(X(B) = 1) = b \tag{9.10}$$
$$P(S(B) = 1) = \sigma(v - \gamma_B) \tag{9.11}$$
$$P(L(B') = 1) = \rho\alpha^t, \tag{9.12}$$

where σ is a Heaviside function as in (8.18);

4. strength γ_B of the single neuron. Whenever B fires, it decides to increment γ_B by a quantity δ provided that the actual value of γ_B is not greater by a fixed τ than the average $\bar\gamma_B$ over all the live d-neurons it is connected to. Namely:

$$\gamma_B = \gamma_B + \delta \text{ if } \gamma_B < \tau\bar\gamma_B; \tag{9.13}$$

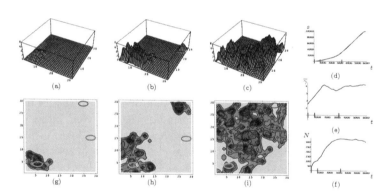

Fig. 9.5. Overlapping scenario of a neuronal evolution. Frames (a-c): spatial distribution of threshold values on a 30×30 grid of neurons; (g-i): spike frequency spatial distribution on the same grid. Color code: a light gray cell denotes an inactive neuron; shading from gray to dark gray denotes a value increase. Bottom left ovals correspond to the first set of awakened neurons; top and median right ovals respectively to the group of neurons awakened by the former, and neurons distant from the previous two which may be activated through signal propagation. Pictures (d-f): course of cumulative number of spikes, mean strength and number of live neurons over time. Gray markers on the horizontal axis show the sampling times of (a-c),(g-i) frames.

5. the localization of the live neurons. It is determined by the connection matrix. Within a given neighborhood we may still maintain the connection uniformly random from upward to downward layers, in close relation to the randomness of the conveyed signal. An analogous randomness on the intra-layer connection between d-neurons will produce radial attraction basins around the neurons that initially start (prime) a thought. Instead, more complex figures may emerge from different settings of the connections. The connection matrix, along with the position of the priming neurons, will generate the spatial process of thought with possible overlapping and alternating phenomena inside the layer.

generating spatial distributions evolving with time.

The simplicity of this paradigm allows for fast computations with the use of dedicated parallel computing languages such as the π-calculus [47]. In Fig. 9.5 is reported the evolution of the life of a grid of 30×30 neurons for a proper setting of the parameters [6].

9.4 Some Practical Design and Implementation

Directing populations offers a number of evident advantages over some other categories of optimization mechanisms. They are general and their conceptual transparency is definitely very much appealing. The population-based style of optimization offers a possibility of a comprehensive exploration of the search space and provides solid assurance of finding a global maximum of the problem. The tools for overcoming local minima are indeed of immediate comprehension and controllability, even in case of deterministic procedure directly pointing at the expected values of the populations. To take full advantage of the potential of evolutionary optimization, one has to exercise prudence in setting up the computing environment. This concerns a number of crucial parameters of the algorithm related to evolutionary operators, size of population, bias-variance balance, stopping criterion, to name the most essential ones. Moreover, what is even more fundamental is the representation of the problem in the genotype space. Here a designer has to exercise his ingenuity and fully capture the essence of domain knowledge about the problem. There is hence no direct solution to the decoding problem. We could come up with a number of alternatives. In many cases, it is not obvious up front which could work the best, viz. lead to the fastest convergence of the algorithm to the global solution, prevent from premature convergence, and help avoid forming an excessively huge genotype space. The scalability aspect of coding

Social computing is not a social event,

needs for careful implementation.

has to be taken into consideration as well. Several examples presented below help emphasize the importance of the development of a suitable genotype space.

The use of directing populations in the development of complex systems can be exercised in many different ways. As we do not envision any direct limitations, one should exercise some caution and make sure that we really take full advantage of these optimization techniques while not being affected by their limitations.

1. Structural optimization of complex systems provided by evolutionary optimization is definitely more profitable than the use of evolutionary methods to their parametric optimization. Unless there are clear recommendations with this regard, we could do better by considering gradient-based methods or exercising particle swarm optimization rather than relying on evolutionary optimization. Another alternative would be to envision a hybrid approach in which we combine evolutionary optimization regarded as a preliminary phase of optimization that becomes helpful in forming some initial and promising solution and then refine it with the aid of gradient-based learning. *(Select proper tools for optimizing,)*

2. The choice of the genotype space is critical to the success of evolutionary optimization; this, however, becomes a matter of a prudent and comprehensive use of existing domain knowledge. Once the specific genotype space has been formed, we need to be cognizant of the nature and role of specific evolutionary operators in the search process. It might not be clear how efficient they could be in the optimization process. *(involve all your knowledge for encoding, and)*

3. The choice of the fitness function must fully capture the nature of the problem. While in directing populations we do not require that such function is differentiable w.r.t. the optimized component, it is imperative, though, that the requirements of the optimization problem are reflected in the form of the fitness function. In many cases we encounter a multiobjective problem, and caution must be exercised so that all the objectives are carefully addressed. In other words, a construction of the suitable fitness function is an essential ingredient to direct populations. *(the problem specifications for defining fitness.)*

One may note that reciprocally, the technology of fuzzy sets could be helpful in structuring domain knowledge which could effectively be used in the organization of population direction. This could result in a series of rules (or metarules, to be specific) that may pertain to the optimization. For instance, we could link the values of the parameters of the evolutionary operators with the performance of the process: if there is high variation of the values of the average fitness of the population, a substantial reduction of mutation rate is advised.

9.5 Conclusions

This chapter hits at very elementary learning mechanisms in a conceptual inverse way of what happened with Turing machine. There we started with a very elementary automaton endowed with a control unit made of a dozen of basic boolean instructions, and with their recursive use we built up wonderfully the complete computational structures of our computers [59]. The typical exercise of designing an algorithm for the sum of two 8 bit numbers basing on these instruction takes a notably student skill to be solved [34]. With time we essentially forgot these algorithms that are hidden in any operational system. *Vice versa*, with granular computing we moved from high level procedures, somehow made jaunty by the abundance of computational power, and went to ever more elementary computational atoms with the awareness of facing an information management problem more than a computational one. On the other side, once defined the elementary behavior of ants or swarm particles, with social computation paradigm we leave them to work by themselves with the general philosophy: if they work, why not!

We pay this softening of our mind work, however, with the nearness of our success. In spite of the NP-completness of the decision problem underlying the traveling salesman problem indeed, we do not find many application in the literature of the algorithms of this chapter to instances quite different from the ones for which they have been assessed. This is why we did not fix an exercise section for this chapter. Rather, you find some design recipes as operational conclusions, to be coupled with a commitment of being so rigorous as imaginative in designing this kind of optimization/learning procedures.

Further Reading

Social computation is a definitely social phenomenon that pervaded the scientific community in last decades [25]. With the new deal of neural networks, in the eighties, it constituted the head of ram to break an axiomatic vision of the sciences where starting from a few truths everything making a sense must be proved. Probably a first jolt to this framework was done earlier, from inside, by the mysticism of quantum mechanics where the wavepacket collapse represents an additional axiom disarranging a monolithic corpus of science [30]. In addition, paradigms like neural network and stochastic optimization proved *in vitro* that Nature does not work optimally but feasibly [56]. Feasibility, indeed, looks as the true key for operationally exploiting the modern Computational Intelligence technology running faster than methodologies. Namely, to employ the large amount of miniaturized logic circuits made available by C-MOS and further technologies, to benefit of terabits of signals pored by any kind of sensors every second, you need simple and quick algorithms [41]. No matter whether their result is optimal, what is important is that they work. The fact that you have a lot of processors each implementing a same simple algorithm connotes your computation as social. Checking whether they really work is the big point. You may partly bypass it bay saying: they imitate algorithms that are well working in Nature [50]. In facts each procedure in this part of the book is a bio-inspired algorithm.

A first lesson we learnt from them is that we primarily want to capture from Nature which kind of information she draws from her environment. Then, the way of processing it plays a minor role, apart from the strategy she adopts to share information between elements of a population moved by the same goal. This is the way a big research effort is devoted today to the feature extraction goal in all its nuances [17]. Meanwhile, very elementary strategies based on numerical computation of the gradient of the cost function prove extremely quick and suitable [21].

A second lesson is connected to the evolution we observe in the design of these algorithms. At the beginning of the neural networks revival, in the eighties, the analogy with human brain led researchers to work with very huge networks deputed to very hard tasks, with the understatement: if natural brain can, then artificial one can too. For instance in [49] a neural network has been devoted to classification of sub-pixels inside the resolutions supplied by remote sensing data, in [5] a network with 5000 neurons decrees the epilepsy class of patients. Today real world applications embed small neural networks (order of 10 neurons) within complex architectures, for instance for image or speech processing, handwritten characters recognition included, within the mentioned philosophy of coupling available knowledge with subsymbolic PEs (see Section 8.4.3). The fact is that we are not yet satisfied by an algorithm working well even though we cannot explain why. Rather we want the steps computed by the elementary processors to be clear and with well understood effects, at least at local level. Moreover, if no formal support is available where to embed these processors we prefer combining them in a statistical way, so that we may elaborate on limited properties of the emerging functionalities rather than remaining fascinated by the outperformance of an exceptional but unexplainable result. Thus huge highly structured neural networks are commonly replaced with ensemble of SVMs [38].

A great drawback is represented by possible misunderstanding of the elementary kind of the processing element computations with naivety. *Per se*, social computation gave room to many practitioners of formalizing and contrasting some rule of thumb that prove suitable in their computations. It opened room, however, to some excess of creativity to many people who daily invent rules and algorithms that they *perceive* could successfully promote the emergence of a goal functionality. Their activity animates the vast universe of Computational Intelligence, rendering difficult to discriminate some notably valuable contributions from spamming. Also at theoretical level we may reckon this drawback, with the emergence of some folklore principle, such as negentropy enhancement on one side [35] and error gaussianity assumptions [42] on the other, that sometime hide the rationale of some successful procedures. As a matter of fact and apart from some meaningful exceptions [44], bio-inspired algorithms act more as case studies inspiring specific steps of operational procedures than procedures themselves capable of fully solving operational problems.

References

1. Agnati, L.F., Zoli, M., Stromberg, I., Fuxe, K.: Intercellular communication in the brain: wiring versus volume transmission. Neuroscience 69(3), 711–726 (1995)
2. Amato, S., Apolloni, B., Caporali, P., Madesani, U., Zanaboni, A.M.: Simulated annealing in back-propagation. Neurocomputing 3, 207–220 (1991)
3. Amit, D., Gutfreund, H., Sommpolinsky, H.: Sping-glass models of neural networks. Physical Review A32, 1007–1018 (1985)
4. Apolloni, B., Armelloni, A., Bollani, G., de Falco, D.: Some experimental results on asymmetric boltzmann machines. In: Garrido, M.S., Vilela Mendes, R. (eds.) Complexity in Physics and Technology, pp. 151–166. World Scientific, Singapore (1992)
5. Apolloni, B., Avanzini, G., Cesa-Bianchi, N., Ronchine, G.: Diagnosis of epilepsy via backpropagation. In: Proceedings of International Joint Conference on Neural Networks, Washington D.C., vol. II, pp. 517–574 (1990)
6. Apolloni, B., Bassis, S.: A feedforward neural logic based on synaptic and volume transmission. Brain Research Reviews 55(1), 108–118 (2007)
7. Apolloni, B., Battini, F., Lucisano, C.: A co-operating neural approach for spacecrafts attitude control. Neurocomputing 16(4), 279–307 (1997)
8. Apolloni, B., Piccolboni, A., Sozio, E.: A hybrid symbolic subsymbolic controller for complex dynamic systems. Neurocomputing 37, 127–163 (2001)
9. Applegate, D.L., Bixby, R.E., Chvàtal, V., Cook, W.J.: The Traveling Salesman Problem: A Computational Study. Princeton University Press, Princeton (2006)
10. Biggs, N.L., LLoyd, E.K., Wilson, R.J.: Graph Theory. Clarendon Press, Oxford (1976)
11. Breiman, L.: Bagging predictors. Machine Learning 24(2), 123–140 (1986)
12. Brownlee, J.: Optimization Algorithm Toolkit (OAT), Swinburne University of Technology (2006), http://optalgtoolkit.sourceforge.net/

13. Cesana, E., Beltrami, S., Laface, A.E., Urthaler, A., Folci, A., Clivio, A.: Current paradigms in immunology. In: Proceedings of Natural and Artificial Immune Systems (WIRN/NAIS) 2005, pp. 244–260 (2005)
14. Condorcet, M.J.A.N., de Caritat, M.: Essai sur l'application de l'analyse à la probabilitè des décisions rendues à la pluralitè des voix. l'Imprimerie Royale, 1–304 (1785)
15. de Castro, L.N., Timmis, J.: Artificial Immune Systems: A New Computational Approach. Springer, London (2002)
16. de Castro, L.N., Von Zuben, F.J.: ainet: An artificial immune network for data analysis. In: Sarker, R.A., Abbass, H.A., Newton, C.S. (eds.) Data Mining: A Heuristic Approach, pp. 231–259. Idea Group Publishing, USA (2001)
17. De Jong, K.A.: Evolutionary Computation: A Unified Approach. Bradford Books. MIT Press, Cambridge (2006)
18. Dorigo, M.: Ant colonies for the traveling salesman problem. IEEE Transactions on Evolutionary Computation 1(1), 53–66 (1997)
19. Dorigo, M., Stutzle, T.: Ant Colony Optimization, Bradford Books. MIT Press, Cambridge (2004)
20. Drucker, H., Schapire, R., Simard, P.: Boosting performance in neural networks. International Journal of Pattern Recognition and Artificial Intelligence 7(4), 705–719 (1993)
21. Duch, W.: Towards comprehensive foundations of computational intelligence. In: Challenges for Computational Intelligence. Springer Studies in Computational Intelligence, vol. 63, pp. 261–316 (2007)
22. Eberhart, R.C., Shi, Y.: Particle swarm optimization: developments, applications and resources. In: Proceedings of Congress on Evolutionary Computation 2001, Piscataway, NJ, Seoul, Korea, IEEE service center, Los Alamitos (2001)
23. Eberhart, R.C., Shi, Y., Kennedy, J.: Swarm Intelligence. The Morgan Kaufmann Series in Artificial Intelligence. Hardcover (2001)
24. Efron, B., Tibshirani, R.: An introduction to the Boostrap. Chapman and Hall, Freeman, New York (1993)
25. Eslick, I.: Scratchtalk and social computation: Towards a natural language scripting model. In: IUI 2008 Workshop on Common Sense Knowledge and Goal-Oriented User Interfaces (CSKGOI 2008) (in press, 2008)
26. Forrest, S., Perelson, A.S., Allen, L., Cherukuri, R.: Self-nonself discrimination in a computer. In: Proceedings of the 1994 IEEE Symposium on Research in Security and Privacy, Oakland, CA, pp. 202–212. IEEE Computer Society Press, Los Alamitos (1994)
27. Freund, Y., Schapire, R.E.: A decision-theoretic generalization of on-line learning and an application to boosting. In: Proc. II European Conference on Computational Learning Theory, Barcellona (March 1995)
28. Freund, Y., Schapire, R.E.: A short introduction to boosting. Journal of Japanese Society for Artificial Intelligence 14(5), 771–780 (1999)

29. Gorman, R.P., Sejnowski, T.J.: Analysis of hidden units in a layered network trained to classify sonar targets. Neural Networks 1, 75–89 (1988)
30. Griffiths, D.J.: Introduction to Quantum Mechanics, 2nd edn. Prentice-Hall, Englewood Cliffs (2004)
31. LaReN Group. Hybrid Learning System, University of Milan (2006), http://laren.usr.dsi.unimi.it/hybrid/index.html
32. Hagan, M.T., Menhaj, M.: Training feedforward networks with the marquardt algorithm. IEEE Transactions on Neural Networks 5(6), 989–993 (1994)
33. Ho, T.K.: The random subspace method for constructing decision forests. IEEE Transactions on Pattern Analysis and Machine Intelligence 20(8), 832–844 (1998)
34. Hopcroft, J., Ullman, J.: Introduction to Automata Theory, Languages and Computation. Addison-Wesley, Reading (1979)
35. Hyvärinen, A., Kahunen, J., Oja, E.: Independent Component Analysis. John Wiley & Sons, Chichester (2001)
36. Johansson, E.M., Dowla, F.U., Goodman, D.M.: Backpropagation learning for multi-layer feed-forward neural networks using the conjugate gradient method. International Journal of Neural Systems 2(4), 291–301 (1992)
37. Kennedy, J., Eberhart, R.C., Shi, Y.: Swarm Intelligence. Morgan Kaufmann Publishers, San Francisco (2001)
38. Kim, H.C., Pang, S., Je, H.M., Kim, D., Bang, S.Y.: Constructing support vector machine ensemble. Pattern Recognition 36(12), 2757–2767 (2003)
39. Kullback, S.: Information theory & statistics. Wiley, Chichester (1959)
40. Lamperti, J.: Stochastic processes: a survey of the mathematical theory. In: Applied mathematical sciences, vol. 23. Springer, New York (1977)
41. Lancaster, D.E.: Cmos Cookbook. Paperback (1997)
42. Levin, E., Tishby, N., Solla, S.A.: A statistical approach to learning and generalization in layered neural networks. In: Rivest, R., Haussler, D., Warmuth, M.K. (eds.) Proceedings of the Second Annual Workshop on Computational Learning Theory, pp. 245–260. Morgan Kaufmann, San Francisco (1989)
43. Littlestone, N., Warmuth, M.K.: The weighted majority algorithm. In: IEEE Symposium on Foundations of Computer Science, pp. 256–261 (1989)
44. Gambardella, L.M., Rizzoli, E., Oliverio, F., Casagrande, N., Donati, A., Montemanni, R., Lucibello, E.: Ant colony optimization for vehicle routing in advanced logistics systems. In: Proceedings of MAS 2003 - International Workshop on Modelling and Applied Simulation, Bergeggi, Italy, pp. 2–4 (2003)
45. Machiavelli, N.: Il Principe 1513
46. Matzinger, P.: Tolerance, danger, and the extended family. Annals Reviews of Immunology 12, 991–1045 (1994)
47. Milner, R.: Communicating and Mobile Systems: the π-Calculus. Cambridge University Press, Cambridge (1999)

48. Mitchell, M.: An Introduction to Genetic Algorithms. MIT Press, Cambridge (1996)
49. Moody, A., Gopal, S., Strahler, A.H.: Artificial neural network response to mixed pixels in coarse-resolution satellite data. Remote Sensing of Environment 58(3), 329–343 (1996)
50. Natural computing, an international journal
51. Novikoff, A.B.: On convergence proofs on perceptrons. In: Symposium on the Mathematical Theory of Automata, Polytechnic Institute of Brooklyn, vol. 12, pp. 615–622 (1962)
52. Parsopoulos, K.E., Vrahatis, M.N.: On the computation of all global minimizers through particle swarm optimization. IEEE Transactions on Evolutionary Computation 8(3), 211–224 (2004)
53. Pollack, J.B.: Recursive distributed representation. Artificial Intelligence 46, 77–105 (1990)
54. Reinelt, G.: Traveling Salesman Problem Library, Heidenberg Univ. (1995),
http://www.iwr.uni-heidelberg.de/groups/comopt/software/TSPLIB95/
55. Roger, H.: Theory of recursive functions and effective computability. McGraw-Hill, New York (1967)
56. Rumelhart, D.E. (ed.): Parallel Distributed Processing, vol. 1. MIT Press, Cambridge (1986)
57. Sahni, S.: Some related problems from network flows, game theory, and integer programming. In: Proceedings of the 13th Annual IEEE Symposium of Switching and Automata Theory, pp. 130–138 (1972)
58. Stuetzle, T.: Ant colony optimization algorithm for symmetric TSP (2004),
http://www.aco-metaheuristic.org/aco-code
59. Turing, A.: On computable numbers, with an application to the entscheidungs problem. Proceedings of the London Mathematical Society 42, 230–265 (1936)
60. Wallace, D.L.: Asymptotic approximations to distributions. Annals of Mathematical Statistics 29, 635–654 (1958)
61. Wilks, S.S.: Mathematical Statistics. Wiley Publications in Statistics. John Wiley, New York (1962)
62. Xie, X.F.: Mini Multiagent Optimization System (MAOS) for solving Combinatorial Optimization Problems, Hakodate, Japan (2006), http://www.adaptivebox.net/research/fields/problem/TSP/index.html

Part V
Granular Constructs

Introduction

Practically almost all problems arising in life are subject to a broad indeterminacy principle: we either find a very accurate solution, but without knowing why this solution works, or we compute a solution exactly on the basis of its requirements, but are unable to stake on its accuracy. With social optimization algorithms of the previous part we follow the first option, that constitutes a really new deal of the past decades. Actually we were trapped by old theoretical constructs capable of generating extremely rigorous results, whose implementation, however, did require a lot of simplifications in the description of the real phenomena and a wide exclusion of boundary cases as outliers. Thus these constructs were abandoned in favor of heuristics whose rationale could be theoretically demonstrated only as a trait, but whose efficiency is numerically validated by a lot of implementations.

Overtaking the dichotomy of accuracy versus explicability.

Nevertheless we are not willing to renounce to the benefits of the above constructs, which mainly consist in their composition into more complex ones, so that, having checked the compliance of the building blocks with the rationale we have about the phenomenon at hand, we may rely on the plausibility of the final results without considering further conceptual checks. The commitments of these constructs are very severe: simple relations on a few variables. In recent years we try to increase their semantical power, hence to reduce the load of simplifications and shrink the set of outliers by evaluating variables through information granules, hence quantizing them at a few levels, and by accurately positioning these granules in the universe of discourse and relations in the functional space. This delineates the way through which in this chapter we manage a complex fuzzy set reflective of a detailed relationship between variables. We will compute its membership function by using a few variables quantized into a few elementary granules and

Simple relationships on a few granular variables.

elementary functions connecting the variables. This leads to design structures between fuzzy sets that denote a fuzzy model. Positioning granules is essentially a matter of fuzzy clustering, so that we will reconsider some of the clustering algorithms, feature representation included, discussed in Part III; refining relationships' parameters is a subsymbolic attitude in the province of social computation expounded in Part IV. Rather these computations will rebound in the whole procedure that leads from features to rules. Although the border line between models and their identification is somewhat unspecified with this matter, we will essentially treat the former in Chapter 10 in terms of *granular constructs*, while in Chapter 11 we will elaborate on the algorithms actually implementing them, thus *identifying fuzzy rules*.

10 Granular Constructs

Consider a highly nonlinear function like the one in Fig. 10.1 as the linear fit of experimental points describing the relationship between variables x and y. Willing to understand this relationship we may decide to express it by a 5-th degree polynomial, whose fitting looks like in Fig. 10.1(a). A better way is to identify three fuzzy sets centered at the points $x_1 = 1, x_2 = 3.5$ and $x_3 = 8$ each with a membership function $\mu_i = e^{-(x-x_i)^2}$ and endowed with its own relation:

A mixture of simple granular functions fits better than a single highly complex function.

$$y = g_i(x) \tag{10.1}$$

that holds between x and y, namely: 1) $y = g_1(x) = 2.4x$; 2) $y = g_2(x) = 0.4 + 1.3x$; 3) $y = g_3(x) = 10.0 - 0.7x$. The mixture of these functions weighted with the above values of μ_i, i.e.

$$y = \sum_{i=1}^{k} g_i(x)\mu_i(x), \tag{10.2}$$

reproduces the function at hand (see Fig. 10.1(b)). This delineates the way through which we manage in this chapter a complex fuzzy set by computing its membership function: using a few variables quantized into a few elementary granules and elementary functions connecting the variables.

10.1 The Architectural Blueprint of Fuzzy Models

From a semantic perspective, in general fuzzy models operate at the level of information granules and in this way constitute highly abstract and flexible constructs. Given the environment of physical variables describing the surrounding world and an

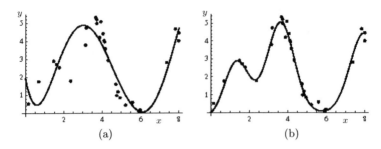

Fig. 10.1. Comparison between fitting methods of points randomly drawn from a complex and noisy function: (a) fifth-degree polynomial fit, and (b) a mixture of 3 fuzzy sets.

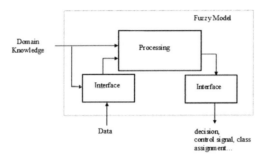

Fig. 10.2. A general view at the architecture underlying fuzzy models showing its three fundamental functional modules.

abstract view of the system under modeling, a very general sketch of the fuzzy model architecture can be portrayed in Fig. 10.2.

We clearly distinguish between three functional components of the model where each of them comes with well-defined objectives. The input interface builds a collection of modalities (fuzzy sets and fuzzy relations) that are required to link the fuzzy model and its processing core with the external world. The latter is represented by a dataset of signals in the form of the finite set of input-output pairs $(\bm{x}_i, y_i)$, $i = 1, 2, \ldots, m$, where $\bm{x}_i = (x_{i1}, x_{i2}, \ldots, x_{in})$. In principle, we combine them into a $(n+1)$-dimensional vectors $\bm{z}_i = (\bm{x}_i, y_i)$. Interfaces manage the mapping from the input variable $\bm{x}$ to fuzzy sets $\{A_1, \ldots, A_n\}$ granulating it and from the resulting fuzzy sets $\{B_1, \ldots, B_r\}$ to the output variable y. In more general term, we may think of a many input-many output (MIMO) model involving a vector of output variables as well, hence $\bm{y} = f(\bm{x})$. For the sake of simplicity, however, we will refer to many input-single output (MISO) models by default, the former being obtained by a collection of latter models, one for each output component, having the same input.

<small>Fuzzy systems: a core processor plus interfaces.</small>

The Architectural Blueprint of Fuzzy Models

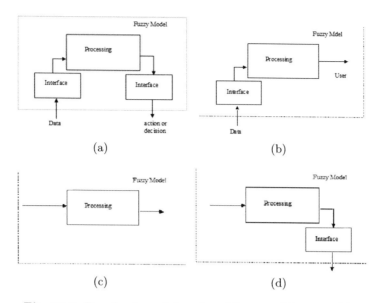

Fig. 10.3. Four fundamental modes of the use of fuzzy models.

The processing core realizes all computing being carried out at the level of fuzzy sets already used in the interfaces. The output interface converts the results of granular processing into the format acceptable by the modeling environment. In particular, this transformation may involve numeric values being the representatives of the fuzzy sets produced by the processing core. The interfaces could be present in different categories of the models yet they may show up to a significant extent. Their presence and relevance of the pertinent functionality depend upon the architecture of the specific fuzzy model and a way in which the model is utilized. The interfaces are also essential when the models are developed on a basis of available numeric experimental evidence as well as some prior knowledge provided by designers and experts.

Referring to Fig. 10.3, we encounter four essential modes of their usage:

(a) the use of numeric data and generation of results in the numeric form (see Fig. 10.3(a)). This mode reflects a large spectrum of modeling scenarios we typically encounter in system modeling. Numeric data available in the problem are transformed through the interfaces and used to construct the processing core of the model. Once developed, the model is then used in a numeric fashion: it accepts numeric entries and produces numeric values of the corresponding output. From the

Sometime is more convenient working with numbers,

perspective of the external "numeric" world, the fuzzy model manifests itself as a multivariable nonlinear input-output mapping. Later on, we discuss the nonlinear character of the mapping in the context of rule-based systems. It will be demonstrated how the form of the mapping depends directly upon the number of the rules, membership functions of fuzzy sets used there, inference schemes and other design parameters. Owing to the number of design parameters, rule-based systems bring in a substantial level of modeling flexibility and this becomes highly advantageous to the design of fuzzy models;

(b) the use of numeric data and a presentation of results in a granular format (through some fuzzy sets), as shown in Fig. 10.3(b). This mode makes the model highly user-centric. The result of modeling comes as a collection of elements with the corresponding degrees of membership and in this way it becomes more informative and comprehensive than a single numeric quantity. The user/decision-maker is provided with preferences (membership degrees) associated with a collection of possible outcomes;

sometime with sets. (c) the use of granular data as inputs and the presentation of fuzzy sets as outcomes of the models (see Fig. 10.3(c)). This scenario is typical for granular modeling in which instead of numeric data we encounter a collection of linguistic observations such as expert's judgments, readings coming from unreliable sensors, outcomes of sensors summarized over some time horizons, etc. The results presented in the form of fuzzy sets are beneficial for the interpretation purposes and support the user-centric facet of fuzzy modeling;

(d) the use of fuzzy sets as inputs of the model and a generation of numeric outputs of modeling (see Fig. 10.3(d)). Here we rely on expert opinions as well as granular data forming aggregates of detailed numeric data. The results of the fuzzy model are then conveyed (through the interface) to the numeric environment in the form of the corresponding numeric output values. While this becomes feasible, we should be cognizant that the nature of the numeric output is not fully reflective of the character of the granular input.

10.2 The Rule System

Since the objects of our relationships are sets we use Boolean algebra to explicit them. Since the sets are fuzzy we embed negation into their membership function and express the remaining monotone part of the relationship through a set of Horn clauses.

Namely, a Horn clause is a clause with at most one positive literal, hence represented by formulas like the following:

$$\overline{A_1} \vee \overline{A_2} \vee \ldots \vee \overline{A_n} \vee B \qquad (10.3)$$

which more suitably reads under our perspective as:

$$A_1 \wedge A_2 \wedge \ldots \wedge A_n \Rightarrow B. \qquad (10.4)$$

Give to each literal in the implication the meaning of indicative function of a set, normally relaxed into a membership function, to n of them in the monomial (left side of the implication) that of antecedents and to the $(n+1)$-th that of consequent of a relation. In this way we express the monomials as conditional "if – then" statements where fuzzy sets occur in their conditions and conclusions. We denote these statements as *rules*, whereas their disjunction constitutes a *rule-based system* as the prototype of any highly modular and easily expandable fuzzy model. Designing a rule system entails two aspects: i) the formulas' layout to completely describe the phenomenon at hand with a suitable set of logic implications, and ii) the rule system evaluation method to draw conclusions with operational instances.

<small>Alternative consequences depending on alternative sets of antecedents,</small>

10.2.1 Designing a Rule System Layout

A common standard format comes in the form:

if x_1 is A_{11} and x_2 is A_{12} and ... and x_n is A_{1n} **then** y is B_1,

if x_1 is A_{21} and x_2 is A_{22} and ... and x_n is A_{2n} **then** y is B_2,

$$\vdots \qquad \vdots \qquad \vdots$$

if x_1 is A_{k1} and x_2 is A_{k2} and ... and x_n is A_{kn} **then** y is B_k, $\qquad (10.5)$

where A_{ij} and B_i, for all $i = 1, \ldots, k$ (k – the number of rules), $j = 1, \ldots, n$ (n – the number of conditions), are fuzzy sets defined in the corresponding input and output spaces, x_i and y variables corresponding to respectively i-th condition and conclusion.

The models support a principle of locality and a distributed nature of modeling as each rule can be interpreted as an individual local descriptor of the data which is invoked by the fuzzy sets defined in the space of conditions. The local nature of the rule is directly expressed through the support of the corresponding fuzzy sets standing in its condition part. The level of generality of the rule depends upon many aspects that could be easily adjusted making use of the available design components associated with the rules. In particular, we could consider fuzzy sets of

<small>to describe a universe at different levels of generality.</small>

350   Granular Constructs

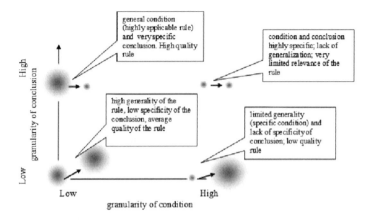

Fig. 10.4. Examples of rules and their characterization with respect to the level of granularity of condition and conclusion parts.

condition and conclusion whose granularity could be adjusted so that we could easily capture the specificity of the problem. By making the fuzzy sets in the condition part very specific (that is being of high granularity) we come up with a rule that is very limited and confined to some small region in the input space. When the granularity of fuzzy sets in the condition part is decreased, the generality of the rule increases. In this way the rule could be applied to more situations. To emphasize a broad spectrum of possibilities emerging in this way, refer to Fig. 10.4 which underlines the nature of the cases discussed above.

The relationship between the clauses could be highly diversified. There are several categories of models where each class of the constructs comes with interesting topologies, functional characteristics, learning capabilities and the mechanisms of knowledge representation. In what follows, we offer a general glimpse at some of the architectures which are the most visible and commonly envisioned in the area of fuzzy modeling.

Let us start with *Tabular fuzzy models* as an archetypal way of describing fuzzy relationships between variables. They are formed as some tables of relationships between the variables of the system granulated by some fuzzy sets, as shown in Table 10.1. For instance, given two input variables with fuzzy sets $A_1, A_2, A_3, B_1, \ldots, B_5$ and the output fuzzy sets C_1, C_2 and C_3, the relationships are articulated by filling in the entries of the table; for each combination of the inputs quantified by fuzzy sets, say A_i and B_j, we associate the corresponding fuzzy set C_d formed in the output space. The tabular models produce a fairly compact suite of transparent relationships represented at the level of information granules. In the case of more than

Table 10.1. An illustrative example of the two-input tabular fuzzy model.

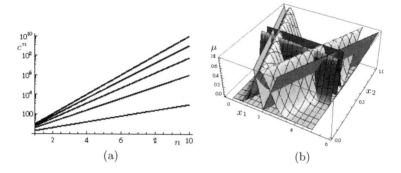

Fig. 10.5. (a) Combinatorial grow (in log-scale) of the number of rulers that can be stated with n input variables; the dependency is illustrated for several variable quantizations c, with $c = 2, 4, 6, 8, 10$ from lower to higher curves. (b) Fuzzy sets centered around hyperplanes in the x space.

two input variables, we end up with multidimensional tables. Let us remind that a complete rule base consists of c^n rules, where c is the number of information granules in each input variable and n denotes the total number of input variables. Even in case of a fairly modest dimensionality of the problem, (say $n = 10$) and very few information granules defined for each variable (say $c = 4$), we end up with a significant number of rules, that is $4^{10} \approx 1.049 \times 10^6$. By keeping the same number of variables and using 8 granular terms we observe a tremendous increase in the size of the rule base; here we end up with 1.074×10^9 different rules which amounts to substantial increase. There is no doubt that such rule bases are not practical. The effect of this combinatorial explosion is clearly visible, as shown in Fig. 10.5(a).

<small>You may consider all possible antecedents,</small>

<small>but it may take too long time.</small>

The evident advantage of the tabular fuzzy models resides with their high readability (transparency). The shortcoming comes with giving up direct mapping mechanisms. This means that we do not have any machinery to transform input (either numeric or granular) into the respective output. Furthermore, the readability of the model could be substantially hampered when dealing with the growing number of variables we consider in this model.

There are several ways of handling the problem of high dimensionality. An immediate one is to acknowledge that we do not need a complete rule base because there are various combinations of conditions that never occur in practice and as such are not supported by any experimental evidence. While this assumptions might be sometimes risky, it also becomes beneficial as leading to a drastic reduction of the number of rules enrolled in our knowledge base.

> To avoid this drawback refer to abstract conditions in place of original variables.

The second more feasible way would be to deal directly with multidimensional fuzzy sets. This means that we gather points of the x space with a limited set of functions describing, with the fuzzy set canons, the possible antecedents of the rule system. For instance, in Fig. 10.5(b) the n-dimensional (3 dimensional for representation simplicity) space is described through the distances of the generic point from one of the three hyperplanes each representing the center of a fuzzy set endowed with Gaussian membership function. In force of the simplicity commitment discussed in the introduction of this part, we look for a number of antecedents that usually is definitely smaller than the cells of a full expansion of individual variables described before. With this approach, rule system descriptions move from the format (10.5) to the following:

$$\begin{array}{c} \text{if } x \text{ is } A_{11} \text{ and } A_{12} \text{ and } \ldots \text{ and } A_{1n} \text{ then } y \text{ is } B_1, \\ \text{if } x \text{ is } A_{21} \text{ and } A_{22} \text{ and } \ldots \text{ and } A_{2n} \text{ then } y \text{ is } B_2, \\ \vdots \qquad \vdots \qquad \vdots \\ \text{if } x \text{ is } A_{k1} \text{ and } A_{k2} \text{ and } \ldots \text{ and } A_{kn} \text{ then } y \text{ is } B_k. \end{array} \quad (10.6)$$

In the opposite direction, i.e. when moving toward rules' increase, we may expect that same cells in the input space represent antecedents of different consequents, which requires a weighting system to get a synthesis on the latter. This denotes a *hierarchical architecture* composed of several levels of knowledge representation where there are collections of rules formed at a few very distinct levels of granularity (generality), as shown in Fig. 10.6. The level of generality of the rules is directly implied by the information granules forming the input and output interfaces.

> Then, a same episode may have different readings as for a consequent.

10.2.2 Evaluating a Rule System

We have various ways of managing the "if-then" logic connectives of the rules which deal with fuzzy sets [15]. In the application

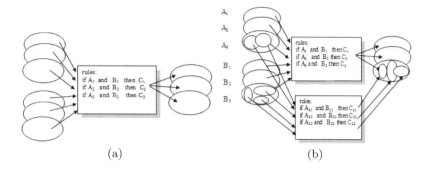

Fig. 10.6. Examples of rule-based systems: (a) single-level architecture with all rules expressed at the same level of generality; (b) rules formed at several levels of granularity of fuzzy sets standing in the condition parts of the rules. A_i and B_j stand for fuzzy sets forming the condition part of the rules; C_d the conclusion.

we will discuss in the book we consider the simplest one focusing on the intersection of the antecedents with the consequent as the result of the rule. Hence the part of B of our interest from a generic rule r as in (10.4) is $A_1 \wedge A_2 \wedge \ldots \wedge A_n \wedge B$, since if any of the antecedents fails we have no reason of considering B within possible results. Add the fact that both antecedents and consequents are fuzzy so that the output of our rule is a membership function as well, and you obtain the following operational companion of the rule: x belongs to all antecedents with a degree $\mu(x)$ that is the T-norm of the membership degrees to each antecedent; the result of the rule r applied to x is a fuzzy set R to which each element y of the universe of discourse $\mathfrak{Y}$ belongs with a degree $\mu_R(y|x)$ given by the T-norm of $\mu_B(y)$ with $\mu(x)$.

Rule system exploitation in terms of T-norms and S-norms of involved membership degrees

Analogously, we use the whole rule system (10.6) focusing on the union of the above sets, that we characterize through its membership function in principle, and any specific value we may draw from it to support our decision as the ultimate goal. In the rest of this chapter we will discuss how to efficiently organize computations to implement the above evaluation according to the functional general architectures in Fig. 10.2 and different strategies of exploiting information.

10.3 The Master Scheme

From a functional perspective, whatever the abstract scheme of the model and the functional relationship between Horn clauses computed by the processing core, the general block diagram of

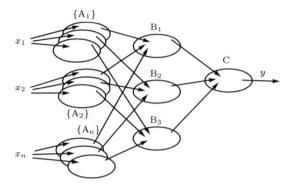

Fig. 10.7. A schematic view of a granular construct; $x_1, \ldots, x_n$: components of the input vector $\boldsymbol{x}$; Ellipses: information granules; y: output. With notation $\{A_i\}$ we mean $\{A_{i1}, A_{i2}, A_{i3}\}$.

Clustering plus mixing,

a granular construct is the one represented in Fig. 10.7 that we may resume with a mixture of information granules like the one mentioned at the beginning of this chapter. You may think of this architecture to be vertically divided into a left part positioned in the province of fuzzy clustering for identifying granules, and a right part located in the province of connectionist methods (for instance a neural network) for mixing them. The frontier between the two parts is somehow fuzzy, since on one side the relation between $\boldsymbol{x}$ and y in (10.1) may be either embedded into a cluster or subsequently *learnt* by the connectionist system. On the other side, the fuzzy annexation of points $\boldsymbol{x}$ into a cluster could be performed by a neural network as well.

The output of the mixture is very variegate as well, depending on the consequences of belonging to the fuzzy set C implied by the set of the fuzzy sets B_i representing the Horn clauses (10.6) emerging from fuzzy clustering. In particular:

through mixing coefficients,

- we may identify C with a single value. With reference to the architecture typologies shown in Fig. 10.3(a) and (d), the overall model computes on an input $\boldsymbol{x}$ a function like:

$$y = \sum_{i=1}^{k} \psi_i \mu_i(\boldsymbol{x}), \qquad (10.7)$$

where ψ_i is the y coordinate of the centroid of cluster B_i, whilst $\mu_i(\boldsymbol{x})$ is the input membership function to the same centroid, as computed by the left part of the architecture in Fig. 10.7. (see Fig. 10.8(a)). Rather,

mixing functions, or

- C may be associated with a function like the one introduced in the first section. In this case we have an output function

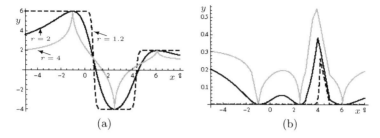

Fig. 10.8. Nonlinear input-output characteristics of the cluster-based models. Fuzzy sets B_is have been identified with method FCM. Dashed line: fuzzification coefficient $r = 1.2$; black plain line $r = 2$; gray plain line $r = 4.0$. (a) Output function: (10.7); point prototypes' coordinates: $(x_1, y_1) = (-1, 6)$, $(x_2, y_2) = (2.5, -4)$, $(x_3, y_3) = (6.1, 2)$. (b) output function (10.8); B_i own membership functions: triangular with vertex coordinates: $(x_1, y_1) = (-1, 1)$, $(x_2, y_2) = (2.5, 1)$, $(x_3, y_3) = (6.1, 1)$ and support $x_i \pm 3$; T-norm and S-norm: Gödel norms.

like (10.2) shown in Fig. 10.1(b). We identify the B_i with a function $g_i(x)$ in place of a single value ψ_i. Furthermore,

- C may represent another fuzzy set to be subsequently processed. Thus, with reference to the architecture typologies shown in Fig. 10.3(b) and (c), we may compute the membership function of C as in Fig. 10.8(b) by computing the following expression:

fuzzy norms.

$$y = \perp_{i=1}^{k} \top (\mu_{B_i}(x), \mu_i(x)), \quad (10.8)$$

where μ_{B_i} is a membership function embedded in fuzzy set B_i *per se*, and products and sums have been substituted by a suitable T-norm and S-norm operators.

In the next two sections, we will enhance the clustering models and the associated connectionist models. Their actual implementation will be the focus of the next chapter in which we will be concerned with the identification of the overall system.

10.4 The Clustering Part

Fuzzy clusters formulate a sound basis for constructing fuzzy models. By forming fuzzy clusters in input and output spaces, we span the fuzzy model on a collection of prototypes. More descriptively, these prototypes are regarded as a structural *skeleton* or a design *blueprint* of the model. While the model could be further refined, it is predominantly aimed at capturing the most essential, numerically dominant features of data by relying on this form of summarization. A single observation is related to these features through membership degrees. Thus each prototype leads to a fuzzy cluster. The more clusters we intend to capture, the more detailed the resulting blueprint becomes. It is important to note that clustering helps to manage the dimensionality problem which is usually a critical issue in rule-based modeling.

Locating data within stakes in the feature space through membership degrees.

In principle we are called to find clusters within an $(n + 1)$-dimensional space $\mathfrak{X}^n \times \mathfrak{Y}$. Its coordinates are constituted by the features to which they refer both the n antecedents and the consequent of the generic Horn clause of the questioned rule system. In order to describe the structures of the clusters, we must consider their projections on both the former n coordinates $(x_1, \ldots, x_n)$, accounting for the input space, and the latter, denote it y, referring to the final result of our computation, and state suitable connections between them. With relationships like (10.7) we often look for a direct mapping of points $\boldsymbol{x}$ of coordinates $(x_1, \ldots, x_n)$ into the fuzzy set projections $\{B_1, \ldots, B_k\}$ on the y axis. There are various ways of expressing these mappings, some of which we will discuss in the next section. With more complex relationships like (10.2) and (10.8) (but even with some instances of the former), we are generally pushed to elaborate on more complex models passing through an accurate evaluation of cluster projection on $\mathfrak{X}$ and a fine tuning of their implications on $\mathfrak{Y}$. This is the second option we will consider for stating rule systems. Its success very often depends on an effectively integrated run of the two parts of the architecture shown in Fig. 10.7.

At beginning data are framed into an $\mathfrak{X}^n \times \mathfrak{Y}$ space,

but which is the true feature space we are interested in?

10.4.1 The Cluster-Based Representation of the Input-Output Mappings

The ideal condition for employing (10.7) is depicted in Fig. 10.9. Its components comprise a monotonic trend (and we are not concerned about its specificity) between $\boldsymbol{x}$ and y and local concentrations around centroids $(\boldsymbol{v}_1, \boldsymbol{v}_2, \ldots, \boldsymbol{v}_k)$. The backbone of the input-output relationship is synthesized by coordinates' coupling in correspondence of the centroids. Then you identify in the most convenient way membership functions that are a function of $\boldsymbol{x}$ but relate to fuzzy clusters distributed around the y coordinates of the centroids. By definition they will be functions $\mu_j(\boldsymbol{x})$ of $\boldsymbol{x}$ indexed by the specific cluster j.

With a substantial monotony between input and output,

Given the mentioned monotonic dependency you may easily work in analogy with the density function of a function of a random variable. Hence you may either compute a membership

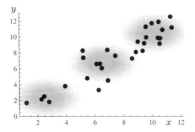

Fig. 10.9. Ideal condition for cluster-based representation through (10.7). Gray disks represent α-cuts of FCM output, computed with $k = 3$ centroids and $r = 2.5$.

function to clusters around the x coordinate of the centroids through the FCM algorithm according to (6.27) and transfer values to the corresponding y projections, or consider directly these projections and compute membership functions on that. In any case $\mu_j(x)$ denotes the attitude of x to be represented by a coordinate ψ_j in the y axis, so that a synthesis of these representations is given by (10.7). The reader familiar with radial basis function (RBF) neural networks [16] can easily recognize that the structure of (10.7) closely resembles the architecture of RBF neural networks, but the function computed through the membership mixture is far more flexible than the commonly encountered RBFs, such as Gaussian receptive fields.

we may easily transfer fuzziness from the former to the latter,

and take numbers as a synthesis.

With a specific reference to the FCM algorithm in Section 6.4.1 to identify the clusters, it is instructive to visualize the nonlinear characteristics of the model, as a function of the fuzzification factor r employed by the former. In analogy with Fig. 6.19, Fig. 10.8 illustrates input-output characteristics for the fixed values of the prototypes and varying values of the fuzzification factor. The commonly chosen value of r equal to 2.0 is also included. Undoubtedly, this design parameter exhibits a significant impact on the character of nonlinearity being developed. The values of r close to 1 produce a stepwise character of the mapping; we observe significant jumps located at the points where we switch between the individual clusters in input space. In this manner the impact coming from each rule is very clearly delineated. The typical value of the fuzzification coefficient set up to 2.0 yields a gradual transition between the rules and this shows up through smooth nonlinearities of the input-output relationships of the model. The increase in the values of r, as shown in Fig. 10.8, yields quite *spiky* characteristics: we quickly reach some modal values when moving close to the prototypes in the input space, while in the remaining cases the characteristics switch

Most depends from the prototypes.

between them in a relatively abrupt manner positioning close to the averages of the modes.

The relationship between centroids coordinates may be captured in a more efficient way than a simple table of input-output pairs. This delineates extended regression problems where the identification of the regression curve is either a follow-out or a premise of granulation, as will be discussed in the following two sections.

Granular regression

Starting from a set of observed (x, y) points, the streamline of this method is constituted by three steps:

1. Gathering points into fuzzy sets described by bell-shaped membership functions, that you read in terms of *relevance score profiles* of the universe of discourse elements. Namely, with each element we associate an S-norm of its membership degree to the above fuzzy sets. With the score distribution we give a different weight to the example contributions in the identification of the regression curve — a functionality that will be resumed by the concept of *context* in a next section.
2. Granulating the observed points by locating new bells centered in these points and having height equal to the above scores (see Fig. 10.10(a)), and building a relevance landscape through the norm of these bells.
3. Assessing an optimization procedure to identify the curve maximizing the line integral of this norm (see Fig. 10.10(b)). In this way, we look for a curve regressing phenomena (say, hyperpoints) rather than single examples; thus, the optimal regression curve should get closest to the points that are mostly representative of them.

In case of linear regression, with the second step we look for the line ℓ maximizing the sum of the integrals I_i of the curves obtained by intersecting the membership functions $\mu_i(x, y)$ with the plane which contains ℓ and is orthogonal to the plane $\mathfrak{X} \times \mathfrak{Y}$. If we refer to the points of ℓ through the equation $a + bx + y = 0$, i.e. ℓ has slope and intercept, respectively, equal to $-b$ and $-a$, the above integral will depend on the latter quantities, thus we write $I_i(a, b)$. Its optimization is particularly convenient if we express the bell membership function as a bidimensional Gaussian density function in the plane having as axes ℓ and any line orthogonal to it, say having coordinates ξ and ψ:

$$\mu_i(\xi, \psi) = h_i e^{-\pi h_i((\xi - \xi_i)^2 + (\psi - \psi_i)^2)}, \qquad (10.9)$$

where ξ_i and ψ_i are the analogous of x_i and y_i in the new space. The bells around our points look like in Fig 10.11, where their

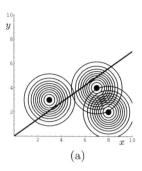
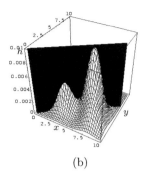

(a) (b)

Fig. 10.10. Fitting the granules' information with a line. (a) Relevance score contour lines; x independent variable, y dependent variable. (b) Crossing landscape with the regression line; h: relevance score.

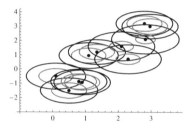

Fig. 10.11. Contours of the fuzzy membership to the granules (centered on bold points) at level 0.1 (light gray ellipses) and 0.005 (dark gray ellipses) suggested by FCM algorithm.

contours at level values 0.1 and 0.005 – squeezed because of the picture aspect ratio – are reported. In this case, the integral $I_i(a,b)$ corresponding to the i-th granule reads:

$$I_i(a,b) = h_i^{1/2} e^{-\pi h_i \frac{(bx_i+y_i+a)^2}{1+b^2}}. \tag{10.10}$$

Therefore, the optimal regression line has parameters:

$$(a^*, b^*) = \arg\max_{a,b} \sum_{i=1}^{m} h_i^{1/2} e^{-\pi h_i \frac{(bx_i+y_i+a)^2}{1+b^2}}. \tag{10.11}$$

since conditional membership functions are Gaussian as well.

Example 10.1. Consider the age adjusted mortality specifications (M) as a function of a demographic index (%NW, the non-white percentage) in the SMSA dataset. To gather the points into fuzzy sets we used FCM with $r = 2$ and $k = 3$. Fig. 10.12(a) shows the output of this procedure together with the emerging clusters (represented by α-cuts in order to visually enhance their fuzziness).

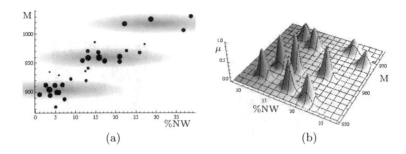

(a) (b)

Fig. 10.12. (a) Output of the scoring procedure with 3 clusters applied to the SMSA dataset. Gray disks: cluster α-cuts. Black circles have a radius proportional to the relevance of the points they refer to. (b) Bell membership functions obtained by applying (10.9) to the points gathered in the central α-cut.

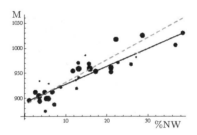

Fig. 10.13. Comparison between optimal regression line (10.11) (black line) and MLE line (dashed line).

The relevance of the point (x_i, y_i) is equal to the highest membership degrees of the point to any of the three clusters. Fig. 10.12(b) shows the bells membership functions corresponding to points gathered by the α-cut of central cluster where $\alpha = 0.8$. These images show a two-layer bell texture, the lower spreading information of the single questioned points, the higher locally linking data into a topological structure. We expect that the global trend of the examples goes linear. After a few thousand iterations of the gradient descent algorithm minimizing (10.10) we obtained the optimal regression line shown in Fig. 10.13.

Incremental granular models

We can take another, less commonly adopted principle of fuzzy modeling whose essence could be succinctly captured as follows:

Splitting bulk from minor details. Adopting a construct of a linear regression as a *first-principle global* model, adjust it through a series of local fuzzy rules which capture remaining and more localized nonlinearities of the system.

More schematically, we could articulate the essence of the resulting fuzzy model by stressing the existence of the two essential modeling structures that are combined together using the following symbolic relationship:

fuzzy model = linear regression + local granular models.

By endorsing this principle, we emphasize the tendency that in system modeling we always proceed with the simplest possible model (Occam's principle), assess its performance and afterwards complete a series of necessary refinements. The local granular models handling the residual part of the model are conveniently captured through some rules.

Let us proceed with some illustrative examples, shown in Fig. 10.14, that help underline and exemplify the above development principle. In the first case, Fig. 10.14(a), the data are predominantly governed by a linear relationship while there is only a certain cluster of points that are fairly dispersed within some region. In the second one, Fig. 10.14(b), the linearity is visible yet there are two localized clusters of data that contribute to the local nonlinear character of the relationship. In Fig. 10.14(c) there is a nonlinear function yet it exhibits quite dominant regions of linearity. This is quite noticeable when completing a linear approximation; the linear regression exhibits a pretty good match with a clear exception of the two very much compact regions. Within such regions, one could accommodate two rules that capture the experimental relationships present there. The nonlinearity and the character of data vary from case to case. In the first two examples, we note that the data are quite dispersed and the regions of departure from the otherwise linear relationship could be modeled in terms of some rules. In the third one the data are very concentrated and with no scattering yet the nonlinear nature of the relationship is predominantly visible.

Various kinds of granules around the bulk,

The fundamental scheme of the construction of the incremental granular model is covered as illustrated in Fig. 10.15. There are two essential phases: the development of the linear regression being followed by the construction of the local granular rule-based constructs that attempt to eliminate errors (residuals) produced by the regression part of the model. The first principle linear regression model comes in the standard form of $\ell_i = \boldsymbol{a}^T \boldsymbol{x}_i + b$, where the values of the coefficients of the regression plane, denoted here by b and $\boldsymbol{a}$, are determined through the methods dicussed in Section 3.2.1. The enhancement of the model at which the granular part comes into the play is based on

same general processing scheme,

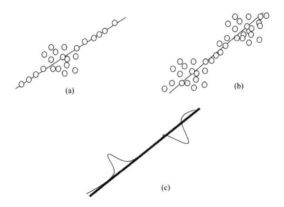

Fig. 10.14. Examples of nonlinear relationships and their modeling through a combination of linear models of global character and a collection of local rule-based models.

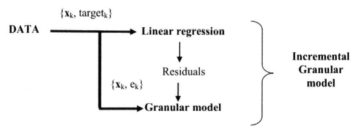

Fig. 10.15. A general flow of the development of the incremental granular models.

rooted on residuals w.r.t. backbone function. the transformed data $\{\boldsymbol{x}_i, e_i\}$, $i = 1, 2, \ldots, m$, where the residual part manifests through the expression $e_i = y_i - \ell_i$, which denotes the error of the linear model. In the sequel, those data pairs are used to develop the incremental and granular rule-based part of the model. Given the character of the data, this rule-based augmentation of the model associates input data with the error produced by the linear regression model in the form of the rules "if input – then error".

Example 10.2. The one-dimensional spiky function $s(x)$ used in this experiment is a linear relationship augmented by two Gaussian membership functions $g_i(x)$, for $i = 1, 2$, described by their modal values ν_i and spreads σ_i (see Appendix C). This leads to the overall expression of $s(x)$ to be in the form:

$$s(x) = \begin{cases} \max(x, g_1(x)) & \text{if } 0 \leq x \leq 1 \\ \min(x, -g_2(x) + 2) & \text{if } 1 < x \leq 2, \end{cases} \quad (10.12)$$

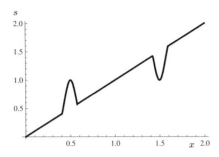

Fig. 10.16. Spiky function (10.12) with $\nu_1 = 0.5$, $\nu_2 = 1.5$, and $\sigma_1 = \sigma_2 = 0.1$.

with $\nu_1 = 0.5$, $\nu_2 = 1.5$ and $\sigma_1 = \sigma_2 = 0.1$, as shown in Fig. 10.16. Each training and test dataset consists of 100 pairs of input-output data. As could have been anticipated, the linear regression is suitable for some quite large regions of the input variable but becomes quite insufficient in the regions where these two spikes are positioned. As a result of this evident departure from the linear dependency, the linear regression produces a high approximation error of 0.154 ± 0.014 and 0.160 ± 0.008 for the training and testing set, respectively. The granular modification of the model was realized by an augmented FCM procedure that we will discuss in the next section. We will come back to this example therein.

10.4.2 The Context-Based Clustering in the Development of Granular Models

In absence of a substantial monotonicity between x and y we are pushed to discover more complex structures between these coordinates. Consider the dataset positioned in the $(n+1)$-dimensional space by vectors $z_i = (x_i, y_i)$ and cluster them with the use of FCM or any other fuzzy clustering technique which leads to the collection of k clusters prototypes $v_1, v_2, \ldots, v_k$, and the partition matrix $U = [u_{ci}], c = 1, 2, \ldots, k$ in the product space $\mathfrak{X}^n \times \mathfrak{Y}$. Fig. 10.17(a) shows an example where we identify four clusters in the dataset. We preliminary discover that there are different clusters having either a same projection on x subspace but different projection on y axis, or different projection on the former and same projection on the latter. Both require a deeper structure to disambiguate the y dependence on x. A main notion to deal with these and intermediate situations is represented by the *context-based* clustering that we may implement in various ways.

With more entangled relationships more structure in the data must be investigated.

364   Granular Constructs

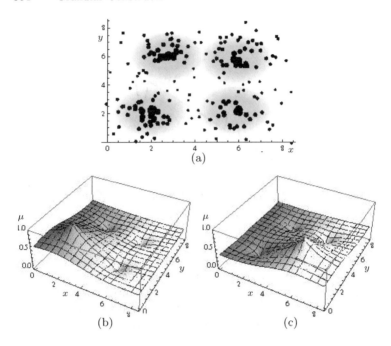

Fig. 10.17. Clustering data in the product space with the use of FCM with $k = 4$. (a) Clusters' representation, the same notation as in Fig. 10.12. (b) and (c) Examples of membership functions obtained.

Context-based clustering is naturally geared towards dealing with direction-sensitive clustering. The context variables are those being the output variables used in the modeling problem. With respect to the general rule:

if x is A_1 and A_2 and ... and A_n **then** y is B,   (10.13)

The output entails input contexts.

y represents a context variable in respect to the x space to be clustered. Given the context, we focus the pursuits of fuzzy clustering on the pertinent portion of data in the input space and reveal a conditional structure therein. By changing the context we continue the search by focusing on some other parts of the data. In essence, the result produced in this manner can be visualized as a web of information granules developed conditionally upon the assumed collection of the contexts. Hence the directional aspects of the model we want to develop on the basis of the information granules become evident. The design of contexts is quite intuitive. The goal is to identify meaningful clusters in the x subspace first, and then recover y from some meaningful relationships between those clusters and meaningful clusters on y axis. The strategy is to use y information from the early stage

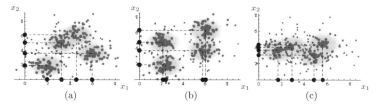

Fig. 10.18. Projection of prototypes on the coordinates of the input space.

of this process, by managing it as a hidden $(n+1)$-th feature in the clustering process implemented on the n-dimensional space where x points are located. In the following sections we graduate the y_i role from a simple additional coordinate on which to run the FCM to a conditional variable weighting the u_{ci} contribution in the FCM cost function.

Images of orthogonal projections

With pairs $\{(x_1, y_1), \ldots, (x_m, y_m)\}$ in the $(n+1)$-dimensional space we may decide of either clustering the first n components and then relate them to the y axis – a strategy considered in Section 10.4.1, or clustering all components and projecting the clustered points on the first n coordinates. Fig. 10.18 shows variants of Fig. 10.17 with x split in x_1 and x_2, y hidden by the projection, and different locations of the 4 clusters. Shown are various distributions of the projections: (a) well delineated clusters in each subspace, (b) substantial overlap between some prototypes observed for the projection on x_1, (c) all prototypes projected on x_2 are almost indistinguishable. The projection scheme brings insight into the nature of the data and resulting rules. In Fig. 10.18(a) we can envision a situation where the projected prototypes are well and visibly distributed. It might be cases, as illustrated in Fig. 10.18(b) that some projections are almost the same. This suggests that we may consider collapsing of the fuzzy sets as being semantically the same. Fig. 10.18(c) shows another example where with the collapse of all prototypes we may consider dropping the corresponding variable from the rules. Through this post-processing of the clustering results in which some fuzzy sets could be combined, the resulting rules may involve different numbers of fuzzy sets. Similarly, we may expect a situation of dimensionality reduction in which some variables could be dropped from the rules.

From simple inspection

many possibilities around really separated clusters or not,

The most critical observation versus this technique concerns a distinction between relational aspects of clustering and directional features of models. By their nature, unless properly endowed, clustering looks at multivariable data as relational constructs so the final product of cluster analysis results in a collection of clusters as concise descriptors of data where each variable is treated in the same manner irrespectively where it is positioned as a modeling entity. For instance the same pictures in Fig. 10.18 could refer to (x_1, y) axes. This stands in sharp contrast with what we observe in system modeling. The role of the variable is critical as most practical models are directional constructs, viz. they represent a certain mapping from independent to dependent variables. In the previous approach we have ignored the directionality aspect. Thus we just concatenate with z the input with output signals and carry our clustering in such augmented feature space. By doing that we have concentrated on the relational aspects of data completely ignoring the possible mapping component that is of interest. To alleviate the problem, we may like to emphasize a role of the output variables by assigning to them higher values of weights. An easy realization of the concept would be to admit that the distance function has to be computed differently depending upon the coordinates of z by using a certain positive weight factor γ. For instance, the Euclidean distance between z_i and z_1 would read as $l^2(x_i - x_1) + \gamma l^2(y_i - y_1)$ (see Appendix A.2), with $\gamma > 0$. The higher the value of γ, the more attention is focused on the distance between the output variables. As usually the dimensionality of the input space is far higher than the output one, so that the value of γ needs to be properly selected to reach a suitable balance. Even though the approach might look sound, the choice of the weight factor becomes a matter of intensive experimentation.

Conditional clustering

This clustering is a variant of the FCM that is realized for individual contexts $W_1, W_2, \ldots, W_q$. The design is quite intuitive. First, contexts are fuzzy sets whose semantics is well defined. We may use terms such as *low*, *medium* and *large* output. In own turn, within a context we can choose fuzzy sets so that they reflect the nature of the problem and our perception of it. For instance, if for some reason we are interested in modeling a phenomenon of a *slow* traffic on a highway, we would define a number of fuzzy sets of context focused on low values of speed (see Fig. 10.19(a)). To model highway traffic with focus on *high* speed we would be inclined to locate a number of fuzzy sets at the high end of the output space (see Fig. 10.19(b)). The customization of the model by identifying its major focus is thus feasible through setting the clustering activities in a suitable context.

In our constructs contexts encompass the $(n+1)$-th feature y. Let us consider a certain fixed context W_j described by some

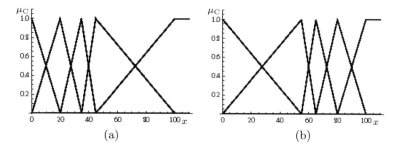

Fig. 10.19. Examples of fuzzy sets within contexts reflecting a certain focus of the intended model: (a) *low* speed traffic, (b) *high* speed traffic.

membership function. A point z_i with projection y_i in the output space is associated with the corresponding membership value of $W_j, w_{ji} = \mu_{W_j}(y_i)$. We introduce a family of partition matrices induced by the j-th context and denote it by $U(W_j)$:

$$U(W_j) = \left\{ u_{ci} \in [0,1] : \sum_{c=1}^{k} u_{ci} = w_{ji} \; \forall i \text{ and } 0 < \sum_{i=1}^{m} u_{ci} < m \; \forall c \right\}, \quad (10.14)$$

where we omit affecting the membership degrees by the additional index j of the context within which it has been defined. What differentiate this family from a context-free one is the sum of the membership degrees of any point z_i over the cluster. It is equal to 1 in absence of context and w_{ji} when context j is defined.

Minimizing the FCM cost with this new constraint moves the fuzziness of the structure one level ahead, meanwhile recalling some probabilistic features in the original level. Namely, recalling (6.22) and (6.23) and noting that asking $\sum_{c=1}^{k} u_{ci} = w_{ji}$ introduces a scale factor depending on i in the context j, we see that:

The rest as usual,

$$C = \sum_{c=1}^{k} \sum_{i=1}^{m} w_{ji} u_{ci}^r \nu(\boldsymbol{x}_i - \boldsymbol{v}_c) = \sum_{i=1}^{m} w_{ji} \sum_{c=1}^{k} u_{ci}^r \nu(\boldsymbol{x}_i - \boldsymbol{v}_c) =$$
$$\sum_{i=1}^{m} \sum_{c=1}^{k} f_{\boldsymbol{X}/c}(\boldsymbol{x}_i) f_C(c) \nu(\boldsymbol{x}_i - \boldsymbol{v}_c) = \widetilde{H}[\boldsymbol{X}|C]. \quad (10.15)$$

Thus, with some caveats about r that is necessarily different from 1, (10.15) denotes an optimization target to be pursued in a probabilistic framework, where y offers a unifying perspective of the whole dataset. What remains definitely fuzzy is the context membership function W_j requiring fuzzy sets instruments to be merged with other contexts'. More specifically, the partition matrix is computed as follows:

Table 10.2. RMSE values (mean ± standard deviation) - training data.

		No. of Contexts q			
		3	4	5	6
No. of clusters k per context	2	0.148 ± 0.013	0.142 ± 0.018	0.136 ± 0.005	0.106 ± 0.006
	3	0.141 ± 0.012	0.131 ± 0.008	0.106 ± 0.008	0.187 ± 0.006
	4	0.143 ± 0.006	0.124 ± 0.007	0.095 ± 0.007	0.078 ± 0.005
	5	0.131 ± 0.012	0.111 ± 0.007	0.077 ± 0.008	0.073 ± 0.006
	6	0.126 ± 0.011	0.105 ± 0.005	0.072 ± 0.007	0.061 ± 0.007

Table 10.3. RMSE values (mean and standard deviation) - testing data.

		No. of Contexts q			
		3	4	5	6
No. of clusters k per context	2	0.142 ± 0.016	0.139 ± 0.028	0.139 ± 0.012	0.114 ± 0.007
	3	0.131 ± 0.007	0.125 ± 0.017	0.115 ± 0.009	0.096 ± 0.009
	4	0.129 ± 0.014	0.126 ± 0.014	0.101 ± 0.009	0.078 ± 0.012
	5	0.123 ± 0.005	0.119 ± 0.016	0.097 ± 0.008	0.082 ± 0.010
	6	0.119 ± 0.016	0.114 + 0.015	0.082 ± 0.011	0.069 ± 0.007

$$u_{ci} = \frac{w_{ji}}{\sum_{d=1}^{k} \left(\frac{\nu(\boldsymbol{x}_i - \boldsymbol{v}_c)}{\nu(\boldsymbol{x}_i - \boldsymbol{v}_d)} \right)^{\frac{1}{r-1}}}, \quad (10.16)$$

where j ranges over the clusters and i over the data points. Let us emphasize here that the values of u_{ci} pertain here to the partition matrix induced by the j-th context. The prototypes $\boldsymbol{v}_c$, $c = 1, 2, \ldots, k$, are calculated as usual in the form of the weighted average:

$$\boldsymbol{v}_c = \frac{\sum_{i=1}^{m} u_{ci}^r \boldsymbol{x}_i}{\sum_{i=1}^{m} u_{ci}^r}. \quad (10.17)$$

but with the further problem of synthesizing output sets.

Once we have obtained $U(\mathbf{W}_j)$ for each context j we must merge them in a suitable way in order to obtain y for a given $\boldsymbol{x}$. The simplest way is to consider so many replicas of rules as many contexts are considered. In the following example we apply this strategy with the incremental granular model introduced in Section 10.4.1.

Example 10.3. Continuing Example 10.2, the augmented granular modification of the model was realized by experimenting with the two essential parameters controlling the granularity of the construct in the input and output space, that is number of context q and number of clusters per context k. The corresponding results are summarized in Table 10.2 and 10.3 in terms of

Table 10.4. Optimal values of the fuzzification coefficient r for selected number of contexts and clusters.

		No. of Contexts q			
		3	4	5	6
No. of clusters k per context	2	3.5	4.0	3.8	3.1
	3	3.2	3.9	3.5	3.1
	4	3.0	2.7	2.6	2.6
	5	3.1	2.8	2.2	2.4
	6	3.0	2.5	2.2	2.0

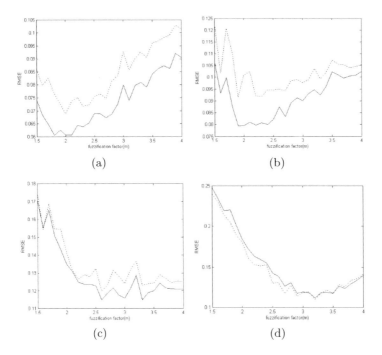

Fig. 10.20. Performance index (RMSE) versus values of the fuzzification coefficient r for some selected combinations of q and k: (a) $(k,q) = (6,6)$, (b) $(k,q) = (5,5)$, (c) $(k,q) = (4,5)$, (d) $(k,q) = (3,6)$, solid line: training data, dotted line: testing data.

mean value ±3 standard deviations over 10 trials. All of them are reported for the optimal values of the fuzzification coefficient as listed in Table 10.4, viz. its values for which the root mean squared error (RMSE) attained its minimum,

$$\text{RMSE} = \sqrt{\frac{1}{m} \sum_{i=1}^{m} (y_i - \psi_i)^2} \qquad (10.18)$$

370   Granular Constructs

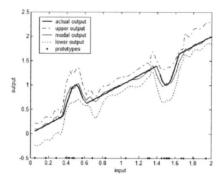

Fig. 10.21. Modeling results for $k = 5$ and $q = 5$ ($r = 2.2$); shown is also a distribution of the prototypes in the input space. Note that most of them are located around the spikes which is quite legitimate as they tend to capture the nonlinearities existing in these regions.

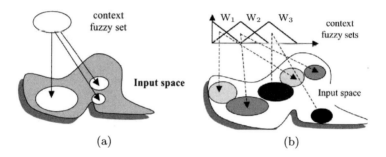

Fig. 10.22. A blueprint of a granular model with contexts: (a) a set of three fuzzy sets within a specific context; (b) a detailed view at the model in case of three contexts and two clusters per context.

where ψ_i is the modal value of Y produced for input x_i. As Fig. 10.20 illustrates, the increase in the number of contexts and clusters leads to higher optimal values of the fuzzification factor. The optimal results along with the visualization of the prototypes when $k = 5$ and $q = 5$ are displayed in Fig. 10.21. The plot shows the modal values as well as the lower and upper bound of the resulting fuzzy number produced by the incremental model. Here we have used the optimal value $r = 2.2$ of the fuzzification factor.

We can move on with some further refinements of context fuzzy sets and, if required, introduce a larger number of granular categories. Their relevance could be assessed with regard to the underlying experimental evidence [21]. To assure full coverage of the output space, it is advisable that fuzzy sets of context form a fuzzy partition. Obviously, we can carry out clustering of data in the output space and arrive at some membership functions being

Conditional clustering is a technique open to various improvements.

generated in a fully automated fashion. This option is particularly attractive in case of many output variables to be treated together where the manual definition of the context fuzzy relations could be too tedious or even impractical. The blueprint of the model, as shown in Fig. 10.22, has to be further formalized to explicitly capture the mapping between the information granules.

10.5 The Connectionist Part

We are left with a set of fuzzy clusters alternatively representing either the information granule each questioned point is endowed with, or expressly the same information but related to the output of a Horn clause on the point. In both cases we must merge the information connected to the various granules in order to provide an output of our information processing in the alternative forms sketched in Fig. 10.3. At the basic level we will use norms as defined in Appendix A.3 to combine the related membership functions, mainly Gödel norms. An efficient way of organizing their composition is through an extended neural network paradigm based on an extended definition of the functions computed by the single neurons, now called *granular neurons*.

Use a neural network to merge granules,

10.5.1 Granular Neuron

This structure consists of a neuron with granular connections and fuzzy norm operators. Hence its input-output relationship is the following:

$$Y = h(u_1, u_2, \ldots, u_k, W_1, W_2, \ldots, W_k) = \perp_{i=1}^{k}(W_i \top u_i), \quad (10.19)$$

with $W_1, W_2, \ldots, W_k$ denoting granular connections weights, hence fuzzy sets, see Fig. 10.23, and $u_1, \ldots, u_k$ numeric inputs,

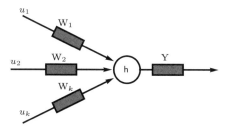

Fig. 10.23. Computational model of a granular neuron; u_i: input signal; W_i: fuzzy weights; Y: fuzzy output; h: fuzzy norm activating the output.

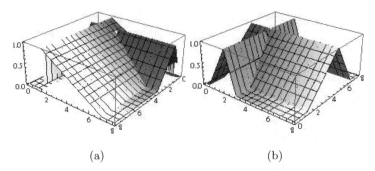

Fig. 10.24. The output of a granular neuron when connection weights are triangular fuzzy numbers and activation function is (10.19); (a) $W_{1\{0.3,0.5,3.0\}}, W_{2\{1.4,1.5,7.0\}}$; (b) $W_{1\{0.3,2.0,3.0\}}, W_{2\{1.4,5.0,7.0\}}$.

generally confined to the unit interval (because of their role of membership functions).

endowed with specially featured neurons

With the notation of Chapter 4 they synthesize second order fuzzy sets composed of first order fuzzy sets W_i and modifiers u_i. In place of maintaining the spread representation of these fuzzy sets as in Fig. 4.11 here we need to compute a unique first order fuzzy set as an operational representative of them. We have ex-

composing fuzzy sets

tended the norm's notation so that the norm between two fuzzy sets is a fuzzy set described in each point by the norm of the membership function of the original fuzzy sets, and analogously for the norm between a fuzzy set and a single value. Apart from degenerate cases, the output of a granular neuron is an information granule as well. If we confine ourselves to triangular fuzzy sets for the connections, we note that the multiplication of W_i by a positive constant scales the fuzzy number, yet retains the piecewise character of the membership function. The plot of the output of the neuron for u_1 and u_2 defined as above is shown in Fig. 10.24.

As for the activation function, the T-conorm operator gives rise to nonlinear functions with features depending on the specific norm adopted. In principle you have no special commitment on the form of the activation function, apart from those deserved to the norms (see Appendix A.3). For instance the dual of (10.19) is:

$$Y = \top_{i=1}^{k}(W_i \bot u_i), \qquad (10.20)$$

with a variety of norm operators

where the activation function looks like in Fig. 10.25. Also this activation function is widely used. Actually, (10.19) translates in fuzzy terms the net input operator (8.16), while (10.20) reflects a Horn clause in its standard form (10.5). In any case the granular neuron may be interpreted as the wiring of a rule having the information carried by the input connections as antecedents

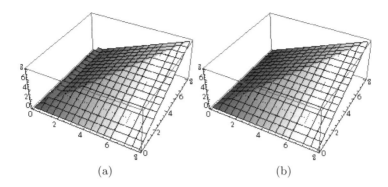

Fig. 10.25. Same as in Fig. 10.24 yet with the activation function (10.20).

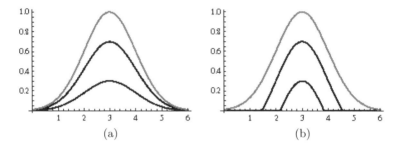

Fig. 10.26. An example of the fuzzy set specification. Gray curve: original Gaussian membership function. Dark curves: their specification through connection weights $w_1 = 0.7$ and $w_2 = 0.3$ and T-norm implemented through (a) Gödel and (b) Lukasiewicz T-norm.

and the neuron output as the consequent. The norm of W_i with u_i is the way of moulding the former. Thus, $W_i \top u_i$ is more specific than the original fuzzy set W_i. Lower values of u_i make the membership function of the associated fuzzy set closer to 0. In the limit case when the connection is equal to 0, the associated fuzzy set becomes *masked*. Practically, in this way the antecedent of the rule has been dropped. Hence by carrying out the same absorption process for each condition in the rule we arrive at the network whose connections have been eliminated. The plots of the modified fuzzy sets for selected values of the connection are shown in Fig. 10.26. Analogous considerations are carried out for $W_i \bot u_i$ as in (10.20), where $W_i \bot u_i$ goes close to 1 when u_i does the same. This masks the contribution of the i-th input connection when all contributions are gathered by a T-norm, as in (10.20). In many applications we will set u_is to 1. In many others we will modify the output of the granular neuron

with the general goal of moulding membership functions.

in order to obtain a single value. The granular neuron exhibits several interesting properties that generalize the characteristics of conventional (numeric) neurons.

10.6 The Neural Architecture

In this chapter, we will adopt the standard architecture shown in Fig. 10.27 with the following convention. Each neuron is in principle a granular neuron defined through (10.19) and represented in Fig. 10.23. Hence neurons implement norm operations while fuzzy sets are located in the connections. Of the three layers of the architecture, the former is devoted to receive inputs, that in general are specific numbers. Thus the functionality of its neurons consists just in giving a name to the signals. The neurons in the second layer are devoted to combine fuzzy sets into rules, while the unique neuron in the third layer synthesizes rules into a single, possibly fuzzy, result. Connections from first to second layer identify the fuzzy sets involved in the rule they lead to. The signal they pipe may be either a signal value x_i or a membership degree $\mu_{W_i}(x_i)$ of the input signal to the fuzzy set W_i embedded into the connection, or a product of the latter with a modifier u_i when we want to take into account the context of the input signal. Connections from second to third layer identify membership degrees of the rules to the final result. They convey either entire fuzzy sets in the form of α-cuts determined by the input signals or simple coefficients, or the product of both when we want to underline the relevance of the single rule in the final result. The output of the third layer neuron is either a fuzzy set or a number.

By definition, having a set of neurons you may combine them in any way, provided you succeed in reproducing the goal input-output function. This is true also with functionality we expect by

A three-layer input, rules, output architecture, with fuzzy sets in the role of connection weights,

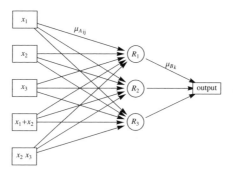

Fig. 10.27. Standard neuro-fuzzy architecture.

the network of granular neurons of suitably mixing the clusters implementing rules as explained in Section 10.2. From an operational point of view all four schemes in Fig. 10.3 may be implemented through a neural architecture, given the high flexibility of the granular neuron definition. With the perspective of training the architecture, however, both interfaces and processing units must be modified from their original form, in order to comply with the learning mechanism. Apart from disingenuous learning rules, this mechanism indeed bases on the continuity of the goal function in respect to the parameters to be learnt. Hence norms and conorms are generally replaced by suitable linear combinations or functions of them, idem for interfaces. Normally these architectures aim to cover the whole trip from input signals to output signals, thus implementing the variant (a) in Fig. 10.3, whereas the remaining variants may represent a relaxed task. In the following we will discuss two extremal architectures: a former postponing the synthesis of fuzzy sets into output signal to the very last module of the system, the latter moving from fuzzy sets to representative values as soon as possible. Of course, a plenty of intermediate strategies may be conceivable and in fact are implemented.

and smooth norm operator to manage their derivatives.

10.6.1 Defuzzification at the Last

The archetype of the former strategy is the Mamdani architecture [14]. It simply implements rules through a T-norm of the antecedents with the consequent, their union through an S-norm of the consequents, getting a final fuzzy set as a true result. Then a variety of methods for synthesizing the set in a single value – an operation that is called *defuzzifying* – are available. Using Gödel or product norms the typical picture illustrating the method is included in Fig. 10.28. Using the former norm, see picture (a), the T-norms between antecedents and, successively, with the consequent is implemented, according to Section 10.2.2, giving the output fuzzy set a membership function $\mu_R(y|x)$, returning for each point y the minimum of the membership degree $\mu_{A_i}(x)$ of x to the input fuzzy sets A_is and the membership degree $\mu_B(y)$ of y to the consequent B. This corresponds to the α-cut in the right part of the pictures in Fig. 10.28(a). *Vice versa*, the union of the rules implies a new fuzzy set in $\mathfrak{Y}$ whose membership function is the maximum in each point of the single rules membership functions. The analogous procedure with product norm is sketched in Fig. 10.28(b). There is no problem for implementing the norms through granular neurons (see Fig. 10.29). The general statement is as follows: fuzzy sets correspond to fuzzy connections, numeric values correspond to the piped signal. Then you may

Use T-norm of antecedents and consequent to obtain a conditional consequent membership function,

merge rules with S-norms.

376   Granular Constructs

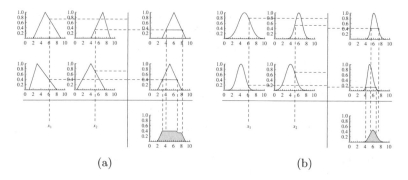

Fig. 10.28. The Mamdani fuzzy inference system using: (a) Gödel norm and (b) product norm.

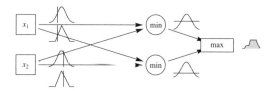

Fig. 10.29. Neural network architecture for a Mamdani fuzzy system with Gödel norm.

Map computations on neural network at your glance, formally relate computations to these items in different evoking ways. On one side you evaluate the membership degree of the signal to the antecedent by evaluating the membership function of the corresponding connection exactly in the value of the signal (the vertical bars in Fig. 10.29). In turn this reads as the involvement degree of the antecedent to the connected rule. Analogously, the connection of the rule neuron to the synthesis neuron is endowed with the α-cuts in output to the single rules, in turn denoting the contributions of the rules to the final fuzzy set (the horizontal bars in Fig. 10.29). Thus you may assume that membership degree is the weight of the input to rule connection.

possibly introducing further modifiers. As a matter of fact, we have in general a surplus of variables. Indeed we may use numeric coefficients u, ω as modifiers to affect either granules or rules through a relevance weight, so that the final rule base may read:

if $u_{11}(x_1$ is $A_{11})$ and ... and $u_{1n}(x_n$ is $A_{1n})$ then $\omega_1(y$ is $B_1)$,

if $u_{21}(x_1$ is $A_{21})$ and ... and $u_{2n}(x_n$ is $A_{2n})$ then $\omega_2(y$ is $B_2)$,

$$\vdots \qquad\qquad \vdots \qquad\qquad \vdots$$

if $u_{k1}(x_1$ is $A_{k1})$ and ... and $u_{kn}(x_n$ is $A_{kn})$ then $\omega_k(y$ is $B_k)$.

(10.21)

Table 10.5. Most used defuzzifiers of a fuzzy set $C = \bot(B_1, \ldots, B_k)$, where $\widetilde{y}_{\max} = \arg\max_{\widetilde{y} \in B_i}\{\mu_{B_i}(\widetilde{y}), i = 1, \ldots, k\}$, ψ_i is the centroid of the fuzzy set B_i, and $\overline{\widetilde{y}_{\max}}$ is the empirical mean value of the set $\{\widetilde{y}_{\max}\}$.

Defuzzifier	Description	Formula
max	maximum	$y = \arg\max_{\widetilde{y} \in C}\{\mu_C(\widetilde{y})\}$
mom	mean of maximum	$y = \overline{\widetilde{y}_{\max}}$
coa	centroid of area	$y = \dfrac{\int_{\widetilde{y} \in C} \widetilde{y}\mu_C(\widetilde{y})d\widetilde{y}}{\int_{\widetilde{y} \in C} \mu_C(\widetilde{y})d\widetilde{y}}$
height	mean centroid	$y = \dfrac{\sum_{i=1}^{k} \psi_i \mu_{B_i}(\psi_i)}{\sum_{i=1}^{k} \mu_{B_i}(\psi_i)}$

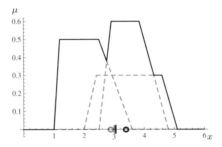

Fig. 10.30. Defuzzifiers applied to the Gödel S-norm of three α-cuts of level 0.5, 0.3 and 0.6 of fuzzy sets $A_1 = \{1, 1.4, 3.6\}$, $A_2 = \{2, 3.4, 5.1\}$ and $A_3 = \{2.5, 3.2, 4.8\}$, respectively from left to right. Black circle: max = 3.38; gray circle: coa = 2.89; black line: height = 3.05; and gray line: mom = 3.01.

The output of our rule system is a fuzzy set C like the ones shown in Fig. 10.28, whose analytical form in the case of Gödel norms reads:

$$C = \max_{i=1,\ldots,k}\left\{\omega_i B_{i\left(\min_{j=1,\ldots,n}\{u_{ij}\mu_{A_{ij}}(x_j)\}\right)}\right\}, \qquad (10.22)$$

where with A_α we denote the α-cut of A. In the common case that we need a single value as an answer to the values put in input to the system, we must draw a synthesis from the fuzzy set. This looks like an *inverse* operation of the process through which we assign input signals to fuzzy sets which is called *defuzzification*. We have no significant hints from the theory on how to perform it. Rather we are often biased by analogous practices with p.d.f. in the probabilistic framework. A list of commonly encountered defuzzifiers is reported in Table 10.5.

> Deduce a central value from the resulting fuzzy set.

378   Granular Constructs

Namely, we may directly point to the modal value of C membership function with max. Rather, if we have many modal points, as in Fig. 10.28(a), then we may mediate their values with mom. Same cases occur if we identify the output of the rule system either with the center of mass of the output fuzzy set C computed with coa, or by mediating through height the B_i centroids whenever the latter operation is computationally simpler than the former. Fig. 10.30 emphasizes the different role played by the above defuzzifiers when applied to a simple fuzzy set produced by the Gödel S-norm of three α-cuts of triangular membership functions.

10.6.2 Use Numbers as Soon as Possible

Antecedent as before, but conditional consequent is a number.

The archetype now is the Sugeno architecture [22] (see Figs 10.31 and 10.32). The first layer (from signal to rules) is the same as in Mamdani's. The weights of the second layer are not yet fuzzy but numeric, while an additional input is suppled in terms of *rule activation function*. With these terms also the output computed by the synthesis neuron is numeric, consisting of a weighted sum. You may think of the system as a mixture of experts, where each rule, in the role of an expert, outputs a single value weighted by a rule relevance, represented by the membership of the input to the antecedents' set of the rule. This value is variously merged with the analogous responses of the other experts through a function outputting a single value in turn. In detail, denoting with (x, y) an input-output couple and with w_i the relevance of the i-th rule computed as:

$$w_i = \mathsf{T}_{j=1}^n \mu_{A_{ij}}(x_j), \qquad (10.23)$$

we obtain y as:

$$y = \sum_{i=1}^k \overline{w}_i \mathsf{h}_i(x) = \frac{\sum_{i=1}^k w_i \mathsf{h}_i(x)}{\sum_{i=1}^k w_i}, \qquad (10.24)$$

The whole boils around a weighted sum.

where h_i defines the activation function of the i-rule, whose shape and arguments depend on the chosen model. It may be simply a costant value $\mathsf{h}_i = c_i$ in case of order-0 Takagi-Sugeno model, or a linear function of the input:

$$\mathsf{h}_i(x) = x \cdot a_i + b_i = \sum_{j=1}^n x_j a_{ij} + b_i, \qquad (10.25)$$

You may compute the conditional consequent either as a function of input, or

for suitable a_i and b_i, defining a Sugeno model of first order, where the order of the model is related to the degree of the activation function.

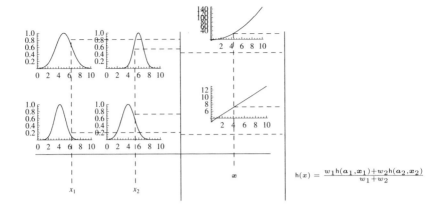

Fig. 10.31. The Sugeno fuzzy inference system.

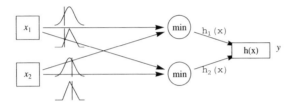

Fig. 10.32. The Sugeno neural network architecture.

An example of intermediate model is represented by the Tsukamoto model [23] (see Fig. 10.33), where the rule activation function h_i does not depend directly on the input, rather on the rule relevance w_i computed as in (10.23): of its membership degrees to antecedents.

$$z = h_i(w_i) = h_i^{-1}(w_i), \qquad (10.26)$$

where h_i is a monotonically increasing function of z. Though in principle any functions satisfying the above monotone requirement could be used, it is common practice to choose as h_i a sigmoid-like function as:

$$h_i(z) = \frac{1}{1 + e^{a_i z + b_i}}, \qquad (10.27)$$

or any piecewise linear approximation of the latter, which despite simplifying computations has the drawback of not being differentiable on the overall domain.

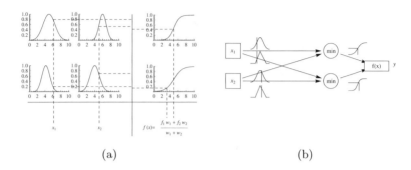

Fig. 10.33. The Tsukamoto fuzzy inference system and its neural network companion.

10.7 Conclusions

In the economy of the research monograph, last chapters combine concepts and methods cumulated in the previous parts. This does not imply they cover more completely the universe of the applications or their more complex instances. In this chapter, in particular, we deal with the common balance between information we draw from both source data and their usage. Even though the general understatement of this book is that you have always in mind an application when you analyze data, where analytical tools are strictly connected to the goal you pursue with their processing, it is undoubted that the goal broadness changes from one method to another. Thus, with clustering methods supporting the left part of the architecture in Fig. 10.7 you have in mind the general class of functions you want to compute with the whole procedure. Rather, with the right part, you focus on the specificities of the function that produces the sampled output you have available. Thus you expect a preliminary insight from the clustering that you will refine on a specific training set with the connectionist architecture in the right part. We may note the fuzziness of the frontier between the two functionalities and architecture features as well. Possibly you go back with the connectionist procedure up to the partition function learnt with FCM in order to refine it. On the opposite direction, you may determine the final output directly with supervised variants of fuzzy clustering algorithms.

All this falls in the province of inductive learning, whereas the product of deductive inference, hence the available symbolic knowledge, is involved in the design of the architecture layout, hence of the gross structure underlying data. From a logical perspective it materializes in terms of fuzzy rules systems as a granular counterpart of the DNF representation of Boolean reasoning in crispy scenarios. They actually represent an immediate way of formalizing our experience on the questioned decisional instance, moving – but not *removing* – all our uncertainty and doubts into the fuzzy characterization of the boiling up variables. They actually will represent the backbone of the chapter avoiding the neuro-fuzzy systems to become a *pout pourry* of algorithms in place of a well meditated synthesis of methods.

Another coupling of symbolic and subsymbolic tools resides in the possibly hybrid feature of the layout, so as for granular regression and incremental models, but also for the mixing function of Sugeno architecture.

This chapter is rather long, given the high number of wedding methodologies between the dichotomies we are involved with. Architectures in Fig. 10.3 denote that they are at the root of many real world instances.

10.8 Exercises

1. Suggest a practical application for each of the four fundamental modes of the use of fuzzy sets described in Fig. 10.3.
2. Describe three suitable families of fuzzy sets $\{A_i\}$, $\{B_j\}$ and $\{C_k\}$ modeling, respectively, weather conditions, speed and safe drive. Use these families in order to write a two-dimensional relationship table as in Table 10.1.
3. Reconsider the tabular model found as a solution of the previous exercise, and rewrite it as a rule-based system.
4. Elaborate on the scatterplot of Guinea pig dataset in order to alternatively fix an incremental granule model and a granular regression model.
5. With Cox dataset we have four blood measurements per individual. Argue from scatterplots if some of these measurements result non influent for the carriers-normal discrimination.
6. Suggest a fuzzy model of decision-making concerning a purchase of a house. In particular, consider the following components that are essential to the model:
    a) variables and their possible granulation realized in the form of some fuzzy sets;
    b) optimization criteria that could be viewed meaningful in the problem formulation;
    c) type of the fuzzy model.

    Justify your choice.
7. Transform the decision tree in Fig. 10.34 into a collection of rules. The information granules of the attributes are given

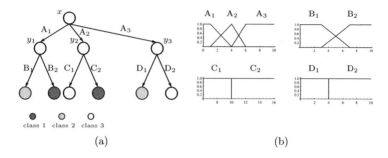

Fig. 10.34. Decision tree (a) and information granules of the attributes (b) referred to Exercise 7.

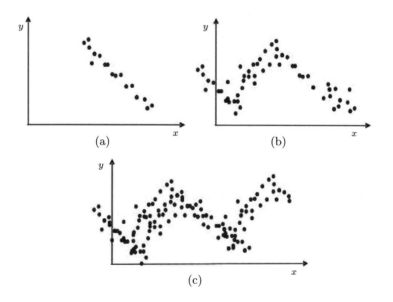

Fig. 10.35. Experimental data referred to Exercise 9.

in the form of fuzzy sets or intervals as illustrated in the picture. What would be the result (class membership) for the input $x = 1.5$ and $y = 2.0$?

8. Continuing Exercise 4 in Chapter 7 on SMSA dataset, now consider the city of the inviduals and attach it to the records of the most significant components found therein. Attribute a weight $w_{ji} = F_\Pi(\pi_j)$ to the i-th pattern in respect to the j-th city, where F_Π is the empirical c.d.f. of the city populations and π_j is the population of the j-th city. With these weights compute a conditional clustering of the data.

9. The experimental data for a single input-single output relationship are shown in Fig. 10.35. What form of rules and how many would you suggest in each case? Justify your choice.
10. In a granular neuron with three inputs, the corresponding connection weights are described by parabolic membership functions $\mu_{P_{\{1,3\}}}(x_1)$, $\mu_{P_{\{4,1\}}}(x_2)$, and $\mu_{P_{\{6,3\}}}(x_3)$ where $\mu_{P_{\{\nu,\sigma\}}}(x)$ is expressed in the form:

$$\mu_{P_{\{\nu,\sigma\}}}(x) = \begin{cases} 1 - \left(\frac{x-\nu}{\sigma}\right)^2 & \text{if } x \in [\nu - \sigma, \nu + \sigma] \\ 0 & \text{otherwise.} \end{cases}$$

What are the outputs of the granular neuron in Fig. 10.23 if the modifiers are $u = (0.5, 1.5, 0.1)$ and the activation rules are either (10.19) or (10.20)? Draw the resulting membership function. What if $u = (1, 1, 1)$ and the input x is $(0.5, 1.5, 0.1)$?
11. Continuing Exercise 8, use a granular neuron to merge the conditional fuzzy sets compute therein. Try to identify the entire structure with one of the three neurofuzzy paradigms enunciated in Section 10.6.

11 Identifying Fuzzy Rules

In Chapters 6 and 8 we discussed separately symbolic rules for identifying membership functions to cluster and subsymbolic rules to learn functions that suitably translate a set of inputs into an output variable. Here we look for complete procedures performing both tasks having the final aim of computing suitable functions from input to output. The general direction is forming these functions as a composition of information granules with mixing functions. Hence we look for a corresponding composition of cluster identification with function learning methods, possibly producing additional benefits in respect to the union of the advantages of the separate methods. Within a wealth shell where any method is admissible provided it is based on meaningful properties of granules and feasible computations, here we focus on three families of procedures: i) the training of a single granular neuron; ii) the partially supervised identification of fuzzy clusters; and iii) the overall training of a neuro-fuzzy system.

A synergy between symbolic and subsymbolic learning methods.

11.1 Training of a Granular Neuron

As mentioned in Chapter 8 we have no problem in principle to train any PE into a general neural network, provided a sensitivity of the function g_θ computed by the PE to its parameter θ becomes available. Now the structure of the input-output function computed by a granular neuron is made up of a net input constituted by a fuzzy norm of the information conveyed by the connections to the neuron, where the latter is the product of a constant by a membership function. Let us denote this function with h_θ where, with respect to the notation used in Chapter 8, we explicitly indicate the parameters, only seldom encompassed by connection weights w. Thus (10.19) reads:

Fuzzy weights plus fuzzy norms to build up a granular neuron:

$$Y = \mathsf{h}_{\boldsymbol{W}}(\boldsymbol{u}) = \perp_{i=1}^{k}(\mathrm{W}_i \top u_i). \qquad (11.1)$$

386   Identifying Fuzzy Rules

derivatives to toss parameter sensitivity,

If we locate the parameters within the membership functions, by the chain rule the optimal situation occurs when we may compute on one side the derivative of the norm w.r.t. its inputs and on the other the derivative of the membership function w.r.t. its parameters. For instance, in case of the Gödel norms we have:

$$\frac{\partial \top_{\min}(a,b)}{\partial a} = \begin{cases} 1 & \text{if } a < b \\ 0 & \text{elsewhere} \end{cases} \qquad \frac{\partial \bot_{\max}(a,b)}{\partial a} = \begin{cases} 1 & \text{if } a > b \\ 0 & \text{elsewhere,} \end{cases} \qquad (11.2)$$

while with a product norm we simply have:

$$\frac{\partial \top_{\text{prod}}(a,b)}{\partial a} = b, \qquad (11.3)$$

and an analogous result holds for the partial derivative w.r.t. b. Specific derivatives may be computed with some other norms reported in Appendix A.3, with a wise management of isolated discontinuity points.

As for the membership functions we encounter similar situations. For instance with a triangular membership function $\mu_{A_{\{a,b,c\}}}(x)$ we have:

$$\frac{\partial \mu_{A_{\{a,b,c\}}}(x)}{\partial a} = \begin{cases} \frac{(x-b)}{(a-b)^2} & \text{if } a \leq x \leq b \\ 0 & \text{elsewhere,} \end{cases}$$

$$\frac{\partial \mu_{A_{\{a,b,c\}}}(x)}{\partial b} = \begin{cases} -\frac{x-a}{(b-a)^2} & \text{if } a \leq x \leq b \\ \frac{c-x}{(c-b)^2} & \text{if } b < x \leq c \\ 0 & \text{elsewhere,} \end{cases} \qquad (11.4)$$

$$\frac{\partial \mu_{A_{\{a,b,c\}}}(x)}{\partial c} = \begin{cases} \frac{x-b}{(c-b)^2} & \text{if } b \leq x \leq c \\ 0 & \text{elsewhere;} \end{cases}$$

while with symmetric Gaussian membership functions $\mu_{A_{\{\nu,\sigma\}}}(x)$ we obtain:

$$\frac{\partial \mu_{A_{\{\nu,\sigma\}}}(x)}{\partial \nu} = \frac{2e^{-\frac{(x-\nu)^2}{\sigma^2}}(x-\nu)}{\sigma^2}, \quad \frac{\partial \mu_{A_{\{\nu,\sigma\}}}(x)}{\partial \sigma} = \frac{2e^{-\frac{(x-\nu)^2}{\sigma^2}}(x-\nu)^2}{\sigma^3}. \qquad (11.5)$$

The derivatives of $h_{\boldsymbol{\theta}}(\boldsymbol{x})$ are carried out as for conventional neurons, refer to Chapter 8. For instance, in case of product norm and triangular membership functions, hence with $h_{\boldsymbol{\theta}}(\boldsymbol{x}) = \prod_{j=1}^{k} \mu_{A_{\{a_j,b_j,c_j\}}}(x_j)$, where $\boldsymbol{\theta} = (\{a_1,b_1,c_1\},\ldots,\{a_k,b_k,c_k\})$, we have:

$$\frac{\partial h_\theta(x)}{\partial a_i} = \begin{cases} \prod_{j \neq i} \mu_{A_{\{a_j,b_j,c_j\}}}(x_j) \frac{(x_i-b_i)}{(a_i-b_i)^2} & \text{if } a_i \leq x_i \leq b_i \\ 0 & \text{elsewhere,} \end{cases}$$

$$\frac{\partial h_\theta(x)}{\partial b_i} = \begin{cases} -\prod_{j \neq i} \mu_{A_{\{a_j,b_j,c_j\}}}(x_j) \frac{x_i-a_i}{(b_i-a_i)^2} & \text{if } a_i \leq x_i \leq b_i \\ \prod_{j \neq i} \mu_{A_{\{a_j,b_j,c_j\}}}(x_j) \frac{c_i-x_i}{(c_i-b_i)^2} & \text{if } b_i < x_i \leq c_i \\ 0 & \text{elsewhere,} \end{cases} \quad (11.6)$$

$$\frac{\partial h_\theta(x)}{\partial c_i} = \begin{cases} \prod_{j \neq i} \mu_{A_{\{a_j,b_j,c_j\}}}(x_j) \frac{x_i-b_i}{(c_i-b_i)^2} & \text{if } b_i \leq x_i \leq c_i \\ 0 & \text{elsewhere,} \end{cases}$$

whose analytical form required, beyond (11.4),the use of chain rule:

chaining rule,

$$\frac{\partial h(x)}{\partial a_i} = \frac{\partial h(x)}{\partial \mu_{A_{\{a_j,b_j,c_j\}}}(x_i)} \frac{\partial \mu_{A_{\{a_j,b_j,c_j\}}}(x_i)}{\partial a_i}, \quad (11.7)$$

and similarly for b_i and c_i.

You are free of using these derivatives with any of the training methods employed with conventional neurons, for instance to navigate in the direction of the gradient descent of the quadratic error function C along its landscape in the θ space, as customary with perceptrons. Hence you have:

cost function,

$$C = \frac{1}{2} \sum_{i=1}^m (t_i - y_i)^2, \quad (11.8)$$

where $y_i = h_\theta(x_i)$ denotes the output of the granular neuron and t_i the observed target value on a sample of size m, and move on the landscape with steps $\Delta \theta_j$ computed by:

and training is served.

$$\Delta \theta_j = -\eta \frac{\partial C}{\partial \theta_j} = -\sum_{i=1}^m (h_\theta(x_i) - t_i) \frac{\partial h_\theta(x_i)}{\partial \theta_j}, \quad (11.9)$$

with a learning rate η selected with an analogous wisdom as in Chapter 8.

Example 11.1. Let us consider a simple case study where we aim at inferring with a granular neuron (as in Section 10.5) the product of two triangular membership functions $\mu_{A_1\{-0.1,0.5,1.1\}}$, $\mu_{A_2\{-0.2,0.55,1.2\}}$ on the basis of a training set of size $m = 100$: each item in this set is a triple (x_1, x_2, t) with (x_1, x_2) randomly drawn from the unit square, and $t = h(x_1, x_2) = \mu_{A_1}(x_1)\mu_{A_2}(x_2))$. Fig. 11.1(a) shows training points together with the learnt h_θ. Fig. 11.1(b) shows the course of the error function with iterations. We adopted an on-line strategy; hence, on each iteration we randomly draw a point from the training set, compute the square error through (11.8) with $m = 1$ and update parameters through (11.9).

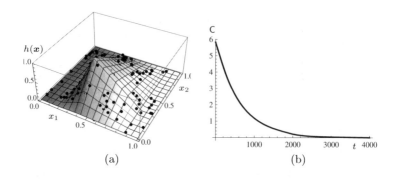

Fig. 11.1. (a) Activation function $h(x)$ of the granular neuron obtained after iterating (11.9) on the sample points as in Example 11.1; (b) course of the MSE (11.8) with successive iterations.

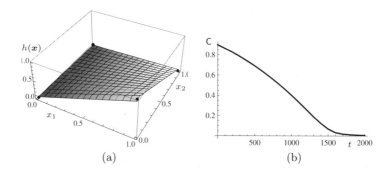

Fig. 11.2. Same as in Fig. 11.1 with OR dataset.

Example 11.2. Following Example 11.1, Figs. 11.2 and 11.3 show the behavior of a granular neuron trained to learn OR and XOR function respectively (see Table 8.1 in Example 8.1 for the latter). The second task performs badly as expected because of the impossibility of the input points to be (even fuzzy) separated by a single neuron.

11.2 Partially Supervised Identification of Fuzzy Clusters

Typically fuzzy clustering is carried out on a numeric data basis. Algorithms such as FCM produce a local minimum of the given objective function which leads to the formation of a collection of information granules. If there is an input from a human which is taken into consideration as a part of the clustering activities, these pursuits become human-centric. In turn, such

So fuzzy to consider user's suggestions,

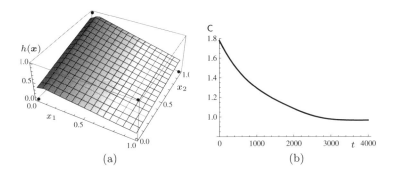

Fig. 11.3. Same as in Fig. 11.1 with XOR dataset.

Fig. 11.4. The principle of fuzzy clustering versus human-centric fuzzy clustering.

clustering produces information granules that are reflective of the human-driven customization. The crux of these clustering activities relies on a seamless incorporation of auxiliary knowledge about the data structure and problem at hand that is taken into consideration when running the clustering algorithm. In this manner we can effectively navigate the clustering activities and impact the geometry of the resulting clusters. In Fig. 11.4 we contrast human-centric clustering with the (data-driven) clustering by pointing at the role of knowledge tidbits. There are several fundamental ways the knowledge tidbits could be incorporated into the generic clustering technique. Here we elaborate on a well-motivated approach.

11.2.1 Using Data Labels When They Are Known

Partially supervised fuzzy clustering is concerned with a subset of patterns (data) whose labels have been provided. A mixture of labeled and unlabeled patterns could be easily encountered in many practical situations.

As an example, consider the clustering problem we faced with the MNIST dataset in Chapter 6. It consists of a huge bench of handwritten characters, some of which are well written, others definitely bad. With pure clustering algorithms we must rely on

When data are incomprehensible,

390   Identifying Fuzzy Rules

Fig. 11.5. Subset of "challenging" digits drawn from the MNIST dataset.

the inner structure of these digits in order to group them correctly. The task is not unbearable *per se*, but we may expect that some items like those in Fig. 11.5 will escape to even sophisticated automatic tools. Thus we move to a framework where we are provided by an expert with some knowledge-based hints
<small>ask user for hints.</small> about a small subset of "challenging" characters. They may be provided either with the plain label of the digit they represent or, in case of very hard scripts, with a set of membership degrees to the various labels. More descriptively, the role of the so labeled patterns would be to serve as *anchor* points when launching clustering: we expect that the structure discovered in the data will conform to the class membership of the labeled patterns. As this example illustrates, we are provided with a mixture of labeled and unlabeled data.

Another scenario in which partial supervision could play an important role originates at the conceptual end. Consider that the patterns have been labeled so on surface we can view that they imply the use of full supervision and call for the standard supervised mechanisms of classifier design. However the labeling
<small>at least for a part of the overall data if a complete advising is too costly.</small> process could have been very unreliable and therefore our confidence in the already assigned labels could be relatively low. Then we resort to the clustering mode and accept only a small fraction of patterns that we deem to be labeled quite reliably. The design scenarios similar to those presented above could occur quite often. We need to remember that labeling is a time-consuming process and labeling comes with extra cost. The clustering, on the other hand, could be far more effective. There is a spectrum of learning spread between "pure" models of unsupervised and supervised learning that could be schematically visualized in Fig. 11.6.

The effect of partial supervision involves a subset of labeled patterns, which come with their class membership. With the perspective that in general there is no dichotomy for supervised and unsupervised learning, to achieve an efficient use of such

Fig. 11.6. A schematic visualization of (infinite) possibilities of partial supervision quantified by the percentage of labeled patterns.

knowledge tidbits we include them as a part of an objective function. During its optimization we anticipate that the structure to be discovered conforms to the membership grades already conveyed by these selected patterns. Hence, we consider the following additive form of the objective function:

A wise balance

$$C = \sum_{c=1}^{k}\sum_{i=1}^{m} u_{ci}^{r} \nu(\boldsymbol{y}_i - \boldsymbol{v}_c) + \alpha \sum_{c=1}^{k}\sum_{i=1}^{m} (u_{ci} - \pi_{ci} b_i)^2 \nu(\boldsymbol{y}_i - \boldsymbol{v}_c).$$
(11.10)

The first term in (11.10) is aimed at the discovery of the structure in data and is the same as used in the generic version of the FCM algorithm. The second term (weighted by the positive coefficient α) captures an effect of partial supervision. Its interpretation becomes clear once we identify the two essential data structures containing information about the labeled data:

between what is already known and what must be compliant with it,

1. the vector of labels, denoted by $\boldsymbol{b} = (b_1, b_2, \ldots, b_m)^T$. Each pattern $\boldsymbol{x}_i$ comes with a Boolean indicator function. We assign b_i equal to 1 if the pattern has been already labeled as a part of the available knowledge tidbits. Otherwise we consider the value of b_i equal to zero;
2. the partition matrix $\Pi = [\pi_{ci}]$, $c = 1, 2, \ldots, k$; $i = 1, 2, \ldots, m$ which contains membership grades assigned to the selected patterns (already identified by the nonzero values of $\boldsymbol{b}$). If $b_k = 1$ then the corresponding column shows the provided membership grades. If $b_i = 0$ then the entries of the corresponding i-th column of Π do not matter; technically we could set up all of them to zero. Let $b_i = 1$. The optimization of the membership grades u_{ci} is aimed at making them close to π_{ik}.

The nonnegative weight factor α helps set up a suitable balance between the supervised and unsupervised mode of learning. Apparently when $\alpha = 0$ we end up with the standard FCM. Likewise if there are no labeled patterns ($\boldsymbol{b} = \boldsymbol{0}$) then the objective function reads as:

$$\mathsf{C} = (1+\alpha)\sum_{c=1}^{k}\sum_{i=1}^{m} u_{ci}^{r}\nu(\boldsymbol{y}_{i} - \boldsymbol{v}_{c}), \qquad (11.11)$$

and becomes nothing but a scaled version of the standard objective function guiding the FCM optimization process. If the values of α increase significantly, we start discounting any structural aspect of optimization (where properly developed clusters tend to minimize the objective function) and rely primarily on the information contained in the labels of the patterns. Subsequently, any departure from the values in Π would be heavily penalized by significantly increasing the values of the objective function. The choice of a suitable value of the weight factor α could be made by considering the ratio of the number q of data that have been labeled versus the overall data size m. To achieve a sound balance we consider the values of α to be proportional to the ratio q/m.

that may be linked to how many data have been already labeled.

One could also consider a slightly modified version of the objective function:

$$\mathsf{C} = \sum_{c=1}^{k}\sum_{i=1}^{m} u_{ci}^{r}\nu(\boldsymbol{y}_{i} - \boldsymbol{v}_{c}) + \alpha\sum_{c=1}^{k}\sum_{i=1}^{m}(u_{ci} - \pi_{ci})^{2} b_{i}\nu(\boldsymbol{y}_{i} - \boldsymbol{v}_{c}), \qquad (11.12)$$

where the labeling vector $\boldsymbol{b}$ shows up in a slightly different format. For $b_k = 1$, we involve the differences between u_{ik} and π_{ik} to minimize them.

11.2.2 The Development of Human-Centric Clusters

As usual, the optimization of the objective function (11.12) is completed w.r.t. the partition matrix and prototypes of the clusters. The first part of the problem is a constraint-based minimization. To minimize it, we consider Lagrange multipliers to accommodate the constraints imposed on the membership grades. Hence the augmented objective function arises in the following form:

An easy optimization target if we involve only second order functions,

$$\mathsf{C}'_{i} = \sum_{c=1}^{k} u_{ci}^{r}\nu(\boldsymbol{y}_{i} - \boldsymbol{v}_{c}) + \alpha\sum_{c=1}^{k}(u_{ci} - \pi_{ci}b_{i})^{2}\nu(\boldsymbol{y}_{i} - \boldsymbol{v}_{c}) - \beta\left(\sum_{c=1}^{k} u_{ci} - 1\right), \qquad (11.13)$$

for fixed $i = 1, \ldots, m$. To compute the gradient of C'_i w.r.t. the partition matrix U we note that choosing an l^2 norm for ν and a fuzzification factor $r = 2$ we obtain a second order objective

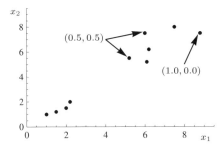

Fig. 11.7. A two-dimensional synthetic dataset. Arrow enhanced points: knowledge tidbits; strings in the brackets: membership grades to the two classes.

function that may be easily optimized. The resulting entries of the partition matrix U assume the form: *leading to an obvious generalization of FCM formulas.*

$$u_{ci} = \frac{1}{1+\alpha}\left(\frac{1+\alpha\left(1 - b_i \sum_{j=1}^{k}\pi_{ji}\right)}{\sum_{j=1}^{k}\left(\frac{\nu(\mathbf{y}_i,\mathbf{v}_c)}{\nu(\mathbf{y}_i,\mathbf{v}_j)}\right)^2} + \alpha\pi_{ci}b_i\right). \quad (11.14)$$

Moving on to the computations of the prototypes, the necessary condition for the minimum of C' w.r.t. the prototypes $\mathbf{v}_c$ comes in the form $\frac{\partial C'}{\partial v_{cj}} = 0$, $c = 1, 2, \ldots, k$; $j = 1, 2, \ldots, n$. By introducing the shortcut notation $\varnothing_{ci} = u_{ci}^2 + (u_{ci} - \pi_{ci}b_i)^2$, this leads to the solution:

$$\mathbf{v}_c = \frac{\sum_{i=1}^{m}\varnothing_{ci}\mathbf{x}_i}{\sum_{i=1}^{m}\varnothing_{ci}}. \quad (11.15)$$

Example 11.3. For illustrative purposes, we consider the small synthetic two-dimensional dataset as shown in Fig. 11.7. The partial supervision comes with the assignments of membership degrees of three patterns to two candidate clusters directly by the user. Namely, in Fig. 11.7 two patterns are assigned the same membership grades $(0.5, 0.5)$ to each cluster. On the contrary a third pattern is assigned definitely to the first cluster; hence its membership degrees are $(1.0, 0.0)$. We also choose a number of clusters $k = 2$ in the FCM algorithm.

The clustering was completed for several increasing values of α and this development gives a detailed view at the impact the classification hints exhibit on the revealed structure of the patterns. This is shown by visualizing the entries of the partition matrices (see Fig. 11.8) w.r.t. the first cluster. We note that by changing α the discovered structure tends to conform to the available classification constraints. For reference, we have shown the results

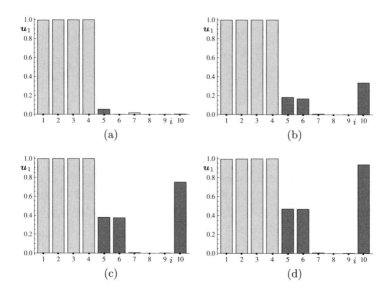

Fig. 11.8. Membership grades of patterns in Fig. 11.7 for selected values of α: (a) $\alpha = 0.0$ (no supervision), (b) $\alpha = 0.5$, (c) $\alpha = 3.0$, (d) $\alpha = 15.0$. Bars: membership grades to the first cluster; bar index: index of points according to the first coordinate. Dark gray bars correspond to the three labeled patterns.

for $\alpha = 0$ so that no supervision effect is taken into consideration. We may see that unsupervised clustering attributes all three questioned points to a cluster different from the one gathering the four leftmost points. However, forcing their labels toward the membership degrees declared before by increasing the value of α changes the verdict of the algorithm.

Remark 11.1. One of the problems arising with this method is that we cannot *a priori* identify the relationship between each cluster and the corresponding labels; consequently we cannot state how to order the rows of the partition matrix Π so that the obtained permutation is consistent with the cluster labeling. All what you realize is that certain patterns group together. However, in case we know the true labeling of each point (since we are in the training phase), a possible solution consists in attributing to each cluster a label equal to the most frequent items', and consequently arranging the rows of Π to satisfy the consistency requirement with the obtained labeling. Such a permutation may be not unique as in the following example where the number of clusters is chosen higher than the item labels in order to obtain better results through an increase in the sparseness of the data versus the clusters.

> A special kind of frequency-based label assignment,

Table 11.1. Confusion matrix for the classification of digits as in Example 11.4 with: (a) $\alpha = 0.0$, and (b) $\alpha = m/(m_e/4) = 27.7$.

	2	4	8
2	66	22	11
4	0	103	2
8	1	6	80

(a)

	2	4	8
2	86	4	8
4	0	103	4
8	2	0	84

(b)

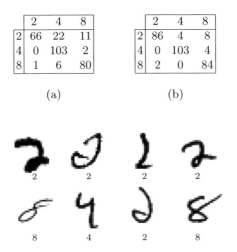

Fig. 11.9. A subset of the 42 digits misclassified by FCM algorithms together with their true labeling.

Example 11.4. Coming back to the MNIST dataset, we extend the methods used in Chapter 6 by applying FCM and its human-centric variant outlined in this section. As we have here only didactical objectives, for the sake of clarity we run the extended FCM algorithm on a small subset of the overall dataset made up of $m = 291$ characters randomly drawn within families of those representing 2, 4 and 8. Namely, with suitable k and α we iterate (11.14-11.15) until we appreciate no significant displacement in the centroids position. Then in order to compute the confusion matrices (see Section 6.3.1) in Table 11.1 we attribute a point to the cluster having maximal membership function. With $k = 9$ and $\alpha = 0.0$ we obtain an overall error of 0.144, corresponding to $m_e = 42$ misclassified items, when chosing as metric the Euclidean distance computed on the first 30 principal components of the bitmaps rendering the characters (see Section 7.1). Going through the misclassified items, we realize that digits like those shown in Fig. 11.9 need more information than the sole topological one in terms of the individual handwriting style to achieve a correct classification. that may render digit recognition more feasible.

Hence we raised α in order to force the true labels of these digits into the fuzzy clustering. Namely, following Remark 11.1 we attributed to each cluster a label equal to the majority of its items' and forced the partition matrix Π to have degree 1 in correspondence of a subset of the badly classified items, and a cluster sharing the same label as the misclassified items. By forcing labels of one quarter of the originally 42 misclassified ones, we obtain only 18 badly classified digits.

11.3 The Overall Training of a Neuro-fuzzy System

We have seen in Section 10.6 that most fuzzy inference systems can be classified into three types, mainly according to the way in which the output of premises is dealt with by the consequence part. Beyond Mamdani rule system – whose fuzzy sets produced in output must be defuzzified according to suitable procedures (see Table 10.5) – we focused also on rule systems computing output directly in a numeric fashion. Its dependence on the input data may derive either from the premises' output (as for Tsukamoto system) or from a possibly nonlinear function of the input data weighted by both premises and some adjustable parameters (as in the case of Takagi-Sugeno rule system). All such paradigms implement a kind of mixture of experts model, where the judgement of the single expert is given in different form depending on the employed scheme, and the influence of the single provision corresponds to a usually normalized computation performed on the premises. In the neuro-fuzzy community, however, there is a recent trend in favor of the second type of systems. The main motivation should not be searched in the increase of the model descriptive power (both are universal approximation algorithms, as classic neural networks are), but rather in the simplicity they offer to the learning process, since skipping the defuzzification procedure is a desirable point. Along the same direction, on one side there is an overall preference toward Takagi-Sugeno rule systems, as judgements of each expert are based on a linear combination of the input data, thus allowing a consistent speed-up in the convergence of the learning algorithm. On the other, to gain the same advantages the inverse sigmoid function used in the Tsukamoto model is often substituted by a linearized version.

A general neuro-fuzzy layered architecture with unitary connection weights and specific layers: In order to implement adaptive inference schemes, a lot of work has been done in transforming fuzzy inference systems into functionally equivalent adaptive networks, whose prototypical class is represented nowadays by variants of the ANFIS architecture [8], developed in 1993. It consists of a layered computation. Moreover, because of the non univocal interpretation of the connection weights functionality, the strategy is adopted of concentrating all computations in the nodes and giving unitary

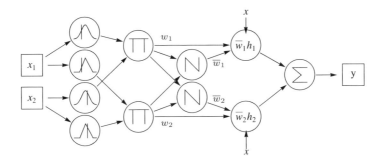

Fig. 11.10. A standard ANFIS architecture.

weight to each connection. In the standard form it is a five layered architecture (see Fig. 11.10) with:

- first layer responsible for fuzzifying input variables. It computes membership functions $\mu_{A_{cj}}$, with c ranging over the different rules, say in number of k, and j over the premises associated with the c-th rule, in any case not exceeding the dimensionality of the input data n; **I. fuzzuify,**
- second layer performing a T-norm of the premises: **II. compute antecedents,**

$$w_c = \prod_{j=1}^{n} \mu_{A_{cj}}(x_j), \quad c = 1, \ldots, k, \quad (11.16)$$

where, to keep notation as simplest as possible, we choose to use the product norm and assumed all rules as involving all the input variables;
- third layer computing the so called *normalized firing strength* through a per-rule normalization, as follows: **III. redistribute them,**

$$\overline{w}_c = \frac{w_c}{\sum_{\kappa=1}^{k} w_\kappa}, \quad c = 1, \ldots, k; \quad (11.17)$$

- fourth layer computing a parametrized *node function* h_c weighted by the corresponding $\overline{w}_c$. It is the form of this function that distinguishes the granular construct implemented; **IV. compute rules**
- fifth layer summing the $\overline{w}_c h_c$s. **V. merge their results.**

Depending on the function h_cs implemented on the fourth layer we encounter different variants of ANFIS. For instance, type-1 ANFIS amounts to:

$$h_c = \frac{a_c - \log(1/w_c - 1)}{b_c}, \quad (11.18)$$

that is useful to implement the Tsukamoto model. Type-3 ANFIS computing:

$$\mathsf{h}_c = \boldsymbol{a}_c \cdot \boldsymbol{x} + b_c = \sum_{j=1}^{d} a_{cj} x_j + b_c, \qquad (11.19)$$

or more simply type-0 ANFIS with h_c equal to a constant κ_c implement Takagi-Sugeno models of degree 1 and 0, respectively. A type-2 ANFIS architecture has been also devised for Mamdani scheme, by replacing the centroid defuzzification procedure with a discrete operator; however, this model is rarely used.

11.3.1 Learning Algorithm for Neuro-fuzzy Systems

Back-propagation forever, Encoding a fuzzy rule system in an adaptive network has the main benefit of helping organizing the available information so that it facilitates the application of most algorithms usually employed in machine learning. In the hypothesis that a set of clustering data have known labels, as depicted in Section 11.2.1, we may apply standard back-propagation schemes like the one sketched with Algorithm 8.4 for computing error gradient and directing error descent algorithms along its landscape in the parameter space. For instance with:

1. a type-3 ANFIS with h_c as in (11.19);
2. the quadratic error function (11.8); and
3. membership functions in the form of generalized Gaussian bells (see Appendix C):

$$\mu_{\mathsf{A}_{cj}}(x_j) = \mathrm{e}^{-\left(\frac{x_j - \mu_{cj}}{\sigma_{cj}}\right)^{2\alpha_{cj}}}, \qquad (11.20)$$

the parameters to be identified are:

- $\boldsymbol{a}_c$ and b_c (for $c = 1, \ldots, k$) in whose respect C derivatives are computed directly as a function of the output. Namely:

$$\frac{\partial \mathsf{C}}{\partial a_{cj}} = (y-t)\overline{w}_c x_j; \quad \frac{\partial \mathsf{C}}{\partial b_c} = (y-t)\overline{w}_c; \qquad (11.21)$$

- $\boldsymbol{\mu}, \boldsymbol{\sigma}$ and $\boldsymbol{\alpha}$, whose derivatives pass through the back-propagation of the error till the first layer. Namely the derivatives have the form:

$$\frac{\partial \mathsf{C}}{\partial \theta} = (y-t)(f_c - y)\overline{w}_c \delta_\theta, \qquad (11.22)$$

with θ equal to a component of any of the above parameter vectors, where the last term specifies as follows:

$$\delta_{\mu_{cj}} = \frac{2\alpha_{cj}\left(\frac{(x_j-\mu_{cj})^2}{\sigma_{cj}^2}\right)^{\alpha_{cj}}}{x_j - \mu_{cj}} \qquad (11.23)$$

$$\delta_{\sigma_{cj}} = \frac{2\alpha_{cj}\left(\frac{(x_j-\mu_{cj})^2}{\sigma_{cj}^2}\right)^{\alpha_{cj}}}{\sigma_{cj}} \qquad (11.24)$$

$$\delta_{\alpha_{cj}} = -\left(\frac{(x_j-\mu_{cj})^2}{\sigma_{cj}^2}\right)^{\alpha_{cj}} \log\left(\frac{(x_j-\mu_{cj})^2}{\sigma_{cj}^2}\right) \quad (11.25)$$

We meet the common problems with the assignment of the learning rates driving the trajectories along the error negative gradient. Moreover, both in on-line and in batch learning modality it is a common practice to alternate the training between two phases until global convergence has been reached, the former being committed to finding a fixed point of the conclusion parameters by iterating (11.21), the latter being focused on fixing the parameters of the membership functions through an iterative update of (11.22).

Actually, node functions (11.19) and their merge in the fifth layer allow for a more direct identification of (a_c, b_c) parameters directly through a minimum MSE procedure once $\overline{w}_c$s have been fixed in the membership function identification phase. Indeed, for given $\overline{w}_c$s the fifth layer output y is a linear combination of input x:

with direct minimum MSE solution whenever is possible,

$$y = \sum_{c=1}^{k}\sum_{j=1}^{n} \overline{w}_c a_{cj} x_j + b_c, \qquad (11.26)$$

admitting elementary derivatives w.r.t. the free parameters and elementary solutions of the system $\partial C/\partial \theta_i = 0$ for all i, with $\boldsymbol{\theta}$ representing the vector of free parameters.

Even simpler procedures may be implemented for type-0 ANFIS. As for type-2 ANFIS, inverse sigmoid functions as (11.18) are usually linearized to allow a similar speed-up, as shown in Fig. 11.11.

plus further simplifications, allow for:

Example 11.5. Consider the case study of inferring a regression curve from the 100 data points in Fig. 11.12(a). Each point is a triplet consisting of two values (x_1, x_2) uniformly drawn in the unit square $[0,1] \times [0,1]$ and a third value y computed as a sum of $\tilde{y} + \epsilon$, where the former comes from the function:

$$\tilde{y} = \frac{1}{20}(\sin(4x_1) + \cos^2(4x_2) + 1)^3, \qquad (11.27)$$

400   Identifying Fuzzy Rules

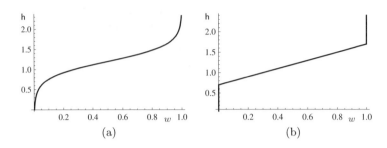

Fig. 11.11. (a) Type-1 ANFIS node function (11.18). (b) Its piecewise linear approximation.

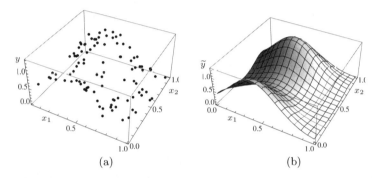

Fig. 11.12. An ANFIS learning instance: (a) training data, and (b) underlying function (11.27) originating them.

Recovering of a complex function.
and the latter is a specification of a Gaussian random variable with the zero mean and the 0.1 standard deviation. We use a *minimalist* model made up of 2 rules and 2 antecedents for each rule, each referring to a single variable. Namely, the antecedents are fuzzy sets described by bell membership functions (11.20) with initial parameters: $\mu_{1j} = 0.25, \sigma_{1j} = 0.3, \alpha_{1j} = 1.5, \mu_{2j} = 0.75, \sigma_{2j} = 0.3, \alpha_{2j} = 1.5$, for $j = 1, 2$. We developed both a type-0 and type-3 ANFIS. The training algorithm run the mentioned two-phases, where node function parameters are estimated through least square fitting and membership functions' via standard back-propagation. The stopping rule triggers when no significant improvement in the error reduction occurs. Results are shown in Fig. 11.13. Lines refer to the two ANFIS types. First column reports the inferred functions that we may contrast with the $\widetilde{y}$ graph in Fig. 11.12(b). The second column reports the course of the square error with iterations. Both columns denote a better performance of type-3, as expected because of the larger number of free parameters this model is provided with. In

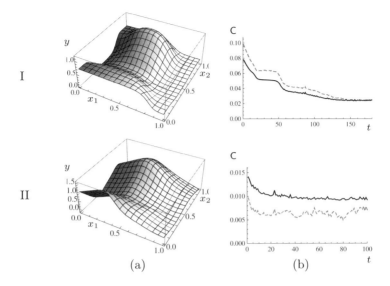

Fig. 11.13. Inferring a function from points in Fig. 11.12 using different ANFIS types. Line I: type-0; line II: type-3. Column (a) the inferred function; column (b) graph of MSE C vs. iteration t with training data (black curve) and test data (gray dashed curve).

particular, the second column highlights that type-3 ANFIS reaches an optimal solution within the first few iterations. With Fig. 11.14 we may go through the data structure realizing that: a) the parameter initialization generates a function far different from the target, and b) the mixture of antecedent fuzzy sets generates a function that is non strictly monotone with the membership functions of these sets, indeed the centers of these sets are located far from the maxima of the consequent.

Example 11.6. Let us infer the relation existing between age adjusted mortality (M) vs. the percentage of non whites in the population (%NW) and sulfur dioxide concentration (SO_2) in the SMSA dataset. We used the same architectures as those described in Example 11.5 with the sole difference in the number of antecedents. Here we ran FCM clustering algorithm with $k = 5$ clusters, whose memberships will be adapted by the whole training procedure as shown in Fig. 11.15(a). As clear from the remaining columns of this figure, also in this case type-3 ANFIS out-perfomed type-0 model, though this time both algorithms fail to converge at the stopping rule triggering toward a configuration with a low MSE as in the previous case study. This might be caused either by a wrong choice of the rules number, or by a bad selection of learning parameters.

Regressing pollution phenomena,

402   Identifying Fuzzy Rules

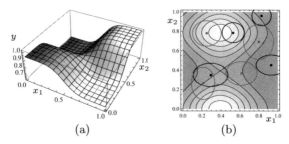

Fig. 11.14. Going through the data structure. (a) the type-0 ANFIS model before training; (b) contour plot of (11.27) together with a rough visualization of membership functions of type-0 (gray) and type-3 ANFIS (black) after training, where the elliptic shape are α-cuts reflecting the bell parameter σ.

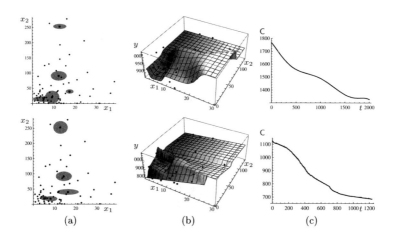

Fig. 11.15. Inferring a function from SMSA dataset. The same notation as in Fig. 11.13, but column (a) shows the SMSA input data and the location of the antecendent bells; column (b) and (c) like columns (a) and (b) in the former figure.

Table 11.2. Confusion matrices for the classification of digits as in Example 11.7 referred to: (a) type-0, and (b) type-3 ANFIS model.

	2	4	8
2	81	4	1
4	16	101	6
8	2	0	80

	2	4	8
2	83	3	0
4	16	102	7
8	0	0	80

(a)   (b)

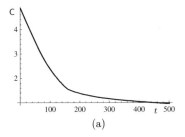 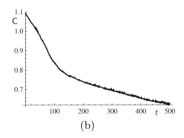

Fig. 11.16. Course of cost C with the number of iterations t for: (a) type-0, and (b) type-3 ANFIS models described in Example 11.7.

Example 11.7. Continuing with Example 11.4 we try to learn the same subset of $m = 291$ digits with both type-0 and type-3 ANFIS architecture. Considering the additional labeling information, we limit here to represent each digit with its first 6 principal components and determine the shape of initial premises membership functions through a FCM clustering algorithm with $k = 9$ centroids, obtaining consequently a total of $9 \times 6 \times 3 = 162$ parameters for the premises and $(6 + 1) \times 9 = 63$ for the consequences (reduced to 9 in case of type-0 model). Once again, to speed-up learning (even considering the high number of parameters) we adopt the two-phases procedure obtaining an error of 0.099 for type-0 and 0.089 for type-3 architecture (corresponding respectively to 29 and 26 misclassified digits as shown in the confusion matrices in Table 11.2 whose items were counted by adopting a nearest-neighbor strategy); a quite good result if compared to the performance of the naive FCM algorithms here realizing an error of 0.19 (55 misclassified digits). The companion courses of the error C with iterations denote a smaller value achieved at the end of the training (see Fig.11.16), since ANFIS outputs are continuous values that are mapped into the admissible labels 2, 4 and 8.

11.4 Conclusions

From a technical point of view this chapter is devoted to the extension of some inference methods from clustering and neural networks frameworks. Perhaps in this sense, they are not exciting news given that we are mainly interested in what we process rather than how it is processed. However our operational approach imposes that we cannot deliver methods if we do not check in advance their implementability. In this respect we may observe that:

and the damned digits.

1. as mentioned before we have no problem of back-propagating error, provided the derivability of the activation function is guaranteed on any kind of PE;
2. introducing supervision in clustering methods accounts for adding a further term in the algorithm's cost function plus a coefficient wisely balancing new and old terms;
3. training a neuro-fuzzy system is accomplished via a standard connectionist procedure. In addition, once the cost function has been defined any shortcut is allowed, such as the use of linear regression coefficients in the output layer neuron.

11.5 Exercises

1. In the expression (11.1) when carrying out optimization, the derivatives of the Gödel norm is a Boolean function assuming values in $\{0,1\}$. Discuss potential drawbacks of this derivative and better situations with other norms in the process of learning.
2. Derive learning formulas in case of a granular neuron when the performance index is of the following form:

$$C = \sum_{i=1}^{m} |t_i - y_i|.$$

3. In the problem of clustering with partial supervision, how could you go about choosing a value of the scaling α? Hint: think of a problem of achieving a balance between the number of all data and the data that have been labeled.
4. State main correspondences between the hybrid architecture proposed in Fig. 10.27 and Fig. 11.10.
5. Continuing Exercise 8 in Chapter 10 on SMSA dataset, of the 16 variables used for clustering take *Mortality* as a consequence Y and the most relevant components from PCA as antecedents' coordinate. Then describe Y through the three fuzzy sets: *low*, *normal* and *high*, and use the partition matrix coming from FCM to connect antecedents' clusters to consequent clusters. Finally, the consequent clusters are connected to the synthesis neuron deciding whether a given citizen has been damaged by pollution or not. Refine this model by training alternatively Sugeno and Tsukamoto neuro-fuzzy system as implemented by ANFIS.
6. Repeat Exercise 5 with a 4-layer perceptron having in input all original variables but *mortality* that plays the role of target. Compare the results.

Further Reading

Fuzzy clustering comes as a fundamental algorithmic framework providing a synthesis between many of the inferential tools we have run through in the previous parts. The book by Bezdek [5] is the first one in the area; the reader may refer to [20] which offers a thorough treatise of the subject of information granulation in the setting of fuzzy modeling including variants of partial supervision, conditional clustering, directional clustering and others.

Takagi-Sugeno architectures of rule-based fuzzy models are highly visible and broadly used [2, 13, 22, 24]. Their functional flexibility is an outstanding feature as the local models forming conclusions of the rules are essential to modeling complex and nonlinear relationships via a finite family of local and simple models such as linear functions. ANFIS and its relatives [8, 9] offer a great deal of architectural and learning capabilities in neuro-fuzzy systems. Furthermore there were interesting studies devoted to the interpretability of such rule-based models [6, 17]. The issue of exploiting various modes of fuzzy modeling was covered in [19].

With fuzzy rules we are covering the dynamic equivalent of the tools for an essentially static analysis of data discussed in the previous parts. We did something similar with recurrent neural networks. However practitioners found them generally unstable, concluding that it is quite difficult to find an equivalent of stochastic processes when the dynamic of the process is both complex and uncertain. From a theoretical perspective, the template of stochastic processes in Algorithmic Inference approach is represented by fractals [7, 3]: a recursive intensive application of elementary transition rules generating a so hard to invert process [1] that you do no realize whether its starting point is random or deterministic. In this thread, fuzzy rules are at an intermediate level with the fractal rules and the transition matrices of Markov chains (see Definition 8.1) [11] or conditional probabilities (see (2.8)) of belief networks [10] within Bayesian probabilistic framework [4], yet exhibiting notably high expressive power. We may also extend the structure of networks to rules obtaining *cognitive maps* [12]. The structure of the network offers a great deal of flexibility and is far less rigid than the fuzzy neural networks where typically the nodes are organized into some layers. We must declare, however, that we are not able to state similar theoretical results as with random processes. This seems an unavoidable drawback as a cost of the information granularity and complexity of referred phenomena as well. We may state some clever theorems at micro level, for instance concerning perceptrons or SVMs. Then we must rely on bootstrapping techniques, like the ones in Chapter 3, and tools for a correct reading of their outputs in order to draw conclusions. Actually, in a very self-referential perspective, we must *learn* from these simulations the properties of the involved tools. Like for Markov blankets [18] a main attitude is to isolate those variables that are locally involved in a phenomenon and the rules of this involvement. The iterative implementation of these rules will give rise to computational architectures for dealing with the whole phenomenon under question. Whilst the synthesis of the latter seems to be preferentially demanded to subsymbolic tools, or statistical analysis to the best.

References

1. Apolloni, B., Bassis, S.: Identifying elementary iterated systems through algorithmic inference: the Cantor Set example. Chaos, Solitons and Fractals 30, 19–29 (2006)
2. Babuska, R.: Fuzzy Modeling for Control. Kluwer Academic Publishers, Dordrecht (1998)
3. Barnsley, M.F.: Fractals everywhere. Morgan Kaufmann, San Francisco (2000)
4. Berger, J.O.: Statistical Decision Theory and Bayesian Analysis, 2nd edn. Springer, New York (1999)
5. Bezdek: Pattern Recognition with Fuzzy Objective Function algorithms. Plenum Press, New York (1981)
6. Casillas, J., et al.: Interpretability Issues in Fuzzy Modeling. Springer, Berlin (2003)
7. Falconer, K.: Fractal Geometry – Mathematical Foundations and Applications. John Wiley & Sons, New York (1960)
8. Jang, J.S.R.: ANFIS: adaptive-network-based fuzzy inference system. IEEE Trans. Syst. Manag. Cyber. 23(3), 665–685 (1993)
9. Jang, J.S.R., Sun, C., Mizutani, E.: Neuro-Fuzzy and Soft Computing. Prentice Hall, Upper Saddle River (1997)
10. Jensen, F.: An Introduction to Bayesian Networks. Springer, Heidelberg (1996)
11. Lamperti, J.: Stochastic processes: a survey of the mathematical theory. Applied mathematical sciences, vol. 23. Springer, New York (1977)
12. Lee, K.C., Kim, J.S., Chang, N.H., Kwon, S.J.: Fuzzy cognitive map approach to web-mining inference amplification. Expert Systems with Applications 22, 197–211 (2002)
13. Yasukawa, T., Sugeno, M.: A fuzzy-logic-based approach to qualitative modeling. IEEE Trans. on Fuzzy Systems 1, 7–31 (1993)
14. Mamdani, E.H.: Application of fuzzy logic to approximate reasoning using linguistic synthesis. IEEE Transactions on Computers 26(12), 1182–1191 (1977)
15. Mizumoto, M., Zimmermann, H.J.: Comparison of fuzzy reasoning methods. Fuzzy Sets and Systems 8, 253–283 (1982)

16. Mi, G.W., Karyannis, N.B.: Growing radial basis neural networks: merging supervised and unsupervised learning with network growth techniques. IEEE Trans. on Neural Networks 8, 1492–1506 (1997)
17. Paiva, R., Dourado, A.: Interpretability and learning in neuro-fuzzy systems. Fuzzy Sets and Systems 147(1), 17–34 (2004)
18. Pearl, J.: Probabilistic Reasoning in Intelligent Systems. Morgan Kaufmann, San Francisco (1988)
19. Pedrycz, W.: Fuzzy Modelling: Paradigms and Practice. Kluwer Academic Press, Dordrecht (1996)
20. Pedrycz, W.: Knowledge-Based Clustering. J. Wiley, Hoboken (1996)
21. Pedrycz, W.: Conditional fuzzy clustering in the design of radial basis function neural networks. IEEE Trans. on Neural Networks 9, 601–612 (1998)
22. Sugeno, M., Takagi, T.: Fuzzy identification of systems and its applications to modeling and control. IEEE Trans. on Systems, Man, and Cybernetics 15, 116–132 (1985)
23. Tsukamoto, Y.: An approach to fuzzy reasoning method. In: Gupta, M.M., Ragade, R.K., Yager, R.R. (eds.) Advances in Fuzzy Set Theory and Applications North Holland, Amsterdam (1979)
24. Yen, J., Wang, L., Gillespie, C.: Improving the interpretability of TSK fuzzy models by combining global learning and local learning. IEEE Trans. Fuzzy Systems 6(4), 530–537 (1998)

A Conceptual Synthesis

12 Knowledge Engineering

While in Part V we have been engaged in assembling computational tools previously assessed to build a complete system from signals to decisions, in this part we close the book with some examples of how to merge conceptual tools. The aim is to render them adequate to face complex data within elementary tasks. The strategy is to drill some either analytical or logical aspects, to the depth they need, in order to embed additional conceptual facilities in the basic procedures. Once more we plan to avoid embarking on algorithmic sophistication, leaving it to specialized books or web repositories.

A merge of conceptual tools.

12.1 Separating Points through SVM

We have met the problem of separating points as a component of many operational tasks. In its simplest version it is a matter of deciding on the labels y_is of the patterns $\boldsymbol{x}_i$ jointly constituting a sample

$$z = \{(\boldsymbol{x}_i, y_i), i = 1, \ldots, m\} \subset \mathfrak{X}^n \times \mathfrak{Y}. \qquad (12.1)$$

You assume the labels to be already known, so that you face a discrimination problem whose objective is to identify a function h so that $y_i = h(\boldsymbol{x}_i)$ for each i, and $\mathfrak{Y}$ to be discrete. Locating h within a family of functions, the above becomes a learning problem, that, in view of the exploitation of its solution on future data, is referred to as a *classification* problem. With this problem h plays the role of a hypothesis named *classifier* attributing a class label to each pattern. Focusing on h linear on $\boldsymbol{x}$ and splitting $\mathfrak{Y}$ into a set of binary sets (for instance, through a decision tree like the one shown in Fig. 3.20), the most used today computational tool is the SVM introduced in Section 3.2.4,

Learning to binary classifying data

thanks to loose symmetry hypothesis on data and relatively fast algorithms available for its implementation.

12.1.1 Binary Linear Classification

Recalling notation in Section 3.2.4, an SVM computes a separator hyperplane h in the $\mathbb{R}^n$ set, for suitable n, that we identify through the equation $\boldsymbol{w}\cdot\boldsymbol{x}+b = 0$. The learning problem consists in identifying the pair $(\boldsymbol{w}, b) \in \mathbb{R}^{n+1}$ of parameters occurring in this equation.

Assuming that the data at hand are *linearly separable*, so that *consistent* hyperplanes exist dividing the patterns' space into two half-spaces each containing only data belonging to a given class, and denoting with $y = +1$ and $y = -1$ the labels of the half-spaces, a family of them satisfies the following constraints:

Far enough from the constraints, that means:

$$\boldsymbol{w} \cdot \boldsymbol{x}_i + b \geq +1 \quad \text{for each } i \text{ such that } y_i = +1 \quad (12.2)$$
$$\boldsymbol{w} \cdot \boldsymbol{x}_i + b \leq -1 \quad \text{for each } i \text{ such that } y_i = -1. \quad (12.3)$$

Equations (3.88-3.92) come from the following requirements:

for a self-referential scale factor,

1. the bounds in (12.2-12.3) are *tight*, i.e. both of them hold as equalities at least once – a condition that fixes the scale factor of the parameters, and

achieve the maximum distances of closest points

2. h has maximum equal distances from the closest points lying in the two halves of the space – in the anticipation that this arrangement improves the probability that also future points will lie in the correct half-space since they are induced to lie as far as possible from the wrong half part of the space.

As mentioned in Section 3.2.4, the second requirement makes sense under loose hypothesis of symmetric displacements of points w.r.t. the separator. Sharing the same notation as being used in Part IV, we satisfy these requirements by associating each hyperplane filling the conditions (12.2-12.3) a *fitness* to be maximized, known in the literature as *margin*, defined as

$$\rho(\boldsymbol{w},b) = \min_{i:y_i=+1} d_r(\boldsymbol{x}_i,\boldsymbol{w},b) - \max_{i:y_i=-1} d_r(\boldsymbol{x}_i,\boldsymbol{w},b), \quad (12.4)$$

where $d_r(\boldsymbol{x},\boldsymbol{w},b)$ denotes the *relative* distance between the hyperplane $(\boldsymbol{w},b)$ and the point $\boldsymbol{x}$, signed according to the half-space where $\boldsymbol{x}$ is located, namely:

$$d_r(\boldsymbol{x}_i,\boldsymbol{w},b) = \boldsymbol{w}\cdot\boldsymbol{x}_i + b. \quad (12.5)$$

Fig. 12.1(a) shows a set of four patterns in $\mathbb{R}^2$ either associated to a positive or a negative class (the gray and black dots,

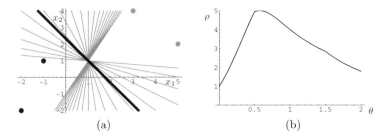

Fig. 12.1. Maximizing the margin for a separating hyperplane. (a) a set of positive and negative points (gray and black dots, respectively), and a sheaf of lines separating positive points from negative ones (gray lines); the black thick line has maximum margin. (b) the margin assumed by the sheaf elements.

respectively), together with a sheaf of lines separating the two classes and crossing the point $(1,1)$. The plot in Fig. 12.1(b) reports the course of the margin of these lines, computed according to (12.4), when we sort them from the one nearest to the negative point in the upper-left part of the graph and following the lines' rotation. The black thick line corresponds to the maximal value for the margin. We deduce from the SVM linearity that its coefficients $\boldsymbol{w}$ are independent of the rotation point.

We have $\rho(\boldsymbol{w}, b) \geq 0$ independently of the patterns. Actually, the margin of $(\boldsymbol{w}, b)$ is independent of b, since a change in this value simply shifts some positive or negative amount from $\min_{i:y_i=+1} d_r(\boldsymbol{x}_i, \boldsymbol{w}, b)$ to $-\max_{i:y_i=-1} d_r(\boldsymbol{x}_i, \boldsymbol{w}, b)$ in the sum (12.4). Thus we may optimize $\boldsymbol{w}$ in advance, obtaining the so-called optimal margin hyperplane, and then decide b in order to have an equal absolute distance of h from the closest positive and negative points. From (12.5) we see that, by chosing $b = 0$ for the sake of simplicity, once you have determined the correct sign of d_r with a given $\boldsymbol{w}$, you decrement its absolute value just by multiplying $\boldsymbol{w}$ by a positive value less than 1, with the possible drawback, however, of falsifying (12.2-12.3). Hence we may write the cost function of our optimization target C as follows, with the help of Lagrangian multipliers $\alpha_1, \ldots, \alpha_m$.

Optimize separately $\boldsymbol{w}$,

balancing its module with constraint satisfaction,

$$\mathsf{C} = \frac{1}{2} l^2(\boldsymbol{w}) - \sum_{i=1}^{m} \alpha_i (y_i(\boldsymbol{w} \cdot \boldsymbol{x}_i + b) - 1). \qquad (12.6)$$

As highlighted in Section 3.2.4, we prefer to solve the dual optimization problem (3.88-3.90) and obtain the optimal $\boldsymbol{w}^*$ through (3.91) as a function of the sole examples in relation to non-null optimal Lagrange multipliers; these examples will be termed

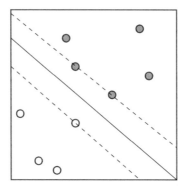

Fig. 12.2. The geometric interpretation of an optimal margin hyperplane.

and locate b, *support vectors*. It can be shown (see Fig. 12.2) that these points are at minimal distance from the hyperplane, so that typically only a small subset of the available data will actually be support vectors. The computation of the optimal b^* as in (3.92) is just a consequence of the fact that the distance of the support vectors from the hyperplane is exactly equal to 1. Once w^* and b^* have been computed, the assignment of a new pattern x' can be computed by determining the sign($w^* \cdot x' + b^*$).

Only a few points are involved.

Example 12.1. The discrimination of one class against the remaining two of the Iris dataset through SVM, discarding the second component having the smallest variance as customary, gives the following results:

- as for class *setosa*, we obtain a classifier based on two support vectors (namely, the 45-th and 99-th pattern) scoring a null error, in line with the dataset description;
- things are worse for class *versicolor*, as in this case the classifier – relying on 4 support vectors – assigns an incorrect label to 60 patterns;
- class *virginica* is handled a little better by a 4-support vector hyperplane misclassifying 20 patterns.

Approximate classifiers

The errors in the previous examples depend on the fact that two classes of the Iris dataset are not linearly separable from the others. Hence the optimization algorithm finds solutions that do not satisfy perfectly the constraints (12.212.3). In these cases it is more convenient to directly control the classification error through the introduction of *slack variables* ξ_i, and modify the constraints in the original problem as follows:

$$w \cdot x_i + b \geq +1 - \xi_i \quad \text{for each } i \text{ such that } y_i = +1 \quad (12.7)$$
$$w \cdot x_i + b \leq -1 + \xi_i \quad \text{for each } i \text{ such that } y_i = -1 \quad (12.8)$$
$$\xi_i \geq 0 \quad \text{for each } i = 1, \ldots, m. \quad (12.9)$$

This amounts to give the hyperplane, that we denote as a *soft-margin* classifier, the possibility of misclassifying the i-th example whenever $\xi_i > 1$. In turn, ξ_is represent relaxation terms that we aim to minimize jointly with previous C. Adopting a further Lagrange multiplier $\grave{\alpha}$ to balance these new terms with the others, the new cost function becomes:

<small>Additional terms to manage inconsistencies,</small>

$$\mathsf{C}' = \frac{1}{2}l^2(w) + \grave{\alpha}\sum_{i=1}^{m}\xi_i - \sum_{i=1}^{m}\alpha_i(y_i(w \cdot x_i + b) - 1 + \xi_i). \quad (12.10)$$

Hiding the last sum that simply reckons the constraints, we are used to consider:

- the second term directly related to the number of misclassification errors, while
- the first one related to the plasticity of the classifier, i.e. its ability to adapt to the provided examples [18].

Their joint minimization, balanced through a choice of the parameter $\grave{\alpha} > 0$ (elsewhere denoted by C so that the method is called *soft C-margin* SVM), allows for an efficient minimization of misclassifications avoiding in the meanwhile the *overfitting* phenomenon, i.e., the useless pursuit of learning even the random drifts of single examples from the general structure, at the expense of not actually learning this structure. This can be viewed as an application of the *structural minimization principle* [18] as a specification of the more general *bias-variance tradeoff* [11].

<small>plus a wisely parameterization to control their effects.</small>

This is the rationale of (3.90′) positioned in place of (3.90) representing the sole change we meet in the reformulation of the problem in terms of dual optimization, hence through (3.88-3.90′).

Example 12.2. Continuing Example 12.1, the introduction of a soft-margin approach does not ameliorate the global classification performance, although it has the effect of balancing the error rate on the partially misclassified *versicolor* and *virginica* classes. Indeed, this approach scores a total number of 49 and 50 errors, respectively, on the above two classes.

Recovering nonlinearities

Moving further in the realm of nonlinear problems, we can consider the kernel trick described in Appendix D.2. Indeed, being all patterns occurrences in (3.88-3.92) expressed in terms of a dot product, we can substitute the latter with a kernel computation. This amounts to nonlinearly mapping patterns in a higher-dimensional space and trying to find a linear separator for the images obtained in this way. The analogon of (3.88-3.90) reads as:

Try with a kernel when the linear classifier fails,

$$\max_{\boldsymbol{\alpha}} \sum_{i=1}^{m} \alpha_i - \frac{1}{2} \sum_{i,j=1}^{m} \alpha_i \alpha_j y_i y_j k(\boldsymbol{x}_i, \boldsymbol{x}_j) \quad (12.11)$$

$$\sum_{i=1}^{m} \alpha_i = 0 \quad (12.12)$$

$$\alpha_i \geq 0 \quad \text{for each } i = 1, \ldots, m, \quad (12.13)$$

while the optimal value for $\boldsymbol{w}$ will be still computed according to (3.91). Things are slightly different for b, as it is necessary to take into account the necessity of mapping the patterns to the new space; this leads to the new formula:

$$b^* = y_i - \sum_{j=1}^{m} \alpha_j^* y_j k(\boldsymbol{x}_j, \boldsymbol{x}_i) \quad \text{for any } i \text{ such that } \alpha_i^* > 0. \quad (12.14)$$

Analogously, a new pattern $\boldsymbol{x}'$ will be associated by the learnt classifier to the class whose label is:

$$y_i^* = \text{sign}\left(\sum_{i=1}^{m} \alpha_i^* y_i k(\boldsymbol{x}_i, \cdot \boldsymbol{x}') + b^*\right). \quad (12.15)$$

Example 12.3. Restricting the discrimination task to the sole classes *versicolor* and *virginica*, each one against the remaining two, the introduction of polynomial kernels of 3-rd degree allows for attaining a zero classification error.

We may relax the constraints on α_is also with this version of the dual problem, thus obtaining approximate (soft-margin) nonlinear separators with controlled classification error.

Combining one-to-many classifiers

The above discrimination of one class versus the remaining ones gives rise to classifiers that, like in a rule system considered in

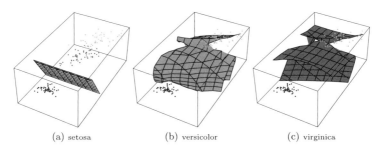

Fig. 12.3. SVM-based classifiers, in a three-dimensional projection for the: (a) *setosa*, (b) *versicolor*, and (c) *virginica* Iris dataset classes, discarding the lowest variance dimension.

Section 10.2, must be merged in order to have the final assignment of a pattern to a single class. If for each class there is a classifier correctly discriminating it, we may assign the pattern to the class of the sole classifier positively labeling it. Otherwise the classifiers looks like fuzzy rules needing mixing functions like those used throughout Chapter 10.

and exploit results to the best.

Example 12.4. If we combine the best results obtained in Examples 12.1-12.3, i.e. we use a linear classifier for discriminating examples from class *setosa* and two cubic classifiers for the remaining classes, we can feed the three classifiers, in turn, on all patterns and verify that each pattern is univocally associated to the corresponding class. Fig. 12.3 shows a three-dimensional projection of the whole dataset missing the smallest variance component and the separator of each class.

The above mentioned approach of dealing with multiple classifiers, each devoted to learn one among a set of classes, is typically criticized as it does not capture possible correlations among the various classes, as each of the single classifiers is trained independently from the remaining ones. Early approaches require to consider a set of additional constraints [18], with the effect of increasing the complexity of the dual formulation. More sophisticated approaches rely on a multi-class generalization of the concept of margin [8], or on the simultaneous training of a classifier for each pair of classes, to be subsequently combined through AND gates [12], Max Win algorithm [10] or direct acyclic graphs [16].

Classifying non homogeneous data

A further complication in the classification problem occurs when the patterns have a different relevance. If we may quantify the

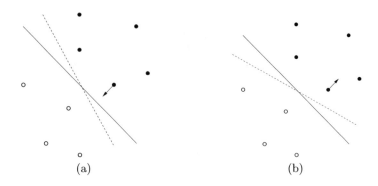

Fig. 12.4. Virtually shifting patterns in order to exploit their relevance: (a) patterns with positive relevance get closer to the separating hyperplane of the original formulation (plain line), with the effect of increasing the distance between the classifier (dashed line) and the original pattern position; (b) when the relevance is negative, the shift is in the opposite direction, thus promoting the pattern misclassification.

relevance, however, it represents an additional information to be injected in the procedure for the SVM identification. The rule of thumb is to give priority to the correct classification of points having higher relevance. Denote by r_i the relevance of the i-th pattern. If we center the r_is around 0, we may realize this rule by virtually moving each pattern $\boldsymbol{x}_i$ in the direction orthogonal to that of the optimal separator hyperplane, the shift being directed toward the latter when the relevance is positive, and in the opposite direction when the relevance is negative. Fig. 12.4 shows how this procedure promotes classifiers whose performance tend to be correct for patterns with high relevance, possibly misclassifying patterns characterized by negative relevance. Indeed:

- patterns with $r_i > 0$ are shifted toward the classifier hyperplane of the standard SVM formulation, with the effect of increasing the distance between the actual classifier and the original pattern position (see Fig. 12.4(a));
- when $r_i < 0$, the shifting occurs in the opposite direction, so that the original pattern is possibly misclassified (see Fig. 12.4(b)).

Starting from the original formulation minimizing $l^2(\boldsymbol{w})$ under the constraints (12.2-12.3) with $r_i = 0$ for each i, its extension with $r_i \neq 0$ for some i and shifts proportional to r_i reads:

$$\min_{\boldsymbol{w},b} \frac{1}{2} l^2(\boldsymbol{w}) \tag{12.16}$$

$$\boldsymbol{w} \cdot \left(\boldsymbol{x}_i - \frac{r_i}{2}\boldsymbol{w}\right) + b \geq +1 \quad \forall i : y_i = +1 \tag{12.17}$$

$$\boldsymbol{w} \cdot \left(\boldsymbol{x}_i + \frac{r_i}{2}\boldsymbol{w}\right) + b \leq -1 \quad \forall i : y_i = -1. \tag{12.18}$$

This leads to the maximize w.r.t. $\alpha_1, \ldots, \alpha_m$ in the dual cost function:

$$\mathscr{L}(\boldsymbol{\alpha}) = \sum_{i=1}^{m} \alpha_i - \frac{1}{2(1 + \sum_{k=1}^{m} \alpha_k r_k)} \sum_{i,j=1}^{m} \alpha_i \alpha_j y_i y_j \boldsymbol{x}_i \cdot \boldsymbol{x}_j, \tag{12.19}$$

subject again to the constraints (3.89-3.90). The optimal $\boldsymbol{w}^*$ has the following expression as a function of the optimal $\alpha_1^*, \ldots, \alpha_m^*$:   with a bit of overhead in computations.

$$\boldsymbol{w}^* = \frac{1}{(1 + \sum_{k=1}^{m} \alpha_k^* r_k)} \sum_{i=1}^{m} \alpha_i^* y_i \boldsymbol{x}_i, \tag{12.20}$$

while the optimal value b^* is still computed through (3.92).

The introduction of kernels in order to deal with the nonlinear case is less straightforward than in the original formulation [3].

Example 12.5. The Cox benchmark can be suitably classified through the above method. We may characterize each individual through a pattern reporting its age a_i and the sample mean $\overline{m}_i$ of the four blood measurements, after a preprocessing phase consisting in whitening the latter (so as to have 0 mean and unitary variance of each measurement) in order to get comparable values. Moreover, we used the linear regression values of each blood measurement w.r.t. the age to get *nominal value* of each observation, missing ones included. The regression was also used in order to attribute each example a relevance: more precisely, for a given observation we computed the mean of the quadratic distance between each blood measurement nominal and observed value (obviously, omitting missing values) as a dispersion index, and then identified relevance with the inverse of this index. Labels $+1$ and -1 are conventionally assigned to the carriers and normal classes, respectively. Fig. 12.5 shows the results obtained by the relevance-based algorithm, contrasted with the usual soft-margin SVM learnt from the same data. The points' representation through bullets with gray level and radius reflecting class and relevance, respectively, enhances the structures discriminated in this way by the algorithm. Indeed, the soft-margin SVM finds a line ℓ that is influenced also by a set of examples having very small relevance (the small gray bullets in the upper-left part of the graph). When we look at the relevance-based SVM, these points are essentially ignored.   Data structure behind their relevance

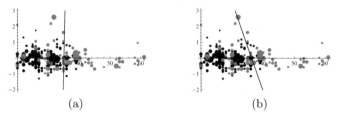

Fig. 12.5. Learnt classifiers for the Cox biomedical dataset: x axis: age a of patients; y axis: mean $\overline{m}_i$ of whitened and rescaled blood measurements. Point: gray levels denote different class labels, the higher the radius and the more relevant is the point. Straight line: classifier in output of the (a) relevance-based and (b) soft-margin SVM.

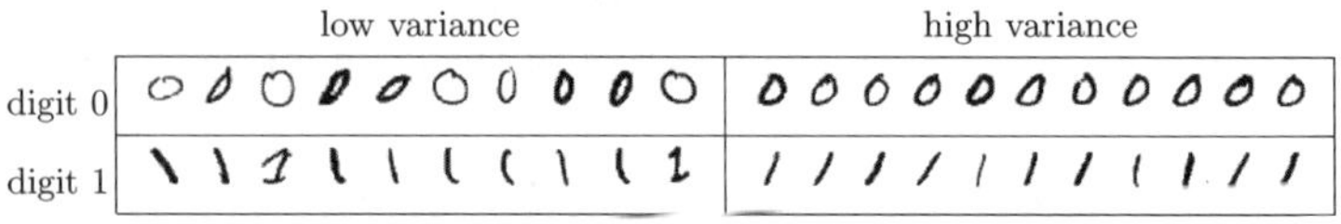

	low variance	high variance
digit 0	○ ◐ ○ ● ● ○ ○ ● ● ○	● ○ ○ ○ ● ○ ○ ○ ● ○
digit 1	╲ ╲ 𝟐 ╲ ╲ ((╲ (𝟐	/ / / / │ / / (│ / /

Fig. 12.6. The 10 sample points having lowest and highest relevance for the class corresponding to the digits 0 and 1, computed according to the FCM algorithm.

Example 12.6. Moving to the first 2000 digits in the MNIST dataset, equally partitioned in training and test sets, the above SVM classifier may be employed when we focus on a pair of two digits at a time. Namely, the FCM algorithm can be used to discover the relevance of the each example through the maximum of its membership grades to the various clusters, as done in Example 10.1. Namely, according to (6.26) and (6.27), with $\nu(\boldsymbol{y}_i - \boldsymbol{v}_j) = l^2(\boldsymbol{y}_i - \boldsymbol{v}_j)$, the relevance h_i is given by:

$$h_i = \max_{c=1,\ldots,k} \left\{ \left(\sum_{j=1}^{k} \left(\frac{l^2(\boldsymbol{y}_i - \boldsymbol{v}_c)}{l^2(\boldsymbol{y}_i - \boldsymbol{v}_j)} \right)^{\frac{1}{r-1}} \right)^{-1} \right\}, \quad (12.21)$$

Relevance as a non ambiguity measure,

Focusing, for instance, on the classes corresponding to digits 0 and 1, Fig. 12.6 shows the sample points having lowest and highest relevance, where a fuzzification factor r was equal to 0.2. A linear relevance-based SVM algorithm scores a 100% accuracy on both training and test sets (having size 213 and 211, respectively). With digits 3 and 8, things get worse. As the clustering procedure does not find sensibly different clusters, relevances have to be assigned manually after visual inspection. Just to check the viability of the procedure, we set to 1 and -1 the relevance of the sample points in Fig. 12.7 denoting particularly well and poorly shaped digits, and set to 0 the relevance of

Separating Points through SVM 421

	highest relevance	lowest relevance
digit 3	3 3 3 3 3 3	8 8 6 8 8 8
digit 8	8 8 8 8 8 8	3 3 3 3 3 3

Fig. 12.7. Sample points having highest and lowest relevance for the class corresponding to the digits 3 and 8, as manually assigned after visual inspection.

Fig. 12.8. Shapes of the misclassified digits for classes 3 and 8.

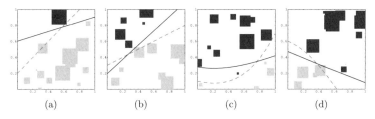

(a)   (b)   (c)   (d)

Fig. 12.9. SVM classifying squares. Gray colors: original classes. Dashed curves: original discriminating functions; plain curves: obtained classifiers.

remaining points. With this assignment the algorithm scored 100% and 90% accuracy in training and test sets, respectively. As it emerges from Fig. 12.8 we have a mix of misclassified digits having ambiguous shapes along with others that appear sufficiently clean. Despite this we have similar values of error rates (ranging from 0.4% to 12%) in comparison with other algorithms available in the literature (see for instance [14]).

Example 12.7. Further insights on the algorithm come from its performance with the problem of learning to discriminate between two classes of square regions, rather than points, in a given space. In this case, each pattern is described by a center x and an area $a > 0$ that identifies with patterns and their relevance, respectively. In this way, regions with larger areas will be assigned higher relevance and the algorithm will be pushed to find a decision surface having higher distance from their centers. Fig. 12.9 shows the results of this method for some instances of this problem, both in the linear and the nonlinear variants. In both cases

or simply a topological feature.

a set of square regions has been randomly generated and subsequently labeled according to a prefixed separation surface (the dashed curve), while the relevance values have been rescaled, in order to obtain also negative values. With points qualified in this manner the SVM algorithm identifies classifiers (the black curves) with a tendency to avoid intersecting the squares, except for regions of very small area.

12.1.2 Regression

Computing regression is a bit more complex.

We move to a *regression* problem when labels of sample z in (12.1) can range in a continuous set $\mathcal{Y}$. In the linear regression instances we face again with the problem of inferring a hyperplane $y = \boldsymbol{w} \cdot \boldsymbol{x} + b$, now related to the sample points as in (3.42) through the relation:

$$y_i = \boldsymbol{w} \cdot \boldsymbol{x}_i + b + \varepsilon_i, \qquad (12.22)$$

where ε_i plays the role of random shift of y_i from its regressed value $\boldsymbol{w} \cdot \boldsymbol{x}_i + b$. In order to reuse the SVM identification methods to infer the regressor hyperplane, we will follow two alternate methods, respectively related to the concepts of ϵ-*tube* and *sliding trick*.

The ϵ-tube

An ϵ wide strip plus slacks for reckoning outliers

We give to the inference the goal of confining all observations $(\boldsymbol{x}_i, y_i)$ at a distance at most ϵ from the hyperplane $(\boldsymbol{w}, b)$ (the ϵ-tube); if this is not possible we introduce non-null values for the *slack variables* ξ_i or $\widehat{\xi}_i$. This translates to the following constrained optimization problem:

$$\min_{\boldsymbol{w},b} \frac{1}{2} l^2(\boldsymbol{w}) + \grave{a} \sum_{i=1}^{m} \left(\xi_i + \widehat{\xi}_i \right) \qquad (12.23)$$

$$y_i \geq \boldsymbol{w} \cdot \boldsymbol{x}_i + b - \epsilon - \xi_i, \quad \forall i = 1, \ldots, m \quad (12.24)$$

$$y_i \leq \boldsymbol{w} \cdot \boldsymbol{x}_i + b + \epsilon + \widehat{\xi}_i, \quad \forall i = 1, \ldots, m \quad (12.25)$$

$$\xi_i, \widehat{\xi}_i \geq 0 \quad \forall i = 1, \ldots, m. \qquad (12.26)$$

In particular:

- the first term in the objective function still allows for controlling the regressor capacity and thus avoids overfitting;
- the constraints are automatically satisfied whenever the label y_i lies within the ϵ-tube, as illustrated for pattern $\boldsymbol{x}_1$ in Fig. 12.10; consequently, $\xi_1 = \widehat{\xi}_1 = 0$;

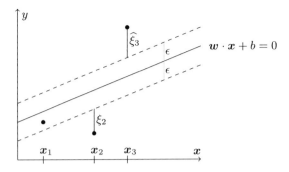

Fig. 12.10. The geometrical interpretation of the problem underlying the ϵ-insensitive SVM approach to regression.

- if previous point cannot be satisfied, either $\xi_i = 0$ and $\widehat{\xi}_i > 0$ will measure the distance between y_i and the ϵ-tube or *vice versa* (see, respectively, patterns $\boldsymbol{x}_2$ and $\boldsymbol{x}_3$ in Fig. 12.10);
- each example falling out of the ϵ-tube is penalized in the objective function minimization through the linear loss function $\xi_i + \widehat{\xi}_i$, that is through the value of its distance from the ϵ-tube, while the remaining points have a null loss.

The parameters ϵ and $\grave{\alpha}$, allowing for tailoring the inference process to the investigated learning instances, are fixed beforehand. Following the same technique adopted for classification involves a larger number of Lagrange multipliers – namely, α_i and $\widehat{\alpha}_i$ for the constraints (12.24-12.25), β_i and $\widehat{\beta}_i$ for (12.26). This leads to the following dual objective that is not affected, however, by β_i and $\widehat{\beta}_i$:

$$\mathscr{L}'(\boldsymbol{\alpha}) = -\frac{1}{2} \sum_{i,j=1}^{m} (\alpha_i - \widehat{\alpha}_i)(\alpha_j - \widehat{\alpha}_j) x_i \cdot x_j$$

$$- \sum_{i=1}^{m} (\alpha_i - \widehat{\alpha}_i) y_i - \epsilon \sum_{i=1}^{m} (\alpha_i + \widehat{\alpha}_i), \quad (12.27)$$

to be maximized under the constraint:

$$\sum_{i=1}^{m}(\alpha_i - \widehat{\alpha}_i) = 0 \qquad (12.28)$$

$$0 \leq \alpha_i, \widehat{\alpha}_i \leq \grave{\alpha}. \qquad (12.29)$$

Once the optimal solution $\alpha_1^*, \ldots, \alpha_m^*, \widehat{\alpha}_1^*, \ldots \widehat{\alpha}_m^*$ has been found, the regressor is obtained through:

$$\boldsymbol{w}^* = -\sum_{i=1}^{m}(\alpha_i^* - \widehat{\alpha}_i^*)\boldsymbol{x}_i \qquad (12.30)$$

$$b^* = \begin{cases} y_i - \boldsymbol{w}^* \cdot x_i - \epsilon & \text{for any } i \text{ such that } 0 < \alpha_i^* < \grave{\alpha} \\ y_i - \boldsymbol{w}^* \cdot x_i + \epsilon & \text{for any } i \text{ such that } 0 < \widehat{\alpha}_i^* < \grave{\alpha}, \end{cases} \qquad (12.31)$$

where the second equation exploits the special relationships:

$$\alpha_i \left(\epsilon + \xi_i + y_i - \boldsymbol{w} \cdot \boldsymbol{x}_i - b\right) = 0 \qquad (12.32)$$

$$\widehat{\alpha}_i \left(\epsilon + \widehat{\xi}_i - y_i + \boldsymbol{w} \cdot \boldsymbol{x}_i + b\right) = 0 \qquad (12.33)$$

$$(\grave{\alpha} - \alpha_i)\xi_i = 0 \qquad (12.34)$$

$$(\grave{\alpha} - \widehat{\alpha}_i)\widehat{\xi}_i = 0. \qquad (12.35)$$

coming from the *Karush-Kuhn-Tucker* (or KKT) conditions involving the optimal values of primal and dual variables [13]. The solution of this problem, which still exhibits a quadratic behavior, has the following properties:

a few support vectors,

- a pattern $\boldsymbol{x}_i$ characterized by either $\alpha_i^* = \grave{\alpha}$ or $\widehat{\alpha}_i^* = \grave{\alpha}$ will lie outside the ϵ-tube centered around the regressor;
- the fact that $0 < \alpha_i^* < \grave{\alpha}$ implies $\xi^* = 0$, and analogously $0 < \widehat{\alpha}_i^* < \grave{\alpha}$ implies $\widehat{\xi}^* = 0$;
- for each $i = 1, \ldots, m$ the product $\alpha_i \widehat{\alpha}_i$ is null, as for each observation only one of the slack variables can assume non-null values.

Moving to nonlinear regression is also in this case pursued through the use of kernels, substituting occurrences of dot products in (12.27-12.29) with a proper mapping accounting for the transformation of patterns in a new space where to compute the dot product.

Example 12.8. Consider the Swiss dataset. If we select the *fertility* and *examination* features and plot the corresponding points, we get the scatterplots of Fig. 12.11. The three graphs in the figure show the regressors learnt from these data from SVM algorithm, using the linear variant and a 2-nd and 3-rd degree polynomial kernel, respectively.

Other SVM approaches to regression

We may solve the regression problem using other objective functions to be fed in the SVM identification procedures. For instance:

As for variants, you may change cost functions,

- ϵ-*insensitive loss* can adopt a quadratic cost instead of the linear one in (12.23) with the effect of adding diagonal terms in the dual objective function, which now becomes:

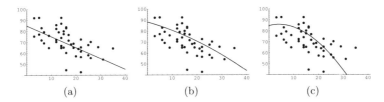

Fig. 12.11. Scatterplots of the *fertility* (x axis) and *examination* (y axis) features in the Swiss dataset, contrasted with the regressors learnt via SVM algorithm using: (a) the linear variant, (b) a quadratic, and (c) a cubic kernel.

$$\mathscr{L}''(\boldsymbol{\alpha}) = -\frac{1}{2}\sum_{i,j=1}^{m}(\alpha_i - \widehat{\alpha}_i)(\alpha_j - \widehat{\alpha}_j)\left(\boldsymbol{x}_i \cdot \boldsymbol{x}_j - \frac{1}{C}\delta_{ij}\right) +$$
$$-\sum_{i=1}^{m}(\alpha_i - \widehat{\alpha}_i)\, y_i - \epsilon \sum_{i=1}^{m}(\alpha_i + \widehat{\alpha}_i), \quad (12.36)$$

where δ_{ij} denotes the Kronecker delta, i.e., $\delta_{ij} = 1$ when $i = j$ and equals zero otherwise.

- *Ridge loss* adopts, in addition to the former, a vanishing ϵ-tube by explicitly setting $\epsilon = 0$. This leads to the minimization of $\lambda l^2(\boldsymbol{w}) + \sum_{i=1}^{m} \xi_i^2$ subject to $y_i - \boldsymbol{w} \cdot \boldsymbol{x}_i - b = \xi_i$ for $i = 1, \ldots, m$, so that variable ξ_i assumes the meaning of drift of the i-th point from the regressor. Introducing kernels as well, this formulation admits the following analytical solution:

$$b^* = \frac{\boldsymbol{u}(K + \lambda I)^{-1}\boldsymbol{y}}{\boldsymbol{u}(K + \lambda I)^{-1}\boldsymbol{u}}, \quad (12.37)$$

$$\boldsymbol{\alpha}^* = 2\lambda(K + \lambda I)^{-1}(\boldsymbol{y} - b^*\boldsymbol{u}), \quad (12.38)$$

$$\boldsymbol{w}^* = \frac{1}{2\lambda}\sum_{i=1}^{m}\alpha_i \boldsymbol{x}_i, \quad (12.39)$$

where $\boldsymbol{u}$ is the vector whose m components are all set to 1, K is the $m \times m$ Gram matrix whose (i,j)-th entry is $k(\boldsymbol{x}_i, \boldsymbol{x}_j)$, $\boldsymbol{y}$ is the vector gathering all examples' labels (see Appendix D.2), and I is the $m \times m$ identity matrix.

Example 12.9. Consider the sinc function, defined as $\operatorname{sinc}(x) = \frac{\sin(\pi x)}{\pi x}$. Fig. 12.12 shows the graph of two regressors learnt through a Gaussian kernel from a sample:

$$\{(x_i, \operatorname{sinc}(x_i) + e_i), \quad i = 1, \ldots, m\}, \quad (12.40)$$

where $m = 100$, x_is are uniformly drawn from $[-6, 6]$, and e_i denotes an error term. More precisely, the black curves in Fig. 12.12(a)

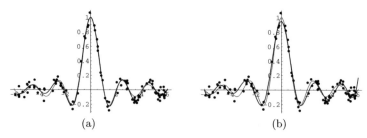

Fig. 12.12. Nonlinear regressors learnt through a Gaussian kernel from a sample of the sinc function, using the SVM algorithm exploiting (a) the ridge and (a) the quadratic ϵ-insensitive loss function. Dots: observations; gray line: sinc function; black line: learnt regressors.

and (b) show the plot of the regressors learnt from the sample via the ridge and the quadratic ϵ-insensitive losses, respectively. In both cases these plots are contrasted with the sinc plot, drawn in gray, and with the sample values.

The sliding trick

Put all points into a moderately wide tube.

We give to inference the goal of minimizing the distance of the farthest points from the hyperplane. This corresponds to a tube with variable ϵ but constant number of patterns inside – all patterns apart from outliers are discarded in advance. Moving to the implicit representation of the goal hyperplane in the form $\boldsymbol{w} \cdot \boldsymbol{z} + q = 0$, with $\boldsymbol{z} = (\boldsymbol{x}, y)$, the optimization problem may be enunciated as follows:

Definition 12.1. *[Original problem] Given a set of points $S = \{\boldsymbol{z}_i, i = 1, \ldots, m\}$, maximize the norm of $\boldsymbol{w}$ under the constraint that all points have functional distance $|\boldsymbol{w} \cdot \boldsymbol{z}_i + q|$ less or equal to 1 from the hyperplane $\boldsymbol{w} \cdot \boldsymbol{z} + q = 0$. In formulas:*

$$\max_{\boldsymbol{w},b} \left\{ \frac{1}{2} l^2(\boldsymbol{w}) \text{ such that } \eta_i(\boldsymbol{w} \cdot \boldsymbol{z}_i + q) \leq 1 \ \forall i \right\}, \quad (12.41)$$

where $\eta_i = \text{Sign}(\boldsymbol{w} \cdot \boldsymbol{x}_i + b)$.
In terms of Lagrangian multipliers $\alpha_i \geq 0$, (12.41) reads:

$$\max_{\boldsymbol{w},b} \left\{ \frac{1}{2} l^2(\boldsymbol{w}) - \sum_{i=1}^{m} \alpha_i \left(\eta_i(\boldsymbol{w} \cdot \boldsymbol{z}_i + q) - 1 \right) \right\}. \quad (12.42)$$

Shift the points to fall in the usual classification problem,

In the present form this problem cannot be dually enunciated because it misses a saddle point in the space $\boldsymbol{w} \times \boldsymbol{\alpha}$. Hence, to fulfill this condition and work with a dual problem with $\boldsymbol{\alpha}$ we consider the equivalent problem which we obtain after having symmetrically slid the points orthogonally to the regression hyperplane, by a fixed quantity that is sufficient to swap the positions w.r.t. the hyperplane of the farthest points [2]. In this way the closest point from one side of it becomes the farthest from the other side (see Fig. 12.13(a)) and the maximization problem in (12.42) translates to a minimization one. Namely:

Definition 12.2 (Modified problem). *For hyperplane, points S and labels as in Definition 12.1 and a suitable instantiation of the hyperplane $\boldsymbol{w} \cdot \boldsymbol{z}_i + b = 0$, denote by $\boldsymbol{w}^*$ the direction orthogonal to it and by $d_{\max} = \max_{\boldsymbol{z} \in S}\{|\boldsymbol{w}\cdot\boldsymbol{z}+b|\}$. Then, map S into S' through the mapping: $\boldsymbol{z}'_i = \boldsymbol{z}_i + \eta_i(d_{\max}+\epsilon)\boldsymbol{w}^*$ with small and positive ϵ. Then find a solution to the following problem:*

$$\min_{\boldsymbol{w},q} \left\{ \frac{1}{2}l^2(\boldsymbol{w}) \text{ such that } \eta_i(\boldsymbol{w} \cdot \boldsymbol{z}'_i + q) \geq 1 \; \forall i \right\} \quad (12.43)$$

i.e.,

$$\min_{\boldsymbol{w},q} \left\{ \frac{1}{2}l^2(\boldsymbol{w}) - \sum_{i=1}^{m} \alpha_i \left(\eta_i(\boldsymbol{w} \cdot \boldsymbol{z}'_i + q) - 1 \right) \right\}. \quad (12.44)$$

The dual formulation of this problem reads:

$$\max_{\boldsymbol{\alpha}} \sum_{i=1}^{m} \alpha_i - \frac{1}{2} \sum_{i,j=1}^{m} \eta_i \eta_j \alpha_i \alpha_j \boldsymbol{z}'_i \cdot \boldsymbol{z}'_j \quad (12.45)$$

$$\sum_{i=1}^{m} \eta_i \alpha_i = 0; \alpha_i \geq 0, \forall i = 1, \ldots, m. \quad (12.46)$$

Note that all we require from the point shifts is that they are orthogonal to the separating hyperplane and to the same extent but in a different direction, depending on whether the point belongs to one or the other of the separated half spaces. This allows to revert the maximization problem (12.42) into the minimization problem (12.44) on the same function but different arguments. Apart from rare pathologies in the hyperplane original instantiation, the procedure that computes y_is translates points according to the running hyperplane and updates the latter on the basis of the above operations has a fixed point in the solution of the problem in Definition 12.1. In particular, continuing our illustrative example we obtain the line in Fig. 10.13(b), where $\boldsymbol{w} = \sum_{i=1}^{m} \alpha_i \eta_i \boldsymbol{z}'_i$ and $q = 1/n_s \sum_{i=1}^{n_s}(\eta_i - \sum_{j=1}^{n_s} \alpha_j \eta_j \boldsymbol{z}'_j \cdot \boldsymbol{z}_i)$, where both sums range on the n_s support vectors corresponding to the non null αs obtained from (12.45). As usual, we find the optimum $\boldsymbol{w}^*$, leaving b locating the hyperplane in the middle of the farthest points of the original S along $\boldsymbol{w}^*$ direction. We synthesize the steps of this procedure in Algorithm 12.1.

then revert to the original configuration.

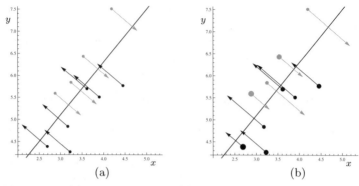

Fig. 12.13. (a) Symmetrical and (b) relevance-based shifts of sample points orthogonally to a tentative regression line.

Algorithm 12.1. Sliding interpolation

Given a training set $z_m = \{(x_i, y_i), x_i \in \mathbb{R}^n, y_i \in \mathbb{R}, i = 1, \ldots, m\}$, and denoting by $w \cdot z + q$ the target hyperplane, fix a norm ν and a threshold θ

---

1. Assign feasible values to w and q
2. **repeat**
   2.1. Slide z_is into z'_is along w^* according to Definition 12.2
   2.2. Compute solution of (12.45), $\widetilde{w} = \sum_{i=1}^{m} \alpha_i \eta_i z'_i$ and $\widetilde{q} = 1/n_s \sum_{i=1}^{n_s}(\eta_i - \sum_{j=1}^{n_s} \alpha_j \eta_j z'_j \cdot z_i)$
   2.3. Set $w = \widetilde{w}$ and $q = \widetilde{q}$
   until $\nu(w - \widetilde{w}) < \vartheta$

As can be seen in Fig. 12.14 in contrast with Fig. 10.13 (and in Fig. 12.14(b) later on) the line computed on the basis of the support vectors lies in an opposite position than the optimal regression line from (10.11) w.r.t. the MLE curve, thus denoting possible drawbacks with the current regression goal proving a simplification of MLE's. To compensate it, we may render the data processing more sophisticated by introducing a nonlinear regressor through kernels as in the previous section.

Exploiting data quality in SVM regression

<small>We may introduce data quality at low expenses,</small> The same motivation illustrated in previous section when dealing with the quality of single observations is valid also for the case of regression. The search for a regressor can exploit the additional information provided by the pattern relevance r_i using

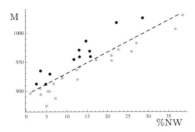

Fig. 12.14. Regression line minimizing the distance of the farthest points from itself, computed from dataset in Fig. 10.12.

an approach similar to that presented in the previous sections. In particular, we may weight with r_i the drifts w.r.t. the regressor.

Focusing, for instance, on ridge regression, this translates into the modification of the objective function to be minimized into $\lambda l^2(\boldsymbol{w}) + \sum_{i=1}^{m} r_i \xi_i^2$, while the constraints are unchanged. The solutions remain the same as (12.37-12.39) apart substituting I with the $m \times m$ diagonal matrix R whose non-null elements are $r_1, \ldots, r_m$. They coincide with:

- the original version solution (12.37-12.39) when $r_i = 1$ for each $i = 1, \ldots, m$;
- the heteroskedastic kernel ridge regression model proposed in [5], where r_i becomes the inverse of the standard deviation for the statistical model associated to the i-th point.

just a change in a diagonal matrix with ridge regeression,

Example 12.10. Consider the so-called motorcycle dataset, describing several accelerometer readings in crash-tests experiments on the efficacy of helmets [17]. It contains 133 pairs (t, g), where t refers to time measurements (not evenly spaced and possibly containing multiple observations for a same time) and g is the related acceleration. These measurements are affected by non-uniform error, as highlighted in the scatter plot of Fig. 12.15, where data are partitioned into three subsets: in the left part of the plot, roughly speaking for $t < 14$, the points are highly concentrated around their mean value; in the middle of the plot, say for $14 \leq t < 30$, the points' variance gets higher; finally, in the remaining part of the plot the points' variance further increases. Subsequently, we define the three relevance values: $r_{\text{low}} = 10^3$, $r_{\text{med}} = 10^2$ and $r_{\text{high}} = 10$. Then using a Gaussian kernel we obtain the results shown in Fig. 12.15, where the regressor is compared to the counterpart in output of the standard SVM algorithm, denoting a less smoothness of the curve computed by the latter than ours.

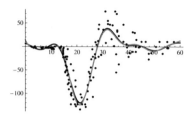

Fig. 12.15. Examples of the learnt regressors for the motorcycle dataset with (gray curve) and without (black curve) data quality exploitation.

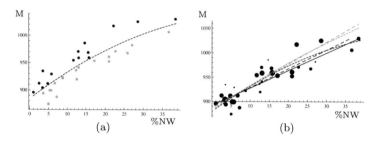

Fig. 12.16. Improving regression in Fig. 12.14 by considering non linearities and points' quality. (a) SVM parabola; (b) MLE line (coarse-grain dashes gray line) and MLE parabola (fine-grain dashes gray curve), optimal regression line (plain black line), SVM regression line (coarse-grain dashes black line) and SVM parabola (fine-grain dashes black curve).

With the sliding trick and the same notation as in Definition 12.2, the solution of the problem taking into account relevance may be achieved after a preliminarily map of S into S^t through the mapping $\boldsymbol{x}_i^t = \boldsymbol{x}_i - \eta_i g(r_i)\boldsymbol{w}^*$, where r_i ranges in $[0, +\infty)$ and g possibly introduces some nonlinearities on the relevance appreciation in order to privilege more/less relevant points – e.g. $g(r_i) = r_i^v$ with $v > 1$ to strengthen the influence of relevant points, and the opposite behavior for $0 < v < 1$. Then we slid S^t into S'^t and proceed with the optimization problem (12.44) with $\boldsymbol{x}_i'$ correspondingly substituted by $\boldsymbol{x}_i'^t$. An analogous formulation holds with hyperplanes in the extended $\boldsymbol{\xi}$ space induced by kernels. While in principle this figures as an extension of the procedure adopted in Fig. 12.13(a), with the sole variant that the point shifts are specifically calibrated, the problem is that the virtual distances depend now not only on the versor but also on the position of the hyperplane (i.e. on the b coefficient). Thus we must find both parameters in the fixed point of the whole procedure. This may require some dumping operator, such as

and a modulation of shifts with the sliding trick.

exponential smoothing, to converge to a fixed point. This occurs in Fig. 12.16, referring the SMSA dataset affected by the relevances computed as in Example 10.1, where we substituted the dot product in the last procedure with a polynomial kernel computing the class of parabolas. Fig. 12.16(a) shows such a curve minimizing the distance (in the feature space) between the farthest points according to the relevance score correction, while Fig. 12.16(b) summarizes the types of forms obtained so far.

12.2 Distributed Architectures for Granular Computing

While there has been a wealth of methodological and algorithmic developments in fuzzy modeling, the subject of distributed and collaborative fuzzy models has not been investigated in great detail. For instance, a lot has been said about rule-based fuzzy models of the form "**if** x is A_c **then** $y = h_c(x, a_c), c = 1, 2, \ldots, k$" as in Section 10.6.2, where h_c denotes a local model endowed with some parameters a_c. What if we encounter individual data sites $\{D(1), \ldots, D(q)\}$ for which such models have to be constructed? Not only they have to be formed on a basis of locally available data $D(i), i = 1, 2, \ldots, q$ but they need to collaborate and exchange their findings, reconcile eventual differences and collectively develop fuzzy constructs. What is visible though, is that the communication dwells on *knowledge* rather than *data*. In communication of this character, we witness a process of knowledge sharing. Formally, the underlying knowledge residing at data site i and being shared between the individual sites can be concisely described as $\chi(i)$. For instance, for the rule based-systems, the shared knowledge assumes the form $\chi(i) = \{A_c(i), c = 1, \ldots, k\}$ where $A_c(i)$s are the information granules formed at $D(i)$. The knowledge of these fuzzy sets is communicated to all other data sites. We may have another format of $\chi(i)$ being a more comprehensive version of the previous knowledge sharing which concerns now both the information granules and the local models, that is $\chi(i) = \{A_c(i), h_c(i), a_c(i)\}$. A schematic, high-end visualization of such machinery of knowledge sharing is presented in Fig. 12.17 where each agent develops and evolves on a basis of experimental data that are locally available and communicates with other agents establishing some interaction at the global level.

Knowledge may be more profitable to exchange than data:

There is a vast array of mechanisms of interaction between individual components of the distributed system. Two of them deserve careful attention. In the first one all nodes operate locally while collaboration can be established between any two of them, see Fig. 12.18(a). The linkages between the nodes are

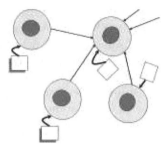

Fig. 12.17. An overview of a multi-agent distributed system. Dark center: computing core; surrounding light gray region: communication layer.

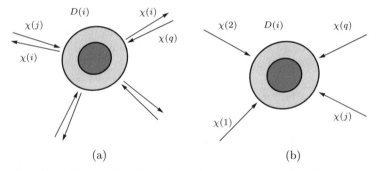

Fig. 12.18. Examples of modes of interaction in a distributed system: (a) collaboration between any two nodes of the system, and (b) experience consistent fuzzy modeling.

it may be either a symmetric or an asymmetric exchange, bidirectional; knowledge sharing is realized in both directions. Node i benefits from knowledge shared with it by node j and *vice versa*: knowledge acquired and conveyed by node j helps to carry out processing completed at node i. The scenario illustrated in Fig. 12.18(b) exhibits a highly asymmetric behavior: a single node benefits from various sources of knowledge, say models, already formed at some other nodes thus augmenting its performance along with the locally available data. This mode of interaction can be referred to as a formation of experience-consistent fuzzy models as the obtained model is not only formed with the aid of some locally available data but, what is more important, takes advantage of some previous experience captured by the already constructed fuzzy models. Note that in this case of interaction, the connections are unidirectional, namely one data site becomes affected by some other models.

There is also a great deal of in-between collaboration scenarios in which the nodes could engage in some selective interaction where strength of interaction itself could vary quite substantially from node to node. There is also no need to have a fully connected network of nodes. In case of bidirectional links they need not be symmetric. There could be stronger impact exerted by node j on node i while a far weaker connection could be established for interaction realized in the opposite direction.

or an intermediate transaction, of course.

12.2.1 Collaborative Clustering

The communication of knowledge involves a structure $\chi(i)$ of node i which embraces a collection of information granules. Considering that such granules have been constructed with the use of the FCM algorithm discussed in Section 6.4.1, they are fully characterized by either their prototypes $\{v_1(i), \ldots, v_{k_i}(i)\}$ or partition matrix $U(i)$. The latter represents the most used operational tool, that we extend in terms of partitioned matrix $U(i|j)$ induced by a node j to reckon communications between the two nodes. By contrast, we use prototypes as comfortable knowledge communication vehicle between the nodes. To ensure consistency, whenever not expressely indicated, we will assume all nodes sharing the same number of prototypes k. With the help of prototypes at node j we construct partition matrix $U(i|j)$ patterns $\{y_1(i), \ldots, y_{m_i}(i)\}$ belonging to data site $D(i)$ of node i. This is done in analogy to (6.27):

You transmit centroids to modify partition matrices;

$$u_{ct}(i|j) = \frac{1}{\sum_{\kappa=1}^{k} \left(\frac{\nu(\boldsymbol{y}_t(i)-\boldsymbol{v}_c(j))}{\nu(\boldsymbol{y}_t(i)-\boldsymbol{v}_\kappa(j))} \right)^{\frac{1}{r-1}}}. \quad (12.47)$$

Refer also to Fig. 12.19 which highlights the essence of this collaboration mechanism by showing how the communication links have been established. Proceeding with all other data sites, $D(1), \ldots, D(i-1), D(i+1), \ldots, D(q)$, we end up with $q-1$ induced partition matrices, $U(i|1), \ldots, U(i|i-1), U(i|i+1), \ldots, U(i|q)$. The minimization of differences between $U(i)$ and $U(i|j)$ is used to establish some collaborative activities occurring between the data sites. At the i-th site, the clustering is guided by the augmented objective function assuming the following form:

the goal is to make them consistent with bordering data sites as well.

$$\mathsf{C}(i) = \sum_{c=1}^{k} \sum_{t=1}^{m_i} u_{ct}(i)^2 \nu(\boldsymbol{y}_t(i) - \boldsymbol{v}_c(i)) +$$
$$+ \alpha \sum_{\substack{j=1 \\ j \neq i}}^{q} \sum_{c=1}^{k} \sum_{t=1}^{m_i} (u_{ct}(i) - u_{ct}(i|j))^2 \nu(\boldsymbol{y}_t(i) - \boldsymbol{v}_c(j)), \quad (12.48)$$

where α is a certain nonnegative number. The objective function $C(i)$ consists of two components. The first one is nothing but a standard sum of weighted distances between the patterns in $D(i)$ and their prototypes. In this sense, it is just the objective function encountered in the standard FCM being applied to $D(i)$ where the customary fuzzification coefficient is $r = 2.0$. The second component reflects an impact coming from the structures formed at all remaining data sites. The distance between the optimized partition matrix and the induced partition matrices is to be minimized – this requirement is captured by this part of the objective function (12.48). The scaling coefficient α strikes a sound balance between the optimization guided by the structure in $D(i)$ and the already developed structures available at the remaining sites. The value of α implies a certain level of intensity of collaboration; the higher its value, the stronger the collaboration. For $\alpha = 0$ no collaboration occurs and the problem reduces to the collection of q independent clustering tasks being confined to the corresponding data sites. In brief, the problem of collaborative clustering can be defined as follows:

> Given a finite number of disjoint data sites with patterns defined in the same feature space, develop a scheme of collective development and reconciliation of a fundamental cluster structure across the sites that is based upon exchange and communication of local findings where the communication needs to be realized at some level of information granularity. The development of the structures at the local level exploits the communicated findings in an active manner through minimization of the corresponding objective function augmented by the structural

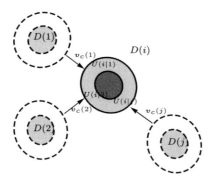

Fig. 12.19. Data sites and communication realized through passing prototypes $v_c(j)$, with $c = 1, \ldots, k$, and the consecutive generation of the induced partition matrices $U(i|j)$.

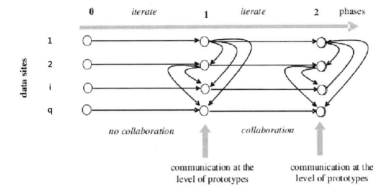

Fig. 12.20. A general functional view at the processing realized in collaborative clustering.

findings developed outside the individual data site. We also allow for retention of key individual (specific) findings that are essential (unique) for the corresponding data site.

Example 12.11. An important and visible category of applications deals with wireless sensor networks. In such communicating systems, we envision a collection of randomly scattered sensors whose communication is established on *ad hoc* basis [6, 1]. Each node (sensor) collects the data available in its neighborhood and realizes their processing leading to the determination of the local characteristics of data (say, formulated as a collection of clusters being observed at this particular local level of the given sensor). At the same time it is recognized that the local processing could benefit from some collective activities established between the sensors. This need for a global and highly collective style of processing is motivated by a limited amount of data available locally and a need to establish a global view at the data collected by the overall network.

A selective communicating protocol

12.2.2 The General Flow of Collaborative Processing

The essence of collaborative clustering pertains to the development of structures at individual data sites on the basis of effective communication of the findings obtained at the level of the individual data sites. There are two phases, namely an optimization of the structures at the individual sites and an interaction between them when exchanging the findings. They intertwine so that these two phases occur in a fixed sequence described by Algorithm 12.2 and visualized in Fig. 12.20.

Algorithm 12.2. Collaborative FCM

Given a set of data sites $\{D(1),\ldots,D(q)\}$, and their patterns $\{\{y_1(i),\ldots,y_{m_i}(i)\}, \ i=1,\ldots,q\}$, a collaboration coefficient α and a termination criterion for the collaboration activity,

---

Inizialization.

1. Choose the number of clusters k to be looked for in the collaborative clustering
2. Carry out FCM clustering to collect prototypes $\{\{v_1(i),\ldots,v_k(i)\}, \ i=1,\ldots,q\}$ for each data site

Collaboration.

3. **repeat**
   3.1. Communicate the results about the structure determined at each data site (*communication step*)
   3.2. **for** each site i, minimize (12.48) through iterative calculations of partition matrices and prototypes as follows (*optimization step*):

$$u_{ct}(i) = \frac{1}{\sum_{\kappa=1}^{k} \frac{l^2(y_t(i)-v_c(j))}{l^2(y_t(i)-v_\kappa(j))}} \left(1 - \sum_{\kappa=1}^{k} \frac{\eta \sum_{\substack{j=1 \\ j \neq i}}^{q} u_{\kappa t}(i|j)}{1+\eta(q-1)}\right)$$

$$+ \frac{\eta \sum_{\substack{j=1 \\ j \neq i}}^{q} u_{ct}(i|j)}{1+\eta(q-1)} \quad (12.49)$$

$$v_{cd}(i) = \frac{\sum_{t=1}^{m_i} u_{ct}^2(i) y_{td} + \eta \sum_{\substack{j=1 \\ j \neq i}}^{q} \sum_{t=1}^{m_i} (u_{ct}(i) - u_{ct}(i|j))^2 y_{td}}{\sum_{t=1}^{m_i} u_{ct}^2(i) + \eta \sum_{\substack{j=1 \\ j \neq i}}^{q} \sum_{t=1}^{m_i} (u_{ct}(i) - u_{ct}(i|j))^2}$$

(12.50)

until termination condition has been satisfied.

A communication-minimization loop

Initially, the FCM algorithm is run independently at each data site (which happens without any communication). After FCM has been terminated at each site, processing stops and the data sites communicate their findings. As already stressed, this communication needs to be realized at some level of information granularity. The effectiveness of the interaction depends on the way in which one data site "talks" to others in terms of what has been discovered so far. Once communication has been established and the nodes are informed about structural findings at other sites, each site proceeds with its optimization pursuits by focusing on the local data while taking into consideration the findings communicated by other data sites. These optimization processes are run independently from each other. Once all of them have declared termination of computing, they are ready to engage in the communication phase. Again they communicate the findings and set up new conditions for the next phase of the FCM optimization. The pair of optimization and communication processes is referred to as a collaboration phase. The overall collaboration takes a finite number of collaboration phases (phases, for short), which terminates once no further significant change in the revealed structure is reported.

As has become clear from this high-end description of the collaboration, there are two important components being crucial to the overall process. First, we have to specify a way of communicating and representing findings at some level of granularity (let us recall that we are not allowed to communicate at the level of individual data but have to establish communication at the higher level of abstraction by engaging the exchange of the granular constructs). Second, we have to come up with an augmented objective function whose minimization embraces both the structures at the local level of the individual data sites and reconciles them with the structures communicated by other data sites.

How and what to communicate

12.2.3 Evaluation of the Quality of Collaboration

The evaluation of the quality of the results of collaboration between the data sites requires a careful assessment. As there are partition matrices associated to each of the $D(i)$s, one could think of computing distance between them and treat it as a measure of quality of the ongoing process. While the idea sounds convincing, its realization requires more attention. We should stress the fact that a direct comparison of two partition matrices could not be feasible as we may not have a direct correspondence between their rows (respective clusters). This is a well-known problem identified in the literature. To get around this shortcoming, we use the concept of proximity and proximity matrix induced by a given partition matrix. Let us recall that for any partition matrix $U = [u_{ct}], c = 1, 2, \ldots, k, t = 1, 2, \ldots, m_i$, an induced proximity matrix, that is $\text{Prox} = [\text{prox}_{ts}], t, s = 1, 2, \ldots, m_i$, comes with entries which satisfy the following properties:

Forming a compromise between global and local characteristics of data

Collaboration is a matter of proximity,

1. symmetry $\text{prox}_{ts} = \text{prox}_{st}$
2. reflexitivity $\text{prox}_{tt} = 1.$

Interestingly enough, here we do not require transitivity (which, albeit nice to have, is always difficult to achieve in practice). The proximity values are based on the corresponding membership degrees present in the partition matrix:

$$\text{prox}_{ts} = \sum_{c=1}^{k} \min\{u_{ct}, u_{cs}\}. \qquad (12.51)$$

It is worth noting that the proximity matrix is more abstract in this form than the original partition matrix it is based upon. It abstracts the clusters themselves and this is what we really need in this construct. Given the proximity matrix, we cannot retrieve the original entries of the partition matrix it was generated from.

Let us consider now the i-th data site with its partition matrix $U(i)$ and the induced partition matrices $U(i|j), j = 1, 2, \ldots, i-1, i+1, \ldots, q$. To quantify the consistency between the structure revealed at the i-th data site with those existing at remaining sites we compute the following expression:

$$W(i) = \frac{2}{m_i^2} \sum_{\substack{j=1 \\ j \neq i}}^{q} \nu\left(\text{Prox}(U(i))\text{Prox}(U(i|j))\right). \qquad (12.52)$$

More specifically, we consider that the distance between the corresponding proximity matrices is realized in the form of the Hamming distance. In other words, we have:

$$\nu\left(\text{Prox}(U(i))\text{Prox}(U(i|j))\right) = \sum_{t=1}^{m_i} \sum_{s>t}^{m_i} |\text{prox}_{ts}(i) - \text{prox}_{ts}(i|j)|, \qquad (12.53)$$

where $\text{prox}_{ts}(i)$ denotes the (t, s) entry of the proximity matrix referred to the partition matrix $U(i)$, and similarly for $\text{prox}_{ts}(i|j)$ when considered w.r.t. the induced partition matrix $U(i|j)$. In a nutshell, rather than working at the level of comparing the individual partition matrices (which requires knowledge of the explicit correspondence between the rows of the partition matrices), we generate their corresponding proximity matrices that allows us to carry out comparison at this more abstract level. Next summing up the values of $W(i)$ over all data sites, we arrive at the global level of consistency of the structure discovered collectively through the collaboration:

$$W = \sum_{i=1}^{q} W(i). \qquad (12.54)$$

The lower the value of W, the higher is the consistency between the q structures. Likewise the values of W being reported during successive phases of the collaboration can serve both as a sound indicator as to the progress and quality of the collaborative process, and as a suitable termination criterion. In particular, when tracing the successive values of W, one could stop the collaboration once no further changes in its values are reported. The use of the above consistency measure is also essential when gauging the intensity of collaboration and adjusting its level through changes of the value of α. Let us recall that this parameter appears in the minimized objective function and shows how much other data sites impact the formation of the clusters at the given site. Higher values of α imply stronger collaborative linkages established between the sites. By reporting the values of W treated as a function of α, that is $W = W(\alpha)$, we can experimentally optimize the intensity of collaboration. One may anticipate that while for low values of α no collaboration occurs and the values of W tend to be high, large values of α might lead to competition and subsequently the values of $W(\alpha)$ may tend to be high. Under some conditions, no convergence of the collaboration process could be reported. There might be some regions of optimal values of α. Obviously, the optimal level (intensity) of collaboration depends upon a number of parameters of the collaborative clustering, in particular the number of clusters and the number of data sites involved in the collaboration. It could also depend upon the data themselves.

that you may improve with a wise dosing of the communicated knowledge influence.

12.2.4 Experience-Consistent Fuzzy Models

In this modeling scenario, it becomes advantageous not only consider currently available data but also actively exploit previously obtained findings. Such observations bring us to the following formulation of the problem:

> Given some experimental data, construct a model which is consistent with the findings produced for some previously available data. Owing to the existing requirements such as data privacy or data security of data as well as some other technical limitations, in the construction of the model an access to these previous data is not available however we can take advantage of the knowledge of the parameters of the existing models.

Considering the need to achieve a certain desired consistency of the proposed model with the previous findings, we refer to the development of such models as *experience-based* or *experience-consistent* fuzzy modeling.

When dealing with experience-consistent models, we may encounter a number of essential constraints which imply a way in which the underlying processing can be realized. For instance, it is common that the currently available data are quite limited in terms of their size (which implies a limited evidence of the dataset) while the previously available datasets could be substantially larger, meaning that relying on the models formed in the past could be beneficial for the development of the current model. There is also another reason in which the experience-driven component plays a pivotal role. The data set D could be quite small and affected by a high level of noise – in this case it becomes highly legitimate to seriously consider any additional experimental evidence available around.

> **Consistent rules must not diverge on the same data;**

In the realization of the consistent-oriented modeling, we consider the following scenario. Given is a data site $D(0)$ for which we intend to construct a fuzzy rule-based model. There is a collection of data sites $D(1), \ldots, D(q)$. For each of them developed is an individual fuzzy model denoted here by $R(j)$. Our assumption is that these models are available when seeking consistency with an analogous model $R(0)$ we are constructing for $D(0)$, while original datasets at their root are not available.

For each data site the model $R(j)$ comes in terms of a rule endowed with a local linear regression function assuming the form:

$$\text{if } \boldsymbol{x} \text{ is B}_c(j) \text{ then } y = \boldsymbol{a}_c(j)^T \boldsymbol{x}, \qquad (12.55)$$

where $\boldsymbol{x} = \{x_1, \ldots, x_n, 1\}$ is an extension of the original input pattern in order to embed in the last component of the coefficient vector $\boldsymbol{a}_c = \{a_{c1}, \ldots, a_{cn}, b_c\}$ the constant term in the regression, and B$_c$ is a fuzzy set.

Alluding to the format of the data at site $D(j)$, it comes in the form of input-output pairs $\{(\boldsymbol{x}_t, y_t), t = 1, 2, \ldots, m_j\}$ which are used to carry out learning in a supervised mode.

The experience-consistent development of the rule-based model

We consider a rule-based model constructed on a basis of data at $D(0)$ where in the construction of the model we are influenced by the models formed with the use of the other data site. However, we assume that data at all sites but $D(0)$ have been previously collected and, due to some technical and non-technical reasons, they cannot be shared with those at $D(0)$. However, to realize a mechanism of experience consistency, the communication between the data sites can be realized at the higher conceptual level such as the one involved in the parameters of the fuzzy models.

Given the architecture of the rule-based system, it is well known that we encounter here two fundamental design phases, that is: i) a formation of the fuzzy sets standing in the conditions of the rules, and ii) the estimation of the corresponding conclusion parts. There are numerous ways of carrying out this

> **hence, share rules,**

construction. Typically, when it comes to the condition parts of the rules, the essence of the design is to granulate data by forming a collection of fuzzy sets via fuzzy clustering. The conclusion part where we encounter local regression models is formed by estimating the parameters a_i. Of the various techniques discussed in Section 3.2, we focus on the squared error criterion's.

The organization of the consistency-driven optimization relies on the reconciliation of the conclusion parts of the rules. We assume that the condition parts, viz. fuzzy sets, are developed independently from each other. In other words, we cluster $D(j)$ data in a same input space and with the same number of clusters k, which results in an a collection of analogous rules. Then the mechanism of experience consistency is realized for the conclusions of the rules. Given the independence of the construction process of the clusters at the individual sites, before moving on with the quantification of the obtained consistency of the conclusion parts of the rules, it becomes necessary to align the information granules obtained at $D(0)$ and the individual data sites $D(j)$. In the ensuing schemes of consistency development we will be relying on the communication of the prototypes of the clusters forming the condition parts of the rules.

and make them to converge.

The consistency-based optimization of local regression models

To make the formulas concise, w.r.t. the rule-based model $R(j)$ pertaining to data site $D(j)$, we use a shorthand notation. In particular, $\text{FM}_j(x)$ is used to denote the output of $R(j)$ with input x. As usual the optimal parameters of the local models occurring in the conclusions of the rules are chosen in such a way so that they minimize the sum of squared errors:

You move from local optimization of local parameters,

$$C_j = \frac{1}{m_j} \sum_{(x_t, y_t) \in D(j)} (\text{FM}_j(x_t) - y_t)^2. \qquad (12.56)$$

Considering the form of the rule-based system, the output $\text{FM}_j(x_t)$ of the fuzzy model is determined as a weighted combination of the local models with the weights being the levels of activation of the individual rules. More specifically we have:

$$\widehat{y}_t = \sum_{c=1}^{k} u_{ct} a_c^T x_t, \qquad (12.57)$$

where u_{ct} is a membership degree of x_t to the c-th cluster being computed on a basis of the already determined prototypes in the input space. In a nutshell (12.57) comes as a convex combination

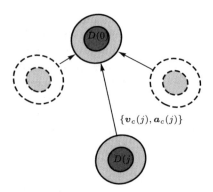

Fig. 12.21. Communication between $D(0)$ and $D(j)$ realized by transferring parameters of the rule-based model available at individual data sites. Same notation as Fig. 12.19.

of the local models which aggregates the local models by taking advantage of the weight factors expressing a contribution of each model based upon the activation reported in the input space.

The essence of the consistency-driven modeling is to form local regression models occurring in the conclusions of the rules on a basis of data $D(0)$ while at the same time making the model perform in a consistent manner (viz. close enough) to the rule-based model formed for the respective $D(j)$'s. The following performance index strikes a sound balance between the two horns of the optimization problem:

to globally conditioned optimization of local parameters,

$$C = \sum_{(\boldsymbol{x}_t, y_t) \in D(0)} (\text{FM}_0(\boldsymbol{x}_t) - y_t)^2 + \alpha \sum_{j=1}^{q} \sum_{(\boldsymbol{x}_t, y_t) \in D(0)} (\text{FM}_0(\boldsymbol{x}_t) - \text{FM}_j(\boldsymbol{x}_t))^2.$$
(12.58)

The calculation of $\text{FM}_0(\boldsymbol{x}_t)$ requires some words of explanation. The model is communicated to $D(0)$ by transferring the prototypes of the clusters and the coefficients of the linear models standing in the conclusions of the rules (refer to Fig. 12.21). Hence $\text{FM}_0(\boldsymbol{x}_t)$ reads $\hat{y}_t$ in (12.57), where the degrees u_{ct} are substituted by degrees $u_{ct}(0|j)$ induced by the communicated $D(j)$ prototypes through (6.27).

The minimization of the cost C for some predefined value of α leads to the optimal vectors of the linear models parameters $\boldsymbol{a}_c^*, c = 1, 2, \ldots, k$ which is reflective of the process of satisfying the consistency constraints. The detailed derivations are a quite standard algebraic exercise. The final result comes in the form:

i.e. to a suitable linear combination of models.

$$\boldsymbol{a}_c^* = \frac{1}{\alpha q + 1} \widehat{X}^\# \left(\boldsymbol{y}_0 + \sum_{j=1}^{q} \alpha \boldsymbol{y}_j \right),$$
(12.59)

where $\boldsymbol{y}_j$ is a vector of the outputs of the j-th fuzzy model (formed on a basis of $D(j)$) whose corresponding coordinate is the output obtained for the corresponding input, that is:

$$\boldsymbol{y}_j = \begin{bmatrix} \mathrm{FM}_j(\boldsymbol{x}_1) \\ \mathrm{FM}_j(\boldsymbol{x}_2) \\ \vdots \\ \mathrm{FM}_j(\boldsymbol{x}_{m_j}) \end{bmatrix} \qquad (12.60)$$

and $\widehat{X}^{\#}$ is a pseudoinverse of the input data matrix.

An overall balance captured by (12.58) is achieved for a certain value of α. An evident tendency of increased impact becomes clearly visible: higher values of α stress higher relevance of other models and their more profound influence on the constructed model. Actually, the second component in C plays a role that is similar to a *regularization* term being typically used in estimation problems; however its origin here has a substantially different format from the one encountered in the literature. Here, we consider other data (and models) rather than focusing on the complexity of the model expressed in terms of its parameters to evaluate the performance of the model. An index quantifying a global behavior of the optimal model arises in the following form:

The optimization of the dosing coefficient is a side task.

$$\mathsf{V} = \sum_{(\boldsymbol{x}_t,y_t)\in D(0)} (\mathrm{FM}_0(\boldsymbol{x}_t) - y_t)^2 + \sum_{j=1}^{q} \frac{1}{m_j} \sum_{(\boldsymbol{x}_t,y_t)\in D(j)} (\mathrm{FM}_0(\boldsymbol{x}_t) - y_t)^2. \qquad (12.61)$$

A schematic view of computing and communication of findings being realized with the aid of (12.61) is illustrated in Fig. 12.22. Note that when the fuzzy model FM_0 is transferred to $D(j)$ as before, we communicate the prototypes obtained at D_0 and the coefficients of the local linear models of the conclusion part of the rules. Likewise as shown in (12.61), the output of the fuzzy model obtained for $\boldsymbol{x}_t \in D(j)$ involves the induced value of membership degree $u_{ct}(j|0)$ and an aggregation of the local regression models.

Apparently the expression of V is a function of α and the optimized level of consistency is such for which V attains its minimal value, namely:

$$\alpha^* = \arg\min_{\alpha} \mathsf{V}(\alpha). \qquad (12.62)$$

The optimization scheme (12.62) along with its evaluation mechanisms governed by (12.61) can be generalized by admitting various levels of impact each data of $D(j)$ might have in the process of achieving consistency.

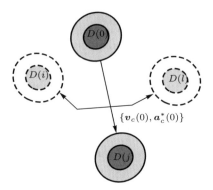

Fig. 12.22. A quantification of the global behavior of the consistency-based fuzzy model.

Characterization of experience-consistent models through its granular parameters

The rules forming each fuzzy model have been formed independently at each data site. If we intend to evaluate a level of consistency of the rules at $D(0)$ vis-à-vis the modeling evidence available at $D(j)$, some alignment of the rules becomes essential. Such an alignment concerns a way of lining up the prototypes forming the condition part of the rules. We consider the models obtained at $D(0)$ and $D(j), j = 1, 2, \ldots, q$ with their prototypes $\{v_1(0), \ldots, v_k(0)\}$ and $\{v_1(j), \ldots, v_k(j)\}$. We say that the c-th rule at $D(0)$ and the κ-th at $D(j)$ are aligned if the prototypes $v_c(0)$ and $v_\kappa(j)$ are the closest within the collections of prototypes produced for $D(0)$ and $D(j)$. The alignment process is realized by successively finding the pairs of prototypes being characterized by the lowest mutual distance. Overall, the alignment process can be described in the following manner.

Once the above loop has been completed, we end up with the list of alignment of the prototypes in the form of pairs $(i_1, j_1), \ldots, (i_k, j_k)$.

We can now look at the characterization of the set of the related parameters of the local regression models. In essence, through the alignment of the prototypes at $D(0)$ and $D(j)$, we obtain the corresponding parameters vectors of the regression models of the conclusion parts. Denote these vectors corresponding to a certain rule by $a_0, a_1, \ldots, a_q$ altogether arriving at $q+1$ of them. If we now consider the i-th coordinate of all of them, we obtain the numeric values $a_{0i}, a_{1i}, \ldots, a_{qi}$. The essence of their aggregation concerns their global representation completed in the form of a single fuzzy set. They employ the same aggregation scheme as presented in Section 4.4. We intend to span a

The maintenance of rules consistency

passes through a suitable alignment process,

followed by rules aggregation,

Algorithm 12.3. Rule alignment

---

1. Form two sets of integers I and J, where $I = J = \{1, 2, \ldots, k\}$
   Start with an empty list of alignments, $L = \emptyset$
2. **repeat**
   2.1. Find a pair of indexes i_0 and j_0 for which the distance attains minimum:
   $$(i_0, j_0) = \arg\min_{c, \kappa} \nu(\boldsymbol{v}_c(0) - \boldsymbol{v}_k(j))$$
   2.2. The pair (i_0, j_0) is added to the list of alignments, $L = L \cup (i_0, j_0)$
   2.3. Reduce the set of indexes I and J by removing the elements that were placed on the list of alignments, $I = I \setminus \{i_0\}$ and $J = J \setminus \{j_0\}$
   until $I = \emptyset$

---

unimodal fuzzy set A over the set of numeric parameters a_j in such a way that A represents these data to the highest possible extent following the principle of justifiable granularity. Consider triangular membership functions for the granular parameters of the fuzzy model. Then the result of the aggregation becomes a triangular fuzzy number of the j-th parameter of the local regression model. Denote it by $A_j = \{a_{j-}, a_j, a_{j+}\}$ with the three parameters denoting the lower, modal, and upper bound of the fuzzy number. Applying the same procedure to all remaining parameters of the vector $\boldsymbol{a}$, we produce the corresponding fuzzy numbers $A_0, A_1, \ldots, A_{j-1}, A_{j+1}, \ldots, A_n$. Given them the rule in $D(0)$ reflects the nature of incorporated evidence offered by the remaining models $D(1), D(2)$, etc. If there is a fairly high level of consistency, this effect is manifested through a fairly concentrated fuzzy number. Increasing inconsistency results in a broader, less specific fuzzy number of the parameters. In summary, a certain fuzzy rule assumes the following format: *(in terms of synthesis of a fuzzy number,)*

if x is B then $Y = A_0 \oplus A_1 \otimes x_1 \oplus A_2 \otimes x_2 \oplus \ldots \oplus A_n \otimes x_n$. (12.63)

The symbols $\oplus$ and $\otimes$ being used above underline the non-numeric nature of the arguments standing in the model over which the multiplication and addition are carried out. For given numeric input $\boldsymbol{x} = (x_1, x_2, \ldots, x_n)$ the resulting output Y of this local regression model is again a triangular fuzzy number $Y = \{y_-, y, y_+\}$ where its parameters are computed as follows: *(throught a possible non-numeric model)*

$$y_- = a_0^- + \sum_{j=1}^{n} \min\{a_{j-} x_j, a_{j+} x_j\} \qquad (12.64)$$

$$y = a_0 + \sum_{j=1}^{n} a_j x_j \qquad (12.65)$$

$$y_+ = a_0^+ + \sum_{j=1}^{n} \max\{a_{j-}x_j, a_{j+}x_j\}. \qquad (12.66)$$

By repeating the above process for all the rules we arrive at the rules of the following forms:

if x **is** B_1 **then** $Y = A_{10} \oplus A_{11} \otimes x_1 \oplus A_{12} \otimes x_2 \oplus \ldots \oplus A_{1n} \otimes x_n$
if x **is** B_2 **then** $Y = A_{20} \oplus A_{21} \otimes x_1 \oplus A_{22} \otimes x_2 \oplus \ldots \oplus A_{2n} \otimes x_n$
$\qquad \vdots \qquad\qquad \vdots \qquad\qquad\qquad\qquad\quad \vdots$
if x **is** B_k **then** $Y = A_{k0} \oplus A_{k1} \otimes x_1 \oplus A_{k2} \otimes x_2 \oplus \ldots \oplus A_{kn} \otimes x_n.$
$\qquad\qquad\qquad\qquad\qquad\qquad\qquad\qquad\qquad\qquad (12.67)$

Given this structure, the input vector x implies the output fuzzy set with the following membership function:

$$Y = \sum_{c=1}^{k} w_c(x) \otimes (A_{c0} \oplus A_{c1} \otimes x_1 \oplus A_{c2} \otimes x_2 \oplus \ldots \oplus A_{cn} \otimes x_n). \qquad (12.68)$$

producing again a fuzzy number. Owing to the fact of having fuzzy sets of the regression model parameters in the conclusion part of the rules, Y becomes a fuzzy number rather than a single numeric value.

12.3 Conclusions

The two veins of results we discuss in this chapter as representatives of the core of knowledge engineering share the common feature of facing the semantic of the problems. It realizes either in terms of quality of available data, mainly with the former, or of consistency of the developed models with experienced data, with the latter. Semantics is a leading perspective of many modern computer science findings. A paradigmatic example is the huge effort paid by researcher into and the strategic value played by the Semantic Web [4]. The results presented in the chapter highlight real advances in this direction concerning the very fundamental problems of discriminating and clustering data. As for the former, the problem of localizing the discrimination surface in the feature space as a function of either the quality or relevance of the questioned data came to be studied as a *per se* topic only recently [7, 3, 19]. As for the latter, the idea of combining experiences to make the outcome model more robust is very popular today, and we elaborated on it in terms of social computing in Chapters 9 and 10. To become more consistent, we have to pursue a different goal, not collapsing with other approaches under the umbrella of distributed clustering [15] as well. Rather, achieving consistency meets the needs of many multi-agents paradigms [9], offering new collaborative schemes where we commonly encounter agents operating quite independently at various levels of specificity, so that it is very likely that the effectiveness of the overall system depends heavily upon a way in which they collaborate and effectively exchange their findings. Within a genuine spectrum of collaborative schemes, we offered here a detailed discussion on the two extremes in which all agents are engaged in collaborative pursuits and when one agent in building its fuzzy model benefits from experience available at other data sites which is made available in the form of the parameters of the models formed there. From a pure information management perspective, the difference concerns the exchangeability of source data between sites (in the former scheme) or only of a structured synthesis of them (in the latter) unavoidably generating asymmetry between agents. A notably value of the discussed methods is the absence of specific assumptions on the data they are processing, with the sole exception concerning homogeneous granularity. As a result, the structure at each data site makes an attempt to reconcile differences, however retaining and quantifying those that are of particular relevance to the given data site, thus overriding higher order fuzzy sets such as type-2 or hierarchical fuzzy sets.

References

1. Apolloni, B., Bassis, S., Gaito, S.: Fitting opportunistic networks data with a pareto distribution. In: Apolloni, B., Howlett, R.J., Jain, L. (eds.) KES 2007, Part III. LNCS (LNAI), vol. 4694, pp. 812–820. Springer, Heidelberg (2007)
2. Apolloni, B., Bassis, S., Malchiodi, D., Pedrycz, W.: Interpolating support information granules. Neurocomputing (in press, 2008)
3. Apolloni, B., Malchiodi, D., Natali, L.: A modified svm classification algorithm for data of variable quality. In: Apolloni, B., Howlett, R.J., Jain, L. (eds.) KES 2007, Part III. LNCS (LNAI), vol. 4694, pp. 131–139. Springer, Heidelberg (2007)
4. Cardoso, J.: Semantic Web Services: Theory, Tools and Applications. Idea Group (2007)
5. Cawley, G.C., Talbot, N.L.C., Foxall, R.J., Dorling, S.R., Mandic, D.P.: Heteroscedastic kernel ridge regression. Neurocomputing 57, 105–124 (2004)
6. Chaintreau, A., Hui, P., Crowcroft, J., Diot, C., Gass, R., Scott, J.: Impact of human mobility on the design of opportunistic forwarding algorithms. In: Proceedings of the 25th IEEE International Conference on Computer Communications (INFOCOM), Barcelona, Spain (2006)
7. Crammer, K., Kearns, M., Wortman, J.: Learning from data of variable quality. In: Weiss, Y., Schoelkopf, B., Platt, J. (eds.) Advances in Neural Information Processing Systems, vol. 18, pp. 219–226. MIT Press, Cambridge (2006)
8. Crammer, K., Singer, Y.: On the algorithmic implementation of multiclass kernel-based vector machines. Journal of Machine Learning Research 2, 265–292 (2001)
9. Ferber, J.: Multi-Agent System: An Introduction to Distributed Artificial Intelligence. Addison Wesley Longman, Harlow (1999)
10. Friedman, J.H.: Another approach to polychotomous classification. Technical report, Stanford University, Department of Statistics (1996),
    http://www-stat.stanford.edu/reports/friedman/poly.ps.Z
11. Geman, S., Bienenstock, E., Doursat, R.: Neural networks and the bias/variance dilemma. Neural Computation 4, 1–58 (1992)

12. Knerr, S., Personnaz, L., Dreyfus, G.: Single-layer learning revisited: A stepwise procedure for building and training a neural network. In: Fogelman-Soulie, Herault (eds.) Neurocomputing: Algorithms, Architectures and Applications. NATO ASI, Springer, Heidelberg (1990)
13. Kuhn, H.W., Tucker, A.W.: Nonlinear programming. In: Proceedings of 2nd Berkeley Symposium, pp. 481–492. University of California Press, Berkeley (1951)
14. LeCun, Y., Bottou, L., Bengio, Y., Haffner, P.: Gradient-based learning applied to document recognition. Proceedings of the IEEE 86(11), 2278–2324 (1998)
15. Pedrycz, W., Lam, P., Rocha, A.F.: Distributed fuzzy modelling. IEEE Transactions on Systems, Man and Cybernetics-B 5, 769–780 (1995)
16. Platt, J., Cristianini, N., Shawe-Taylor, J.: Large margin dags for multiclass classification. In: Solla, S.A., Leen, T.K., Muller, K.R. (eds.) Advances in Neural Information Processing Systems, vol. 12, MIT Press, Cambridge (2000)
17. Schmidt, G., Mattern, R., Schueler, F.: Biomechanical investigation to determine physical and traumatological differentiating criteria for the maximum load capacity of head and vertebral column with and without protective helment under the effects of impact. EEC research program on biomechanics of impact, final report. Technical report, University of Heidelberg (1981)
18. Vapnik, V.: Statitical Learning Theory. John Wiley & Sons, New York (1998)
19. Yang, M.S., Li, C.Y.: Mixture poisson regression models for heterogeneous count data based on latent and fuzzy class analysis. Soft Computing 9(7), 519–524 (2005)

Appendices

A Norms

A.1 Norms

A leitmotiv of this book is that you discover structure in currently available data in order to exploit it in future data, for instance through their probability distribution, for instance through their membership degree to a given fuzzy set. A main tool to synthesize this structure is supplied by a metric on the set $\mathfrak{X}$ where data are defined. To this aim we state here some definitions with the goal of being mostly self contained, sometimes bearing the risk of appearing somehow broad. In particular, since our sets are in principle discrete, possibly using continuous sets for their approximation, we will avoid explicit links to the theory of measure.

A.2 General Norms

Let us start with defining a metric.

Definition A.1. *A distance on a set $\mathfrak{X}$ is a function d (also called* metric*) from pairs of elements of $\mathfrak{X}$ to the set $\mathbb{R}^+$ of non-negative real numbers with the following properties. For all $x_1, x_2, x_3 \in \mathfrak{X}$:*

D1. $d(x_1, x_2) \geq 0$ *(non-negativity)*;
D2. $d(x_1, x_2) = 0$ *if and only if* $x_1 = x_2$ *(highest similarity says identity)*;
D3. $d(x_1, x_2) = d(x_2, x_1)$ *(symmetry)*;
D4. $d(x_1, x_3) \leq d(x_1, x_2) + d(x_2, x_3)$ *(triangle inequality)*.

The pair $(\mathfrak{X}, d)$ is called a metric *space.*

We assume the set of real numbers as codomain of the distance function in order to comply with the general theory, even though the set of rational numbers would be sufficient. Moreover, roughly identifying a *vector space* with a set of elements, called vectors, that can be scaled and added as it happens with vectors in any Euclidean space, most of distances we regularly deal with play the role of *norms* in that they are scalable functions of the vector joining the two objects in the feature space the distances refer to, independently of the vector direction. We resume the connotation of a norm with the following definition.

Definition A.2. *Given a vector space $\mathfrak{Y}$ a norm on $\mathfrak{Y}$ is a function ν from vectors $\boldsymbol{y}$ to $\mathbb{R}^+$ with the following properties. For all scalars $a \in \mathbb{R}$ and all $\boldsymbol{y}, \boldsymbol{w} \in \mathfrak{Y}$, denoting with $|a|$ the absolute value of a,*

N1. $\nu(\boldsymbol{y}) = \nu(-\boldsymbol{y})$ *(symmetry);*
N2. $\nu(a\boldsymbol{y}) = |a|\nu(\boldsymbol{y})$ *(positive scalability);*
N3. $\nu(\boldsymbol{w} + \boldsymbol{y}) \leq \nu(\boldsymbol{w}) + \nu(\boldsymbol{y})$ *(triangle inequality);*
N4. $\nu(\boldsymbol{y}) = 0$ *if and only if $\boldsymbol{y}$ is the zero vector   (positive definiteness).*

We will often refer to l_r-norms and l^r-norms that are defined as follows.

Definition A.3. *For a vector $\boldsymbol{y} = (y_1, \ldots, y_n) \in \mathbb{R}^n$ its l_r-norm, when r is a real number ≥ 1, is:*

$$l_r(\boldsymbol{y}) = \left(\sum_{i=1}^{n} |y_i|^r \right)^{\frac{1}{r}}. \tag{A.1}$$

Slightly drifting from Definition A.2, we also define a power norm l^r as the r-th power of l_r-norm, i.e.:

$$l^r(\boldsymbol{y}) = \sum_{i=1}^{n} |y_i|^r. \tag{A.2}$$

Intuitively l_r-norms provide different notions of length of a vector $\boldsymbol{y}$, starting from different metrics adopted for measuring the distance between two points in $\mathbb{R}^n$ and, definitely, using different *geometries*. To give an example, the most common Euclidean norm l_2 is substituted by the so-called Manhattan (or Taxicab) norm l_1 when the distance between two points is measured as the sum of the (absolute) differences of their coordinates. In general, we will be more interested in the geometry than in the norm *per se*. This is why we introduce the function l^r; it requires obvious modification of scalability property and triangle inequality. As a matter of fact these properties are not crucial in many instances of this monograph, where we use the norm functionality to position data in respect to some reference points, rather than each other directly. Namely, with no foundation pretense and no link to similar notations in the literature, we expressly define here for expository convenience *weak norms* (possibly omitting mentioning the adjective in the text), at an intermediate acceptation between geometrical and fuzzy norms, as follows.

Definition A.4. *Given a vector space $\mathfrak{Y}$ a weak norm on $\mathfrak{Y}$ is a function ν from vectors $\boldsymbol{y}$ to $\mathbb{R}^+$ with the following properties. For all $\boldsymbol{y}, \boldsymbol{w} \in \mathfrak{Y}$,*

W1. $\nu(\boldsymbol{y}) = \nu(-\boldsymbol{y})$ *(symmetry);*
W2. $\nu(\boldsymbol{y}) \geq \nu(\boldsymbol{w})$ *if* $\nu'(\boldsymbol{y}) \geq \nu'(\boldsymbol{w})$, *for all norms ν' as in Definition A.2*   *(monotonicity);*
W3. $\nu(\boldsymbol{y}) = 0$ *if and only if $\boldsymbol{y}$ is the zero vector*   *(positive definiteness).*

In Table A.1 we report some norms (in regular, power or weak acceptation) that are of widespread use and the related distribution laws in the light of (6.9). Note that l_r too, with $r < 1$, is a weak norm. In some cases we are more familiar with the metric, in other ones with the distribution law, still in others we ignore one of the two features.

Table A.1. Common (broad definition) norms $\nu(\boldsymbol{y})$ – expressed when convenient in terms of l_r-norms (and their companion l^r-norms) defined in (A.1) – and corresponding density models $f_Y(\boldsymbol{y})$. $\Gamma(\cdot)$ and $\beta(\cdot)$ are respectively the Gamma and Beta function, and $|x|_\varepsilon$ is defined to be $\max\{|x| - \varepsilon, 0\}$. Polynomial norms have here a quite different acceptation than Bombieri norm.

	norm $\nu(\boldsymbol{y})$	density model $f_Y(\boldsymbol{y})$
ε-insensitive	$l_1(\boldsymbol{y})_\varepsilon$	$\frac{1}{2(1+\varepsilon)} e^{-l_1(\boldsymbol{y})_\varepsilon}$
Laplacian	$l_1(\boldsymbol{y})$	$\frac{1}{2} e^{-l_1(\boldsymbol{y})}$
Gaussian	$\frac{1}{2} l^2(\boldsymbol{y})$	$\frac{1}{\sqrt{2}} e^{-l^2(\boldsymbol{y})/2}$
Huber's robust loss	$\begin{cases} \frac{1}{2\sigma} l^2(\boldsymbol{y}) & \text{if } l_2(\boldsymbol{y}) \leq \sigma \\ l_2(\boldsymbol{y}) - \frac{\sigma}{2} & \text{otherwise} \end{cases}$	$\propto \begin{cases} e^{-\frac{l^2(\boldsymbol{y})}{2\sigma}} & \text{if } l_2(\boldsymbol{y}) \leq \sigma \\ e^{\frac{\sigma}{2} - l_2(\boldsymbol{y})} & \text{otherwise} \end{cases}$
Polynomial	$\frac{1}{p} l^p(\boldsymbol{y})$	$\frac{p}{2\Gamma(1/p)} e^{-l^p(\boldsymbol{y})}$
Piecewise polynomial	$\begin{cases} \frac{1}{p\sigma^{p-1}} l^p(\boldsymbol{y}) & \text{if } l_p(\boldsymbol{y}) \leq \sigma \\ l_p(\boldsymbol{y}) - \sigma \frac{p-1}{p} & \text{otherwise} \end{cases}$	$\propto \begin{cases} e^{-\frac{l^p(\boldsymbol{y})}{p\sigma^{p-1}}} & \text{if } l_p(\boldsymbol{y}) \leq \sigma \\ e^{\sigma \frac{p-1}{p} - l_p(\boldsymbol{y})} & \text{otherwise} \end{cases}$
Cosh	$\sum_{i=1}^n \log \cosh y_i$	$\prod_{i=1}^n \frac{1}{\pi \cosh y_i}$
Cauchy	$\sum_{i=1}^n -\log r^2 + \log(r^2 + y_i^2)$	$\prod_{i=1}^n \frac{r}{\pi(r^2 + y_i^2)}$
Student's t	$\sum_{i=1}^m -\log \frac{\left(\frac{r}{r+y_i^2}\right)^{\frac{1+r}{2}}}{\sqrt{r}\,\beta(\frac{r}{2},\frac{1}{2})}$	$\prod_{i=1}^n \frac{\left(\frac{r}{r+y_i^2}\right)^{\frac{1+r}{2}}}{\sqrt{r}\,\beta(\frac{r}{2},\frac{1}{2})}$

A graphical representation of these norms is provided in the following pictures where, for the sake of visualization, we consider two-dimensional vectors $\boldsymbol{y} = (y_1, y_2)$.

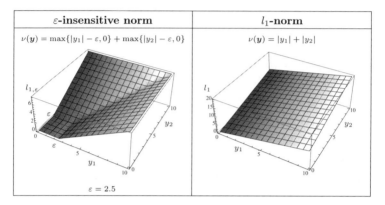

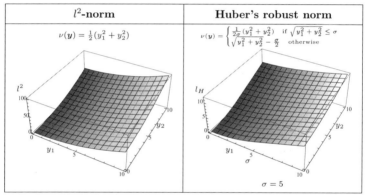

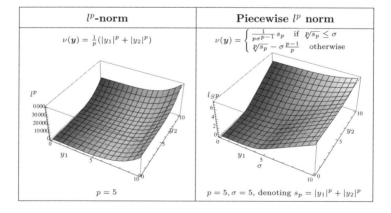

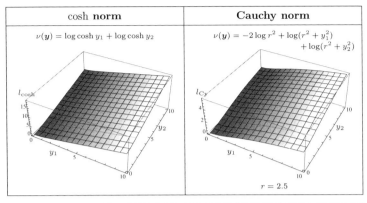

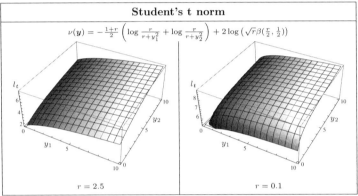

Norm l_2 and the companion Gaussian model are the most employed, hitting the most straightforward notion of distance in the Euclidean space. Norm l_1 looks a simplified version of the former as it substantially preserves the linearity of the space the items belong to. Hence many variants of them have been proposed in order to take into account special rationales. Table A.2 is a typical repertoire of them where the metric is explicated as a function $d_{ij} = d(\boldsymbol{y}_i, \boldsymbol{y}_j)$ of two vectors having one extreme coinciding with the origin of the coordinate axes. Note that, by definition, norms relate each other through a non decreasing monotonic relationship (say: if $l_r(\boldsymbol{y}) > l_r(\boldsymbol{w})$ then $l_p(\boldsymbol{y}) \geq l_p(\boldsymbol{w})$ for every r, p). We speak of similarity measures s when we have an opposite relationship (say: if $d_{ij} > d_{hk}$ then $s_{ij} \leq s_{hk}$). Their rationale is synthesized as follows:

- Minkowski distance enhances the geometrical meaning of the l_r norm: when $r = 1$ it becomes the city-block distance (Manhattan distance) and when $r = 2$ the Euclidean distance;
- Chebyshev distance is a special case of Minkowski distance with $r \to \infty$;

Table A.2. Similarity/Dissimilarity metrics for quantitative variables, denoted respectively with $s_{ij} = s(\boldsymbol{y}_i, \boldsymbol{y}_j)$ and $d_{ij} = d(\boldsymbol{y}_i, \boldsymbol{y}_j)$. As usual y_{ik} denotes the k-th component of the pattern $\boldsymbol{y}_i$.

Euclidean Distance	$d_{ij} = \sqrt{\sum_{k=1}^{n} (y_{ik} - y_{jk})^2}$
Manhattan Distance	$d_{ij} = \sum_{k=1}^{n} |y_{ik} - y_{jk}|$
Chebyshev Distance	$d_{ij} = \max_{k} |y_{ik} - y_{jk}|$
Minkowski Distance	$d_{ij} = \sqrt[p]{\sum_{k=1}^{n} |y_{ik} - y_{jk}|^p}$
Canberra Distance	$d_{ij} = \sum_{k=1}^{n} \frac{|y_{ik} - y_{jk}|}{|y_{ik}| + |y_{jk}|}$
Sorensen Distance	$d_{ij} = \frac{\sum_{k=1}^{n} |y_{ik} - y_{jk}|}{\sum_{k=1}^{n} (y_{ik} y_{jk})}$
Angular Separation	$s_{ij} = \frac{\sum_{k=1}^{n} y_{ik} \cdot y_{jk}}{\sqrt{\sum_{k=1}^{n} y_{ik}^2 \cdot \sum_{k=1}^{n} y_{jk}^2}}$
Correlation Coefficient	$s_{ij} = \frac{\sum_{k=1}^{n} (y_{ik} - \overline{y}_i) \cdot (y_{jk} - \overline{y}_j)}{\sum_{k=1}^{n} (y_{ik} - \overline{y}_i)^2 \cdot \sum_{k=1}^{n} (y_{jk} - \overline{y}_j)^2}$

- Canberra distance examines the sum of *fraction differences* between coordinates of a pair of vectors: each addend has value between 0 and 1. If one coordinate is zero, the term becomes unity regardless the other value, thus the distance will not be affected. This distance is very sensitive to a small change when both coordinates approach zero;
- Sorensen (also known as Bray Curtis) distance has the nice property that if all coordinates are positive, its value is between zero and one;
- angular separation represents the cosine angle between two vectors: it measures similarity rather than distance or dissimilarity. Like cosine, the value of angular separation lies in the interval $[-1, 1]$. If standardized by centering the coordinates to their mean values, it is called correlation coefficient.

Table A.3 concerns categorical data, i.e. data assuming only a small number of values, each corresponding to a specific category value or label. Continuity breaks by definition with these data; thus distances are based on *logical* properties the pair of data owns. Therein an excerpt of these distances is reported. In particular, in the case that each object can be characterized by a pattern of binary variables, similarity between two objects

Table A.3. Similarity/Dissimilarity metrics for categorical data, both for nominal and ordinal variables. Notation of Table A.2 is extended with symbols: p (s) counting the number of variables that are positive (negative) for both $\boldsymbol{y}_i$ and $\boldsymbol{y}_j$, and q (r) counting the number of variables that are positive (negative) for $\boldsymbol{y}_i$ and negative (positive) for $\boldsymbol{y}_j$.

Binary variables	Simple Matching Coefficient	$s_{ij} = (p+s)/n$
	Jaccard's coefficient	$s_{ij} = p/(p+q+r)$
	Hamming distance	$d_{ij} = q + r$
Ordinal variables	Spearman distance	$d_{ij} = \sum_{k=1}^{n}(y_{ik} - y_{jk})^2$
	Footrule distance	$d_{ij} = \sum_{k=1}^{n}|y_{ik} - y_{jk}|$
	Kendall distance	minimum number of transpositions of discordant pairs
	Cailey distance	minimum number of transpositions of any pair of objects
	Ulam distance	minimum number of "Delete-Shift-Insert" operations

is generally based on the functions: p (s) counting the number of variables that are positive (negative) for both i-th and j-th objects, q (r) the number of variables that are positive (negative) for the i-th object and negative (positive) for the j-th object, and n the total number of variables. With these functions we may measure similarity/dissimilarity between the objects using: i) simple matching coefficients, when both positive and negative values carry equal information, ii) Jaccard's coefficient, when the negative value has no meaningful contribution to the similarity, and iii) Hamming distance just counting the number of different bits between the two patterns. If the number of categories is higher than two, we need to transform these categories into a set of dummy binary variables (either by associating at each attribute a single bit denoting the presence/absence of that attribute, or by describing each category through a set of bits by exploiting the binary representation).

If an order relation can be stated between the values of the categorical data, we may use ordinal variables. Beyond simple adaptations of the metrics developed for quantitative variables (for instance the Spearman and Footrule distance, extending the Euclidean and Manhattan ones), other notions of distance may be introduced which measure the *disorder* of ordinal variables.

They may be thought as the minimum number of allowed steps to equate the two compared objects. So:

i) with Kendall distance we can interchange the order of *adjacent pair* of variables when at least one digit does not match the corresponding one in the other vector;
ii) with Cailey distance the transposition may involve any pair of non-corresponding objects (not requiring the adjacency); and
iii) with Ulam distance the allowed operations are: a) the deletion of single digits; b) the shift of the remaining digits; and c) the insertion of a deleted digit in an empty space. Of course also Hamming Distance may be used for ordinal variables, counting the number of unmatched items.

A.3 Fuzzy Norms

A fuzzy norm shares with a norm the properties of symmetry (N1), positive scalability (N2) and triangle inequality (N3) *but not* positive definiteness (N4). In its place it identifies a null and unitary operators that may have two alternatives definitions, in turn specifying either a T-norm or an S-norm (also called co-norm). To give them clearer operational meaning we will resume their properties as functions of the extremes of the vector they refer to. Namely:

Definition A.5. *A T-norm is a function* $\top : [0,1] \times [0,1] \rightarrow [0,1]$ *with the following properties. For each* $a, b \in [0,1]$:

T1. $\top(a,b) = \top(b,a)$ *(commutativity);*
T2. $\top(a,b) \leq \top(c,d)$ *if* $a \leq c$ *and* $b \leq d$ *(monotonicity);*
T3. $\top(a, \top(b,c)) = \top(\top(a,b), c)$ *(associativity);*
T4. $\top(a,0) = 0$ *(null element);*
T5. $\top(a,1) = a$ *(identity element).*

S-norms are in a certain sense dual to T-norms. Given a T-norm, the complementary S-norm is defined by:

$$\bot(a,b) = 1 - \top(1-a, 1-b). \quad (A.3)$$

It follows that an S-norm satisfies the following relations. For each $a, b \in [0,1]$:

S1. $\bot(a,b) = \bot(b,a)$ *(commutativity);*
S2. $\bot(a,b) \leq \bot(c,d)$ *if* $a \leq c$ *and* $b \leq d$ *(monotonicity);*
S3. $\bot(a, \bot(b,c)) = \bot(\bot(a,b), c)$ *(associativity);*
S4. $\bot(a,1) = 1$ *(null element);*
S5. $\bot(a,0) = a$ *(identity element).*

In the following tables we report both the analytical expression and the graphical representation of some among most used fuzzy norms.

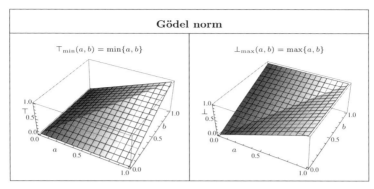

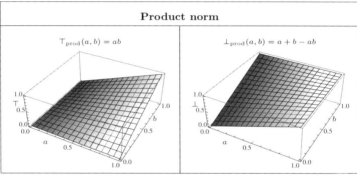

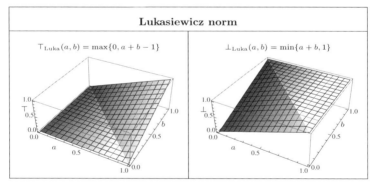

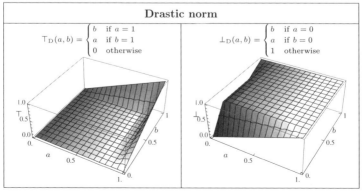

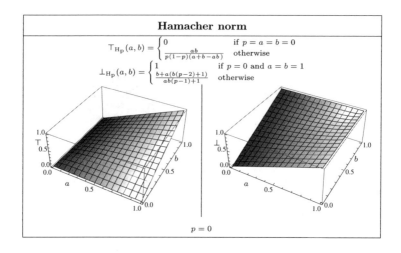

Hamacher norm

$$\top_{H_p}(a,b) = \begin{cases} 0 & \text{if } p=a=b=0 \\ \frac{ab}{p(1-p)(a+b-ab)} & \text{otherwise} \end{cases}$$

$$\bot_{H_p}(a,b) = \begin{cases} 1 & \text{if } p=0 \text{ and } a=b=1 \\ \frac{b+a(b(p-2)+1)}{ab(p-1)+1} & \text{otherwise} \end{cases}$$

$p = 0$

B Some Statistical Distribution

A general philosophy of this book is to analyze the information we may take from data in advance; then we possibly synthesize them into probabilistic models. This book does not offer an exhaustive panorama of these models. Here we just describe with a few formulas and graphs the most used probability distributions inside the book itself. They refer to one-dimensional variables, leaving the reader with the charge of building up multidimensional models by assembling conditional distributions. The sole exception concerns the multidimensional Gaussian variable which plays a special role in the modeling of linear dependency between random data. The interested reader may find more models in common handbooks.

B.1 Some One-Dimensional Variables

Fact B.1. *A cumulative distribution function (c.d.f.) of a one-dimensional random variable X is a function $F_X : \mathbb{R} \to [0,1]$ such that:*

1. $\lim_{x \to -\infty} F_X(x) = 0;\quad \lim_{x \to +\infty} F_X(x) = 1;$
2. $x_1 < x_2 \Rightarrow F_X(x_1) \leq F_X(x_2)$, *i.e. F_X is a non decreasing monotone function;*
3. $\lim_{h \to 0^+} F_X(x+h) = F_X(x)$, *i.e., F_X is continuous from the right.*

Fact B.2. *A probability function (p.f., also denoted d.f.) of a one-dimensional discrete random variable X is a function $f_X : \mathbb{R} \to [0,1]$ such that for enumerable I (using the symbol $\forall$ for denoting the universal quantifier "for each"):*

1. $f_X(x_i) > 0 \quad \forall i \in I;$
2. $f_X(x) = 0 \quad \forall x \neq x_i, i \in I;$
3. $\sum_{i \in I} f_X(x_i) = 1.$

Fact B.3. *A density function (d.f.) of a one dimensional continuous random variable X is a function $f_X : \mathbb{R} \to \mathbb{R}^+$ such that:*

1. $\int_{-\infty}^{+\infty} f_X(x)\mathrm{d}x = 1$.

It follows an excerpt of some probability distributions.

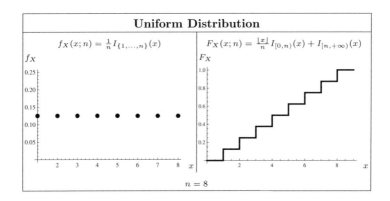

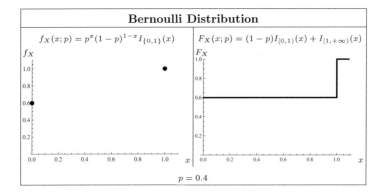

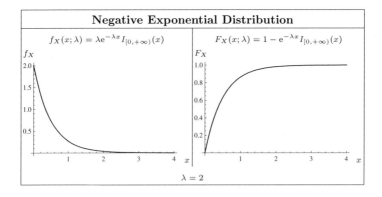

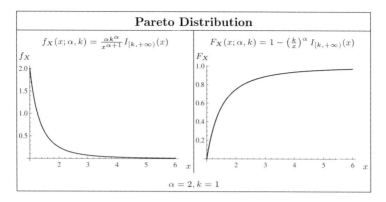

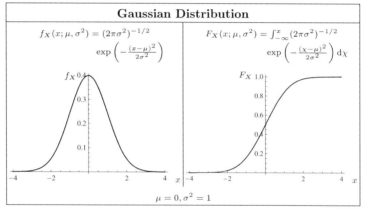

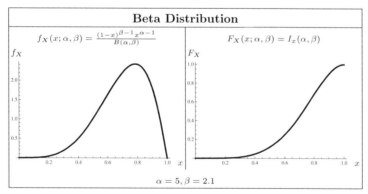

In the last picture we refer to $B(\alpha, \beta)$ as the *beta function* (also called the Euler integral of the first kind) of parameters α and β, defined as:

$$B(\alpha, \beta) = \int_0^1 x^{\alpha-1}(1-x)^{\beta-1} dx, \tag{B.1}$$

and to $I_x(\alpha, \beta)$ as the *regularized incomplete beta function*:

$$I_x(\alpha, \beta) = \frac{\int_0^x z^{\alpha-1}(1-z)^{\beta-1}dz}{B(\alpha, \beta)}, \quad (B.2)$$

B.2 Multidimensional Gaussian Variable

Let $\boldsymbol{X}$ be an n-dimensional variable with specifications $\boldsymbol{x} = (x_1, \ldots, x_n) \in \mathbb{R}^n$. Then:

- $\boldsymbol{\mu_X} = \mathrm{E}[\boldsymbol{X}] = (\mathrm{E}[X_1], \ldots, \mathrm{E}[X_n])$;

- $\Sigma_{\boldsymbol{X}} = \mathrm{E}\left[(\boldsymbol{X} - \mathrm{E}[\boldsymbol{X}])(\boldsymbol{X} - \mathrm{E}[\boldsymbol{X}])^T\right] = \begin{pmatrix} \sigma_{11} & \cdots & \sigma_{1n} \\ \vdots & \ddots & \vdots \\ \sigma_{n1} & \cdots & \sigma_{nn} \end{pmatrix}$,

  where $\sigma_{ij} = \mathrm{E}[(X_i - \mathrm{E}[X_i])(X_j - \mathrm{E}[X_j])]$ is the *covariance* of X_i with X_j, and $\Sigma_{\boldsymbol{X}}$ is the *covariance matrix* of $\boldsymbol{X}$.

Let $\boldsymbol{X}$ be an n-dimensional Gaussian variable. Its d.f. assumes the analytical form:

$$f_{\boldsymbol{X}}(\boldsymbol{x}) = \frac{1}{(\sqrt{2\pi})^n \sqrt{|\Sigma_{\boldsymbol{X}}|}} e^{-\frac{1}{2}(\boldsymbol{x}-\boldsymbol{\mu_X})^T \Sigma_{\boldsymbol{X}}^{-1}(\boldsymbol{x}-\boldsymbol{\mu_X})}, \quad (B.3)$$

where $|A|$ denote the determinant of the matrix A. In particular, with two-dimensional variable (X_1, X_2):

$$f_{X_1, X_2}(x_1, x_2) = \frac{1}{(\sqrt{2\pi})^2 \sqrt{|\Sigma_{\boldsymbol{X}}|}} e^{-\frac{1}{2}(x_1 - \mu_{X_1}, x_2 - \mu_{X_2}) \Sigma_{\boldsymbol{X}}^{-1} \binom{x_1 - \mu_{X_1}}{x_2 - \mu_{X_2}}}, \quad (B.4)$$

where:

$$\Sigma_{\boldsymbol{X}} = \begin{pmatrix} \sigma_{11} & \sigma_{12} \\ \sigma_{12} & \sigma_{22} \end{pmatrix}. \quad (B.5)$$

Its 3D-plot and contour plot are reported in Fig. B.1.

Let $\boldsymbol{\chi} = A^T(\boldsymbol{X} - \boldsymbol{\mu_X})$ the representation of $(\boldsymbol{X} - \boldsymbol{\mu_X})$ through its principal components, being A the matrix whose columns are the eigenvectors of $\Sigma_{\boldsymbol{X}}$ corresponding to the eigenvalues $\{\lambda_1, \ldots, \lambda_n\}$. Then:

$$f_{\boldsymbol{\chi}}(\boldsymbol{\chi}) = \frac{1}{(\sqrt{2\pi})^n \sqrt{\prod_{i=1}^n \lambda_i}} e^{-\frac{1}{2} \sum_{i=1}^n \frac{\chi_i^2}{\lambda_i}}. \quad (B.6)$$

Coming to the two-dimensional case we have:

$$f_{\chi_1, \chi_2}(\chi_1, \chi_2) = \frac{1}{2\pi\sqrt{\lambda_1 \lambda_2}} e^{-\frac{1}{2}\left(\frac{\chi_1^2}{\lambda_1} + \frac{\chi_2^2}{\lambda_2}\right)}. \quad (B.7)$$

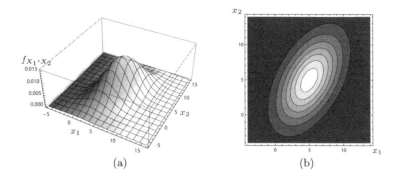

Fig. B.1. Shape of the d.f. of a two-dimensional Gaussian distribution of parameters $\mu_{X_1} = \mu_{X_2} = 5$, $\sigma_{11} = 9$, $\sigma_{22} = 16$ and $\sigma_{12} = 6$: (a) 3D plot, and (b) contour plot of f_{X_1, X_2}.

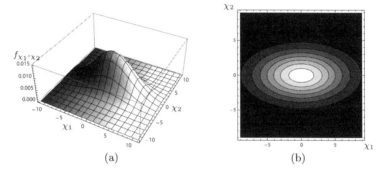

Fig. B.2. Shape of the d.f. of a two-dimensional Gaussian distribution with principal components of parameters $\mu_{\chi_1} = \mu_{\chi_2} = 0$, $\sigma_{11} = 19.45$ and $\sigma_{22} = 5.55$ representing the Σ_X eigenvalues of (X_1, X_2) as in Fig. B.1.

Pictures in Fig. B.2 give the companion of Fig. B.1 when (X_1, X_2) are represented through their principal components (χ_1, χ_2).

Let $\chi' = \Lambda^{1/2}\chi'$ the whitened representation of $(\boldsymbol{X} - \boldsymbol{\mu_X})$, being Λ the diagonal matrix with $\Lambda_{ii} = \lambda_i$. Then:

$$f_{\chi'}(\chi) = \frac{1}{\left(\sqrt{2\pi}\right)^n} e^{-\frac{1}{2}\sum_{i=1}^{n} \chi_i^2}. \tag{B.8}$$

Coming to the two-dimensional case we have:

$$f_{\chi'_1, \chi'_2}(\chi_1, \chi_2) = \frac{1}{2\pi} e^{-\frac{1}{2}\left(\chi_1^2 + \chi_2^2\right)}. \tag{B.9}$$

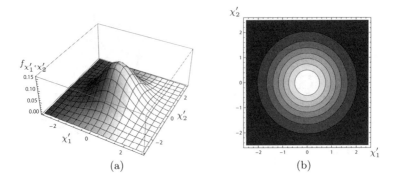

Fig. B.3. Shape of the d.f. of a two-dimensional Gaussian distribution with independent components of parameters $\mu_{\chi_1'} = \mu_{\chi_2'} = 0$, $\sigma_{11}\sigma_{22} = 1$ representing the whitened version of of (X_1, X_2) as in Fig. B.1.

The marginal distribution of X_i w.r.t. $\boldsymbol{X}$ is:

$$f_{X_i}(x) = \int_{-\infty}^{+\infty} f_{\boldsymbol{X}}(x_1, \ldots, x_n) \mathrm{d}(x_1, \ldots, x_{i-1}, x_{i+1}, \ldots, x_n) =$$

$$= \frac{1}{\sqrt{2\pi}\sigma_{X_i}} e^{-\frac{1}{2}\left(\frac{x - \mu_{X_i}}{\sigma_{X_i}}\right)^2}, \quad \text{(B.10)}$$

i.e. that of a (unidimensional) Gaussian distribution of parameters μ_{X_i} and σ_{X_i}.

The conditional density function of one component given the value of the remaining one reads:

$$f_{X_i|\boldsymbol{X}=\boldsymbol{x}}(x_i) = \frac{1}{\sqrt{2\pi}\sigma'_{X_i}} e^{-\frac{1}{2}\left(\frac{x_i - \mu'_{X_i}}{\sigma'_{X_i}}\right)^2}, \quad \text{(B.11)}$$

with:

$$\mu'_{X_i} = \mu_{X_i} + \Sigma_{X_i, \boldsymbol{X}_{-i}} \Sigma^{-1}_{\boldsymbol{X}_{-i}, \boldsymbol{X}_{-i}}(\boldsymbol{x}_{-i} - \mu_{\boldsymbol{X}_{-i}}) \quad \text{(B.12)}$$

$$\sigma'^2_{X_1} = \sigma_{ii} - \Sigma_{X_i, \boldsymbol{X}_{-i}} \Sigma^{-1}_{\boldsymbol{X}_{-i}, \boldsymbol{X}_{-i}} \Sigma^T_{X_i, \boldsymbol{X}_{-i}}, \quad \text{(B.13)}$$

where $\Sigma_{t_0,(t_1,\ldots,t_k)}$ denotes the vector $(\sigma_{01}, \ldots, \sigma_{0k})$ and extensions for $\Sigma_{(r_1,\ldots,r_q),(t_1,\ldots,t_k)}$, and $\boldsymbol{X}_{-i} = (X_1, \ldots, X_{i-1}, X_{i+1}, \ldots, X_n)$. Equation (B.12) identifies a regression line between $(\boldsymbol{X}_{-i}, X_i)$ specifications.

C Some Membership Functions

In line with Appendix B, we report here a set of popular membership functions, remarking that *any function* from $\mathbb{R}$ to $[0,1]$ is a candidate. Thus we rely on its *compliance* with available data and logical suitability in order to get correct results from the employment of the selected function.

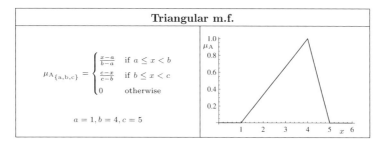

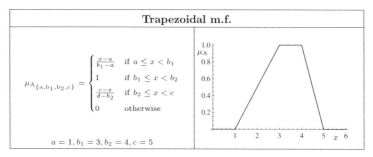

470   Some Membership Functions

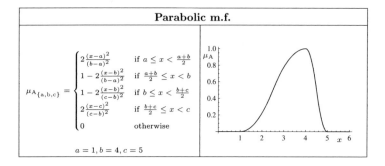

Parabolic m.f.

$$\mu_{A_{\{a,b,c\}}} = \begin{cases} 2\frac{(x-a)^2}{(b-a)^2} & \text{if } a \leq x < \frac{a+b}{2} \\ 1 - 2\frac{(x-b)^2}{(b-a)^2} & \text{if } \frac{a+b}{2} \leq x < b \\ 1 - 2\frac{(x-b)^2}{(c-b)^2} & \text{if } b \leq x < \frac{b+c}{2} \\ 2\frac{(x-c)^2}{(c-b)^2} & \text{if } \frac{b+c}{2} \leq x < c \\ 0 & \text{otherwise} \end{cases}$$

$a = 1, b = 4, c = 5$

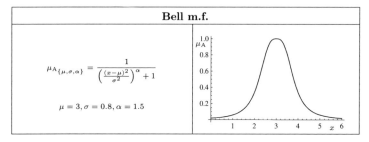

Bell m.f.

$$\mu_{A_{\{\mu,\sigma,\alpha\}}} = \frac{1}{\left(\frac{(x-\mu)^2}{\sigma^2}\right)^\alpha + 1}$$

$\mu = 3, \sigma = 0.8, \alpha = 1.5$

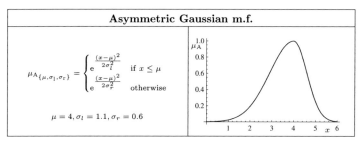

Asymmetric Gaussian m.f.

$$\mu_{A_{\{\mu,\sigma_l,\sigma_r\}}} = \begin{cases} e^{\frac{(x-\mu)^2}{2\sigma_l^2}} & \text{if } x \leq \mu \\ e^{\frac{(x-\mu)^2}{2\sigma_r^2}} & \text{otherwise} \end{cases}$$

$\mu = 4, \sigma_l = 1.1, \sigma_r = 0.6$

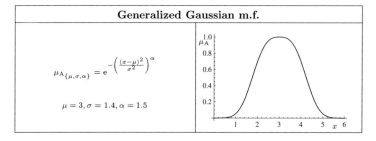

Generalized Gaussian m.f.

$$\mu_{A_{\{\mu,\sigma,\alpha\}}} = e^{-\left(\frac{(x-\mu)^2}{\sigma^2}\right)^\alpha}$$

$\mu = 3, \sigma = 1.4, \alpha = 1.5$

D Some Algebra

We assume the reader to be familiar with the basic operations with vectors. While the norm *per se* is a function of the length of the vector it refers to (see the symmetry property N1), a normed space allows also operations based on the reciprocal comparison of vectors in their directions. In their essence, they come from the definition of *dot product*.

Definition D.1. *Let $\mathfrak{X}$ be a vector space, $\boldsymbol{a}, \boldsymbol{b}$ two vectors and $r \in \mathbb{R}$. The dot product $\boldsymbol{a} \cdot \boldsymbol{b}$ of $\boldsymbol{a}$ and $\boldsymbol{b}$ is an operator $\cdot : \mathfrak{X} \times \mathfrak{X} \to \mathbb{R}$ satisfying the following properties:*

P1. $\boldsymbol{a} \cdot \boldsymbol{b} = \boldsymbol{b} \cdot \boldsymbol{a}$ *(symmetry)*;
P2. $\boldsymbol{a}(\boldsymbol{b} + \boldsymbol{c}) = \boldsymbol{a} \cdot \boldsymbol{b} + \boldsymbol{a} \cdot \boldsymbol{c}$ *(distributivity)*;
P3. $\boldsymbol{a}(r\boldsymbol{b} + \boldsymbol{c}) = r(\boldsymbol{a} \cdot \boldsymbol{b}) + \boldsymbol{a} \cdot \boldsymbol{c}$ *(bilinearity)*.

Represent a vector $\boldsymbol{a}$ through its components $\boldsymbol{a} = (a_1, \ldots, a_n)$ in the Euclidean space $\mathfrak{X} \in \mathbb{R}^n$, then you have:

$$\boldsymbol{a} \cdot \boldsymbol{b} = \sum_{i=1}^{n} a_i b_i. \tag{D.1}$$

Represent a matrix B through a vector of vectors $B = (\boldsymbol{b}_1, \ldots, \boldsymbol{b}_n)$ and you obtain $\boldsymbol{d}$ as a vector whose j-th component is:

$$d_j = \boldsymbol{a} \cdot \boldsymbol{b}_j = \sum_{i=1}^{n} a_i b_{ji}, \tag{D.2}$$

with b_{ji} the i-th component of the vector $\boldsymbol{b}_j$.

You gain a topological visualization of these operation minding at a squared grid where you locate the components of vector $\boldsymbol{a}$ on contiguous vertical cells (you say that a vector is a *column vector* by default) and denote $\boldsymbol{a}^T$ a vector having the same ordered components of $\boldsymbol{a}$ but disposed on horizontal contiguous cells. With this notation you denote equivalently the dot product, also called *inner product*, between $\boldsymbol{a}$ and $\boldsymbol{b}$ as:

$$\boldsymbol{a} \cdot \boldsymbol{b} = \boldsymbol{a}^T \boldsymbol{b}, \tag{D.3}$$

thus adopting the convention of a *row by column product*. Analogously the product of vector $\boldsymbol{a}$ by the above matrix B reads:

$$\boldsymbol{d}^T = \boldsymbol{a}^T B = (\boldsymbol{a}^T \boldsymbol{b}_1, \ldots, \boldsymbol{a}^T \boldsymbol{b}_n). \tag{D.4}$$

In particular, the product $\boldsymbol{a}\boldsymbol{b}^T$ is a matrix whose ij-th element is $a_i b_j$. With the same representation the vector $\boldsymbol{e} = \boldsymbol{a} + \boldsymbol{b}$ is a vector whose i-th component is $e_i = a_i + b_i$.

Of the huge set of features and properties deriving from the algebra of matrices we recall just a couple here below that are specifically employed in the book.

D.1 Eigenvalues and Eigenvectors

Definition D.2. *With reference to the vector space $\mathcal{X} \in \mathbb{R}^n$, a square matrix A with n rows and n columns (of order n, henceforth), a vector $\boldsymbol{x} \in \mathcal{X}$, and the equation:*

$$(A - \lambda I)\boldsymbol{x} = \boldsymbol{0}, \tag{D.5}$$

the elements of solution pair $(\lambda, \boldsymbol{x})$ are called respectively eigenvalue and eigenvector of A. By abuse, we identify the full rank property of A with having n distinct eigenvalues and corresponding eigenvectors.

As it emerges from the definition, the key-property of eigenvectors is to represent pivots of the single rotations of the reference frameworks (more in general, of any linear transformation). Their employment in signal analysis is extremely wide-spread, so that any general-purpose mathematical package contains routines for their computation. As for us, we have that in the case of full rank of A coinciding with $\widehat{\Sigma}_Y$ as in (7.9), the n eigenvectors are orthogonal and identify the n directions of residual maximal dispersion of the data in terms of the sum of quadratic distances from the origin along these directions, where the associate eigenvalues are proportional to the dispersions.

D.2 Kernels

Consider a set of vectors $\{\boldsymbol{y}_1, \ldots, \boldsymbol{y}_m\}$ in a space $\mathcal{Y}$ endowed with dot product. We are interested on symmetric matrices K coming from the application of binary functions k to each pair

($\boldsymbol{y}_i, \boldsymbol{y}_j$) of vectors. We ask the matrices to be constituted of inherently non- negative terms, a property that we identify with the non negativeness of its eigenvalues or equivalent conditions you may find in any textbook. We associate these matrices to broad sense *kernels* functions, preferably restricting the field to Gram matrices so that the involved binary functions are also bilinear. Thus, with some algebra (see Mercer's theorem), we may consider the term $K_{ij} = k(\boldsymbol{y}_i, \boldsymbol{y}_j)$ at the cross of i-th row and j-th column to be equal to the dot product $\boldsymbol{\phi}(\boldsymbol{y}_i) \cdot \boldsymbol{\phi}(\boldsymbol{y}_j)$ in a feature space into which $\boldsymbol{y}$ maps through $\boldsymbol{\phi}$. The trivial kernel is constituted by the dot product $\boldsymbol{y}_i \cdot \boldsymbol{y}_j$ on the above set.

The general suitability of a kernel with our scope comes from its mentioned interpretation in terms of a dot product of a function ϕ applied to $\boldsymbol{y}_i$ and $\boldsymbol{y}_j$, which projects $\boldsymbol{y}$ into a higher dimensional space (called feature space). Typical kernels of wide use, bejond the above mentioned dot product, are listed in the following table.

Inhomogeneous polynomial kernels of degree d	$k_p(\boldsymbol{y}_i, \boldsymbol{y}_j) = (\boldsymbol{y}_i \cdot \boldsymbol{y}_j + c)^d$
Gaussian kernel	$k_G(\boldsymbol{y}_i, \boldsymbol{y}_j) = e^{-hl^2(\boldsymbol{y}_i - \boldsymbol{y}_j)}$
Sigmoid kernel	$k_s(\boldsymbol{y}_i, \boldsymbol{y}_j) = \tanh(\alpha \boldsymbol{y}_i \cdot \boldsymbol{y}_j + \beta)$

Inhomogeneous polynomial kernel of degree d may be regarded as a dot product of the $\binom{n+d}{d}$-dimensional space of all monomials in n variables (where n is the dimensionality of $\mathcal{Y}$), up to degree d with c arbitrary constant. On the contrary, the feature space associated to both Gaussian and sigmoid kernels are of infinite dimensionality and may be reformulated through quite complex $\phi(\boldsymbol{y})$. In particular the shape parameter h in Gaussian kernel relates to the width of radial basis-like functions, while scale and position parameters α and β of sigmoid kernel allow for a stretching/translation of the sigmoid function at the basis of neural network. Note that, despite its large use, sigmoid kernel does not belong to the class of positive semidefinite bilinear functions, thus losing most properties strict sense kernels hold.

Kernels are widely used in machine learning community since, given an algorithm formulated in terms of dot products, one can always construct an alternative algorithm by replacing the dot product · with a kernel k (*kernel trick*). This proves very efficient for extending in a quite natural way and without extra-efforts the strict functionalities of a linear algorithm to handle nonlinear problems.

Pictures reported below point out what happens, in terms of 3D-plot and one contour plot line, when the standard dot product in the function $\ell(x) = \sum_{i=1}^{m} x_i \cdot x$ is substituted with the aforementioned kernels, where $m = 3$ points: $x_1 = (-50, 20)$, $x_2 = (10, 80)$ and $x_3 = (60, -40)$ are used in the above sum.

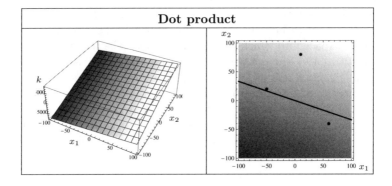

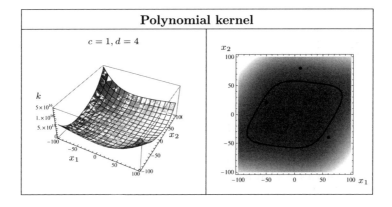

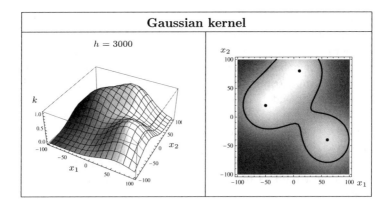

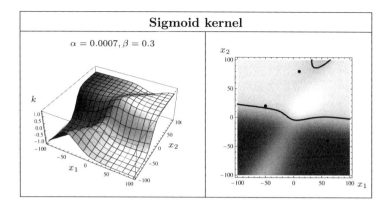

E List of Symbols

E.1 Variables

Notation:

- small caps, any character (ex. $\{x, y, u, \theta\}$): deterministic variables, ex. random variable specification, value assignment;
- italic capital letters, any character (ex. $\{X, Y, U, \Theta\}$): either random variables (with specifications $\{x, y, u, t\}$) or matrices. Greek symbols have often large cap coinciding with roman symbols (for instance large cap of ϵ is E);
- roman capital letters, any character (ex. $\{A, B, C, W\}$): granule, in the form of fuzzy set, rough set, interval-valued set and shadow set.
- gothic letters (ex. $\{\mathfrak{X}, \mathfrak{Y}, \mathfrak{U}, \mathfrak{f}\}$): sets (with elements $\{x, y, u, \theta\}$);
- different fonts may be employed to disambiguate the meaning of a symbol.

Modifiers:

- $\widehat{}$: high quality approximation (ex. $\widehat{\Theta}$: unbiased estimator of Θ; $\widehat{y}$ approximate value of the solution y);
- $*$: optimal solution (ex. $\widehat{\theta}^*$: maximum likelihood estimate of θ; y^*: optimal solution of a problem);
- $\widetilde{}$: either less quality approximation (ex. $\widetilde{\Theta}$: weakly unbiased estimator of Θ; $\widetilde{y}$ rough approximate value of the solution y), or arbitrarily fixed value (ex. $\widetilde{p}$ as a pivot of a twisting argument on P);
- $\check{}$: bootstrap computation of a random parameter (ex. $\check{\theta}$: bootstrap computation of θ) and involved seeds (ex. $\{\check{u}_1, \ldots, \check{u}_m\}$: seeds of a universal sampling mechanisms);
- $\bar{}$: sample average (ex. $\bar{x} = 1/m \sum_{i=1}^{m} x_i$).

478 List of Symbols

Indices:

- indices denote either a position in a set (say $\{x_i, y_j\}$) or a reference framework (for instance μ_X to denote the mean value of random variable X, or $\theta_{\text{dw}}, \theta_{\text{up}}$ to denote extreme values). Indices in round brackets refer to ordered values;
- apices refer to a power exponent (say x^3).

Ensembles:

- indexed variables within braces: set constituted of these elements;
- indexed variables within round brackets: ordered set, i.e. vector having these elements as components;
- bolsymbol: vector having components described by indexed non-bold symbols;
- capital letters (such as W, A) may denote matrices.

Favorite meaning:

- last characters (say $\{x, y, z\}$ for variables, first characters (say $\{a, b, c\}$ for items in a set), intermediate values (say $\{i, m, n\}$) for indices. In particular:
- n, m, N for dimensionality of the sample space (possibly number of bits of a string), cardinality (size) of a sample, cardinality (size) of a population (possibly number of bootstrap replicas), respectively; ν for number of items in a set, q number of contexts, k number of clusters;
- v_i for a boolean variable and $\overline{v}_i$ its negation;
- θ for a parameter (in particular μ and σ for mean and standard deviation), ϵ for a noisy term, δ or γ for confidence level, η for learning rate, λ either eigenvalue or parameter of a negative exponential distribution law;
- $\mathbb{N}, \mathbb{I}, \mathbb{R}$ for the set of natural, integer and real numbers, respectively;
- b_i for the i-th binary output of a function;
- μ_h, t_h, ρ_h for detail of h, number of points misclassified by it and their cumulative probability;
- W, A for matrices supporting linear transforms: in particular W describes connection weights matrix in a neural network, P transition matrix in a Markov process;
- W for fuzzy weight in a granular neuron;
- $\boldsymbol{y}$ for pattern, $\boldsymbol{v}$ for centroid, $\boldsymbol{z}_m$ for sample of size m.
- $\boldsymbol{\tau}, \boldsymbol{w}, \boldsymbol{\gamma}$ for state vector, weight vector, and threshold vector in a neural network, respectively;
- $\boldsymbol{w}, b, \boldsymbol{\alpha}$: support vector machine coefficients and Lagrange multipliers.

E.2 Functions

Functions use symbols with same notations as for variables, but with looser interpretation. The argument of a function is in round brackets. In addition:

- apices refer either to a power exponent, like with variables, or to a functional (for instance G^{-1} to denote the inverse function of G).

Favorite meaning:

- F_X, f_X respectively for cumulative distribution function, and either probability function or density function; $\widetilde{F}_{X_\theta}^{-1}$ for generalized inverse function of F_{X_θ}, L for likelihood function;
- P, ϕ for probability and asymptotic frequency, respectively;
- g_θ for explaining function, s for a statistic, s_θ for a well behaving/sufficient statistic for θ (same management of small caps and capital letters to distinguish randomness of both statistic and parameter);
- μ_A for membership function of the fuzzy set A, $\mu_{A_{\{a,b,c\}}}$ for three-positional parameters fuzzy number;
- c for a goal function, C its random counterpart, and C the class it belongs to; h, H, H analogously for an hypothesis about c;
- $I_\alpha(h_1, h_2)$ for Incomplete Beta function with parameters h_1, h_2, Γ for Gamma function, $\mathrm{B}(a,b)$ for a Beta function with parameters a and b, $I_A(x)$ for indicator function of the set A;
- $|\ |$ for either length of a string (say $|x|$ equal to the number of bits of x) or module of a vector, π shortest program coding a string, K_z conditional Kolmogorov complexity of a string;
- $d(a,b), s(a,b)$ for distance/similarity between a and b, $\nu(\boldsymbol{y})$ norm of $\boldsymbol{y}$;
- U_k probability measure of k blocks, U_c probability measure of concept c, $\mathfrak{U}(s)$ partition induced by statistic s, $\boldsymbol{\mu}_A$ vector of membership grades, U partition matrix;
- $\mathsf{h}_i(\boldsymbol{\tau})$ for general form of an activation function.

E.3 Operations on Sets

- $\cap$ intersection, $\cup$ union, $\div$ symmetric difference;
- $\top$ t-norm, $\bot$ s-norm.

E.4 Operators

- $E[X], V[X], H[X]$: mean, variance and entropy of random variable X;
- $I[X, Y]$: mutual information between X and Y;
- $\ell[D, d_j]$: loss function of decision D w.r.t. environment d_j;
- C: cost function.

Index

ϵ-tube   422
π-calculus   332

absence of prejudice   281, 288, 301
absorption process   373
abstraction   437, 438, 440
acceptance probability   291
activation function   294
  Heaviside   296, 327, 330
  linear   295
  probabilistic   296, 300
  sigmoid   296
  temperature   296
activation mode   295
  asynchronous   295
  delayed   295
  parallel   295
  random   295
adaptation   314
affinity   325
age parameter   328
agglomeration   *see* clustering, algorithm, agglomerative
aggregation   *see* clustering, algorithm, aggregative, 259
aging   319, 327, 328
agnostic approach   194, 195, 201, 203, 221, 228
algorithm
  branch and bound   219
  clustering, *see* clustering
  conjugate gradient   304

  gradient descent   360, 387, 399
  greedy   219, 289
  heuristic   322, 343
  intrusion detection   326
  learning, *see* learning algorithm
  least square fitting   399, 400, 441
  Levenberg-Marquardt   304
  linear programming   255
  optimization, *see* optimization problem
  simulated annealing   219, 266, 304
  steepest descent   302
  tabu search   219
  training, *see* learning algorithm
algorithmic inference   12, 118, 405
anchor point   390
ANFIS   396, 398, 405
  type   397
angular cross-correlation   241
ANOVA   210
anti-entropy   318
antibody   325
antigen   325
  receptor   326
approximation   59
  Gaussian   54
  linear   361

piecewise linear 379, 396, 399
aprioristic approach 194, 201, 228
artificial intelligence 265

Back-Propagation
 algorithm 305
bagging *see* ensemble, bagging
balance 310, 323, 366, 380, 391, 415, 423, 430, 434, 439, 442, 443
benchmark *see* dataset
bias 289
 inductive 24
bias-variance tradeoff 176, 332, 415
BICA 247, 311
binarization 248
biologically inspired optimization 317, 327
 analysis 332
 design 332
boolean
 algebra 348
 assignments 306
 clause 19, 112
  canonical 21
  Horn 348, 372
 formula 111
  CNF 20, 112, 175
  consistent 22
  descriptional length 175
  DNF 20, 112, 175
  focal set 176
  fuzzy, *see* fuzzy set, boolean formula
  fuzzy frontier, *see* fuzzy set, frontier
  monotone 19
  nested domain 176
  support 19
  understandable 175
 function 19, 96, 309
 literal 19, 112
 monomial 19, 112, 175
  canonical 21
  core 176
 propositional variable 19, 112

vector 247, 297
boosting *see* ensemble, boosting
bootstrap 312
 nonparametric 313
 population 71, 229
 accuracy 72
 regression 86

Cartesian product 234
categorical data 204, 458
centroid 196, 200, 228, 249, 290, 356, 378
Chapman-Kolmogorov equation 280
chromosome 287
Church-Turing thesis 57
classification 108, 227
classifier 390, 411
 approximate 414
 merge 416
 multiple 417
 nonlinear 416
cluster 193, 374
 number of 210, 225, 368, 439
cluster analysis *see* clustering
clustering 191, 193, 239, 246, 265
 algorithm 198, 246, 249, 261
 agglomerative 198, 212
 aggregative 198, 212, 218
 classification of 204
 collaborative Fuzzy C-Means 435, 436
 DIANA 214
 divisive 214
 exact relocation 219
 Fuzzy C-Means 221, 344, 357, 359, 363, 366, 388, 391, 393, 403, 420, 433, 436
 Fuzzy C-Means, features 223
 GNAT 212
 greedy 219
 hierarchical 200, 212, 213, 225
 implementation mode 219
 incremental 220
 ISODATA 215

Index 483

k-means   205, 244, 252, 256
k-means, analysis   209
k-means, improvements
   209, 219
multistarting   219
Nearest Neighbor, *see*
   Nearest Neighbor
QT Clust   214
spectral   267
stopping rule   219, 220
collaborative   433, 435
   evaluation   437
conditional   366, 370
context-based   363
data-driven   389
direction-sensitive, *see*
   clustering, context-based
fuzzy   354, 363, 395, 441
human-centric   388, 389,
   392, 404, 420
incremental   249
partially supervised, *see*
   clustering, human-centric
codeword   246
coding   61
coefficient
   cost balancing   366, 391,
      442, 443
   scaling   434
collaboration   431, 433, 434,
   437, 439, 447
communication   437, 440–442
compatibility   34, 65, 118
   measure   7
compatible population   *see*
   population compatible
complexity   56, 443
   conditional   57, 203
   Kolmogorov   56, 202, 230
   Levin   57, 201, 221, 236
   lower bound   58
   prefix   57
   shortest program   56, 202
      approximation   60
   unconditional   57
   upper bound   58
concise description   *see* data
   compression
confidence   390
   interval   78

   Bernoulli   79
   Gaussian   81
   level   78
   misclassification error   111
   two-sided symmetric   79
region   85, 310
   envelope   88
   Gaussian   95
   level   85
   nested   85
   peeling   89
   pivot   85
   score   88
confusion matrix   206, 237,
   244, 250, 251, 395
connection   293, 328, 385, 396
   backward   307
   matrix   297, 330
   weight   294, 295, 396
connectionism   277, 309, 354
connectionist paradigm   *see*
   connectionism
consensus function   218
consistency   394, 433, 438,
   440–442, 446
   level   444, 445
constraint   260, 412, 414, 419,
   423, 429, 439
context   364, 366, 374
   computational   59
   number of   368
controllability   332
cost function   97, 178, 194,
   220, 222, 233, 235, 239,
   247, 249, 292, 301, 321,
   367, 387, 388, 391, 392,
   404, 413, 415, 418, 422,
   424, 429, 433, 437, 441, 442
   dual   419, 423, 424
   estimator   81
covariance matrix   197, 206,
   235, 236, 239, 240, 466
cross-validation   179
crossover   287
curse of dimensionality   313,
   350

danger   326
Danger theory   326

data compression 58, 61, 203, 266, 306, 312, 366
dataset 25–30
  censored 26
  Cox 26, 231, 232, 381, 419
  Guinea pig 26, 381
  Iris 28, 204, 207, 211, 214, 231, 237, 244, 249, 252, 258, 260, 263, 414–417
  Leukemia 25, 35, 91, 115
  MNIST 28, 31, 153, 195, 197, 200, 208, 217, 218, 232, 252, 304, 389, 395, 403, 420
  motorcycle 429
  satellite 307
  sinc function 425
  SMSA 27, 54, 115, 263, 359, 382, 401, 404, 428, 431
  Sonar 29, 248, 306, 310
  Swiss 28, 232, 264, 424
  synthetic 25, 197, 205, 207, 212, 216, 223, 226, 228, 237, 244, 249, 251, 256, 262, 361, 365, 387, 393, 399
    2-spiral 198
    halfring shaped 218
    knapsack 291
    spiky function 362, 368
    square 421
  TSP 321, 324, 326
  Vote 29, 112, 178
  XOR 299, 388
data site 431, 433, 435, 437, 440
  number of 439
death
  functional 329
  physiological 329
decay 330
decision 194
  function, see decision problem
  problem 194, 203, 220, 227
  rule 194, 228, 251
  theory see decision problem
  tree 112, 205
    C4.5 113
decoding 281
defence 329

defuzzification 348, 375, 377, 396, 398
defuzzifier 378
  coa 378
  height 378
  max 378
  mom 378
dendrogram 212, 214
diameter intra-cluster 214
dimensionality 366
  reduction 365
discriminant analysis 191, 227, 248, 259
dispersion 210, 361
  between clusters 194, 210, 237
  residual maximal 236
  within cluster 194, 210, 237, 253
dissimilarity 196, 214, 459
  intercluster, see linkage method
  Kullback 292
  Kullback, sample estimate 292
distance see metric
distribution law 202, 285, 463
  compatible 68, 275
  conditional 177, 468
  cumulative distribution function 6, 48, 463
  empirical 8, 55, 71, 291
  generalized inverse 10, 52
  inference 74
  density function 48, 356, 464
  empirical 247
  environmental 298
  factorization 50
    criterion 68
    lemma 84
  joint 50
  marginal 95, 468
  multimodal 244
  peaked in the maximum 287, 290
  prior 177
  probability function 6, 48, 247, 279, 377, 463
    histogram 290

subgaussian 239
supergaussian 239, 255
support 259
symmetric 237
domain knowledge *see* knowledge base
dot product 235, 257, 260, 416, 424
  properties 471
dumping term 331, 430
dynamics 276, 279, 298

edge pulling function 247, 306
eigenvalue 166, 236, 251, 466, 472
eigenvector 166, 236, 262, 466, 472
ellipsoid 236
elongation 239
emerging behavior 275, 277, 301, 307, 319, 325, 326, 336
encoding 246, 247, 281
  binary 281, 288
  correct 246
  floating point 282
  fuzzy logic hierarchy 282
  fuzzy system 282
  rule-based system 283
  tree representation 284
ensemble 278, 309, 416, 442, 443, 447
  analysis 310
  bagging 313
  base learner 311, 312
  boosting 305, 313
  majority voting 215, 309, 311, 394, 395
  random subspace method 313
  replica 309, 368
  SVM, of 310
  weighted majority algorithm 310, 313
entropy 60, 83, 202, 225, 230, 244, 247
  conditional 221
  sample estimate 202, 247
environment 335
estimator
  convergence 54

cost function 81
loss function 81
  quadratic 82
maximum likelihood 84, 202
minimization problem 82
MMSE 82
moments, method of 83
point 81
risk 82
robust 243
UMVUE 82, 197
unbiased 82
uniform 82
weakly unbiased 82
evolutionary
  algorithm 219, 286, 297, 299, 336
    clonal selection 326, 327
    genetic algorithm 219, 285–292
    genetic programming 284
    immune system 219, 325–327
    negative selection 326
  optimization 281, 285, 318, 332
expected value 49, 60, 201, 378
  empirical 60
experience 319, 439, 440, 447
expert 308, 347, 378, 396
explaining function 10, 298
  dynamic 278
exploitation 288, 323
exploration 288, 323, 332
exponential family 87
exponential smoothing 431

FastICA 241, 243, 262
Feasibility 335
feature 199, 234, 254
  extraction 335
  hidden 364
  moulding 325
  relevant 234, 248, 356
feature space 196, 233, 234, 254, 258–260, 287, 313, 356, 366, 416, 424, 430, 473
fitness 286, 287, 314, 321, 322, 325, 333, 412

486 Index

fixed point   163, 399, 427, 430
fractal   405
frequency   5
　asymptotic   279
fuzzification   348, 397
　transfer of fuzziness   357
fuzzification coefficient   222,
　　223, 357, 369, 392, 420, 434
fuzzy inference system   see
　　fuzzy reasoning
fuzzy model   443, 444
　architecture   355
　blueprint   345, 355, 371
　clustering   355
　collaborative   431
　communication layer   431
　connectionist part   371
　consistency-based optimiza-
　　tion, see fuzzy model,
　　experience-consistent
　distributed architecture   431
　experience-based, see fuzzy
　　model, experience-
　　consistent
　experience-consistent   432,
　　439–441, 444
　hierarchical   352
　incremental   360, 368
　　analysis   365
　interface   346, 375
　　granular   348
　　numeric   347
　local regression   440–442
　master scheme   353
　MIMO   346, 371
　MISO   346
　non-numeric   445
　nonlinear   357, 396
　processing core   346, 375,
　　431
fuzzy norm   161, 230, 353, 358,
　　371, 385, 397, 460
　fuzzy set, between   372
　Gödel norm   17, 161, 162,
　　371, 375, 386, 437
　product norm   375, 386, 397
　properties   460
fuzzy number   168, 169, 171,
　　370, 445, 446
fuzzy partition   370

fuzzy reasoning   396
　Mamdani   375, 396, 398
　Takagi-Sugeno   378, 396, 398
　　order   378
　Tsukamoto   379, 396, 398
fuzzy rule system   see
　　rule-based system
fuzzy set   16, 345, 440
　α-cut   158, 360, 374, 375,
　　377
　boolean formula   176
　calibration   131, 137, 152
　circle   173
　compatible   18
　condensation, see fuzzy set,
　　localization
　core   169, 176
　coverage   168, 170, 370
　degenerate   17, 161
　descriptors   137
　design   185
　distance, of   174
　equalization, see membership
　　function, estimation, fuzzy
　　equalization
　family of, see fuzzy set,
　　vocabulary
　feasibility   131
　first order   372
　frontier   176, 177
　　radius of   177
　fuzzy norm, see fuzzy norm
　generalization   150, 185, 374
　higher order   144
　identification   344
　inference   18, 162
　intersection   17
　interval-valued   143, 160
　linearization   132
　linguistic approximation
　　146
　localization   140–143
　masked   373
　metaelement   144
　multidimensional   352
　norm
　　product   164
　pivot   137, 160, 165, 169,
　　173, 174, 179
　population   18

preference   135
referential, *see* fuzzy set, vocabulary
refinement   131, 225, 361, 370
representative point   162
rough   148
second order   144, 372
semantic   130, 366
  overlap   179
shadowed set, transformation in   140
shift trick   162, 163
specificity   168, 170, 373, 380, 445
support   170, 176
synthesis   161, 375, 445
transform
  nonlinear   137
  optimization   137
  piecewise linear   137
type-2   149
typicality   133
uncertainty   169, 179
union   17
vocabulary   136, 146, 160, 179

gap   246
gene   286
generalization   295, 313, 412
genotype   55–63, 84, 201, 275, 281, 286, 332
Gram matrix   425, 472
granular activation function   372, 378, 385, 399
  constant   378
  linear function   378
  sigmoid function   379, 396, 399
granular connection   371, 375
granular construct   *see* fuzzy model
granularity   78, 84, 265, 334, 343, 350, 352, 406, 434, 436, 437, 447
granular neural network   371
  architecture   374, 375
granular neuron   371, 374, 375
  training   385

granular weight   *see* granular connection
granule   3, 60, 130, 344, 345, 371, 372, 385, 388, 431, 433
  conditional   364
  fuzzy   4, 15, 127, 155
  generation   16
  logical   4, 19, 96
  semantic   152, 181
  statistical   3, 5, 45
graph   294
ground state   329

hazard function   92
  censored data   93
  master equation   92
  sampling mechanism   92
heteroskedastic model   429
Hidden Markov model   231
hidden variable   200
homeostasys   328
hybrid approach   5
hybrid system   294, 307, 333, 381
hyperplane   207
  numerical accuracy   110
  optimal margin   109, 259, 310, 413
  sentry point   109, 110
  separating   108, 109, 411, 412, 427
  symmetric difference   110
hyperpoint   358
hypersphere   259

ICA   200, 237
  local   249
idiotypic network   325
Immune Network theory   325
immune system   *see* evolutionary algorithm, immune system
Incomplete Beta function   33, 79, 103, 465
independence   234, 238, 261
  linear   239, 240
  local   303
  w.r.t. rotation   239
independent component   239, 240, 247, 255, 261, 468

488   Index

boolean, 248
indeterminacy   73, 109, 113, 343
index
   CH   211
   collaboration   438
   dispersion   419
   elbow criterion   210
   equalization performance   181
   experimental evidence   170
   GAP   211
   global behavior   443
   H   211
   inconsistency   167
   KL   211
   performance   141, 227
   quality   437
   separation   224, 225
   Silhouette   211
   transitivity, lack of   167
indicator function   9, 391
inequality
   Jensen   301
   Levin, *see* complexity, Levin
inference   63, 65, 293, 403, 443
   deductive   380
   pivotal quantity method   81
   predictive   117
information   246, 335, 385, 417
information granule   *see* granule
information processing   371

Karush-Kuhn-Tucker condition   424
kernel   256, 257, 259, 260, 416, 419, 424, 425, 429, 430, 473
   Gaussian   259, 425, 429, 473
      width   260
   PCA   258
   polynomial   258, 417, 431, 473
   sigmoid   473
kernelization of metric   257
kernel trick   256, 293, 310, 311, 416, 424, 428, 473
knapsack problem   284, 291
   profit   284
   weight   284

knowledge   389, 439
   base   130, 332, 352
   engineering   411, 446
   prior   347
   sharing   431, 433, 440
      asymmetric   432
      symmetric   431
   symbolic   380
   tidbits   389, 391
Kolmogorov framework   82
Kronecker delta   425
kurtosis   244

label   197, 227, 246, 398, 411
   unreliable   390
   vector   391
labeled sample   *see* sample labeled
Lagrange multiplier   392, 413, 415, 423, 426
latent variable model   258
law of large numbers   313
layer   328, 374, 378, 396
   hidden   303
   input   303
   output   303
learning   23, 62, 63, 65, 118, 317, 396
   algorithm   97, 106, 277, 298, 387
   back-propagation   302, 398, 400
   convergence   396
   learning rate   304, 387, 399
   momentum term   304, 305
   strong   313
   SVM, *see* support vector machine
   two-phase   399, 400, 403
   weak   313
algorithmic expansion   100
batch modality   399
boolean function
   bounds   103, 106
   interval estimate   106
   point estimator   108
   bounded rectangle   104
   complexity index   108
   computational load   107
   detail   101

ensemble, *see* ensemble
error probability   97
frontier   101
function   84, 422
  concept   97
  fairly strongly surjective
    105
  granular regression   358
  hypothesis   97, 411
  linear regression   86, 228,
    358, 361, 419, 422
  nonlinear regression   277,
    301, 345, 358, 361, 399,
    422, 424, 428
  optimal regression line   359
  symmetric difference   97
hypothesis
  approximate   107
  consistent   102, 103, 412
  inductive   380
  learnable concept class   106,
    107
  mislabeled points   107, 111
  nested domain   99
  numerical precision   104
  on-line modality   399
  PAC   96
  relevant points   111
  sample size   107, 439
    bounds   106
  sampling mechanism   97
  sentinel, *see* learning, sentry
    point
  sentry function   100
  sentry point   100, 109
  supervised   259, 440
  symmetric difference   103
  training strategy   305
  twisting argument   103
  uncertain examples, from
    103
  unsupervised   206, 259
  Vapnik-Chervonkis dimension   102
  witnessing   99
linear combination   239, 441,
  443
linear congruental generator
  46
linear dependence   235

linear separator   *see* hyperplane
linguistic variable   130, 348
linkage method   212
locality   349, 361
local optimum   219, 304, 388
log-likelihood   84
loss function   196, 227, 228
  ϵ-insensitive, 425
  estimator   81
  quadratic   82
  ridge loss   424, 429

machine learning   277, 398
majority voting   *see* ensemble,
  majority voting
manifold   258
mapping   *see* transform
margin   412, 427
  generalization   417
Markov chain   279, 288, 297,
  406
  detailed balance   280
  homogeneous   , 279
  master equation   280
  stationarity   280
  stationarity state, asymptotic
    280, 298
mask   239, 241, 256
master equation   12, 67
  double exponential   71
  Gaussian   229
  hazard function   92
  Laplace   87
  Markov chain   280
  system of   94
maturation   325
mean   *see* expected value
median   70, 169
membership degree   *see*
  membership function
membership function   16, 252,
  345, 353, 366, 386, 437,
  446, 469
  bell   358
  bounds   143
  compatibility   144, 161, 164
  definition   131, 133
  discrete   142

estimation   151, 155, 168, 186, 225
  consistency   159
  experimental evidence   170, 179, 180, 225, 347, 352, 366, 370, 392, 440
  Fuzzy C-Means   *see* clustering, algorithm, Fuzzy C-Means
  fuzzy equalization   179–181
  horizontal   156
  maximum compliance   162, 164
  optimization   159, 170
  Saaty's priority method   165
  vertical   158
gap   144
Gaussian   18, 137, 138, 142, 164, 352, 362, 386
generalized Gaussian   398
joint   163
parabolic   142, 383, 470
polynomial   173
strength   144
trapezoidal   169
triangular   137, 138, 142, 149, 150, 160, 169, 171, 179, 372, 386, 387, 445, 469
uniform   137
memory   325
merge
  aggregation   445
  clusters   213–215
  granules   367, 371
metagranule   160
metric   196, 201, 204, 223, 228, 230, 233, 239, 255, 257, 412, 434, 437, 444, 453
  angular separation   250, 458
  Chebyshev   458
  correlation coefficient   250, 458
  Euclidean   196, 321, 366, 395, 458
  Hamming   307, 438, 459
  Hausdorff   174
  Kullback   *see* dissimilarity, Kullback
  Mahalanobis   251
  Manhattan   458
  Mbiknowsi   458
  properties   453
  relative   412
minimization problem   442
  constrained   260, 392, 413, 415, 418, 422, 427
mirroring   306
misclassification error   415
  confidence interval   111
  sampling mechanism   111
mixture
  coefficient   355
  experts, of   378, 396, 447
  function   385
  fuzzy norm   355
  Gaussian model   194, 197, 228, 237, 244, 249, 251
  granule   354
  labeled-unlabeled pattern   389
  linear function   309
  local model   441, 443
  nonlinear function   345, 355
MLE   *see* estimator, maximum likelihood
MLE curve   428
mode   229, 370, 378
modifier   372, 374
  linguistic   146
    concentration   146
    dilution   146
moment
  first order, *see* expected value
  second order, *see* variance
monotonicity   363, 379
morphology   248
mutation   287, 325
mutual information   246

Nearest Neighbor   198, 403
neighboring notion   287
net input   295, 372, 385
network
  ad hoc   435
  wireless sensor   435
neural network   259, 293, 327, 344, 354, 385
  architecture   294, 305
  autoassociative MLP   258

Boltzmann machine 292, 297–301
  awake phase 299
  clamped mode 299
  dream phase 299
  intermediate belt 298
  thermal core 298
  unclamped mode 299
  visible belt 298
granular, *see* granular neural network
hourglass 306
hybrid, *see* hybrid system
MLP 299, 301–308, 328, 387
multi-layer perceptron, *see* neural network, MLP
RBF 357
recurrent 307
SOM 259
unfolding 307
neural spike 329
neuro- fuzzy system
  learning algorithm 398
neuro-fuzzy system 374, 404
  ANFIS, *see* ANFIS
  architecture 396
  training 396
neuron *see* PE
neuron strength 329, 331
node 435
node function *see* activation function, granular activation function
noise 440
nongaussianity 245, 266
norm 202, 204, 220, 230, 234, 249, 392, 453
  l^r-norm 454
  l_r-norm 454
  general 453
  properties 454, 455
normalized firing strength 397
NP-complete problem 319, 334
NP-hard problem 219, 285

Occam's principle 361
optimization problem 326
  constrained 426
  dual 413, 415, 416, 427

order 3
order relation 459
orthonormal matrix 240
outlier 175, 343, 426
overcomplete basis 254
overfitting 415, 422

parallel computation 277
parameter
  free 12, 164, 194, 227, 385, 399, 400, 412, 431, 444
  granular 444, 445
  vector of 93
partition matrix 222, 363, 367, 392, 394, 433, 436, 437
  human-centric 391
  induced by a collaborative node 433, 438, 442
pattern 193, 389, 390, 411, 440
  linearly separable 412
  missing value 419
  nominal value 419
  nonhomogeneous 417
  nonlinearly separable 414, 416
  partitioned 433
  quality 446
pattern recognition 325, 326
PCA 200, 208, 235, 472
  local 249, 252
PE 293, 385
  firing mechanism 327
  subsymbolic 293
  symbolic 293, 307
penalty *see* cost function
permutation
  consistent 394
phenotype 55–63, 275, 281, 287
polytope 256
population 9, 314, 335
  compatible 12, 72, 228, 275, 291
  hidden 293
  moulding 276
  visible 293
possibility and necessity theory 185
postprocessing 218, 365
predator-prey scenario 328

492   Index

prefix function   57
prefix machine
    universal   57
preprocessing   200, 265, 312, 419
principal axis   see eigenvector, 254
principal component   236, 239, 247, 250, 251, 260, 395, 403, 466
principal surface   258
    probabilistic   258
principle
    first-principle global model   360
    justifiable information granularity   168, 186, 445
    structural minimization   415
priority reciprocal matrix   165, 166
probability   5
    conditional   50, 194, 220
    model   43, 155, 201
processing element   see PE
profile graph   212, 214
projection   260, 356, 363, 365, 366, 417
    nonlinear, see transform, nonlinear
prototype   222, 355, 368, 392, 433, 436, 441, 443, 444
    collapse   365
    indistinguishable   365
proximity   437
    matrix   437, 438

quadratic error   301, 306, 369, 387, 398, 403, 441
quantile   78

random function   85
    MLE   89
    moments, method of   86
    pivot   85, 89
    sampling mechanism   86
    score   88
randomness epistemology   117
random number generator   46
random seed   13, 92
    Normal   80

    uniform   30
random variable   6, 47
    Bernoulli   98, 111, 201, 464
        confidence interval   79
        sampling mechanism   11
        sum   51
        twisting argument   73
        well behaving statistic   68
    Beta   33, 79, 465
    Binomial   51, 106, 156
        estimator   83
        sampling mechanism   66
    continuous   48
    discrete   48
    double exponential
        master equation   71
        sampling mechanism   70
        well behaving   70
    function of   49, 356
    Gamma   51
    Gaussian   80, 89, 95, 157, 228, 238, 310, 400, 465
        block   61
        confidence interval   81
        confidence region   95
        master equation   229
        multidimensional   197, 206, 236, 358, 466
        sampling mechanism   80, 95, 228
        sufficient statistic   80, 228
        sum   51
        truncated   290, 291
    independent   51
    Laplace   86
        master equation   87
        sampling mechanism   86
    multidimensional   50
    negative exponential   464
        block   61
        estimator   84
        sampling mechanism   11
        sufficient statistic   69
        sum   51
        twisting argument   77
        well behaving statistic   68
    Normal   80, 245
    Pareto   465
        well behaving statistic   68
    Poisson   51

specification 47
sum 51
uniform 45, 171, 287, 464
   estimator 82
   sampling mechanism 11
   sufficient statistic 69
   well behaving statistic 68
Weibull 92
random walk 297
reflexitivity 437
regression *see* learning, function
regularization 260, 415, 443
relation
   equivalence 162
   order 230, 290, 394
relevance 358, 360, 365, 370, 374, 378, 379, 417, 420, 428, 430, 443, 447
representation 199, 233, 266, 281, 286, 304, 311, 332, 344, 357, 437, 445
   binary 247
   knowledge 281, 350, 352
   linear 233
reproduction 329
residual 361
ridge regression 429
risk 194, 227
   estimator, of 82
rotation
   invariance, w.r.t. 241
   matrix 235
rough set 147
   fuzzy 148
   lower bound 148
   upper bound 148
   vocabulary 148
rule 198, 348, 349, 353, 362, 364, 374, 375, 378, 431, 440, 441, 443
   agglomerative 198
   aggregative 198
   alignment 444, 445
   assignment 194
   subsymbolic 385
   symbolic 385
   synthesis 374, 378
   universal 266

rule-based system 348, 352, 362, 376, 398, 431, 440, 446
   antecedent 349, 373, 375, 396, 440, 444
   consequent 349, 373, 375, 396, 440, 442, 446
   evaluation 352
   identification 385
   layout 349
   local 361, 431, 440, 441, 443, 444
   premise, *see* rule-based system, antecedent
   tabular fuzzy model 350
   weight 378, 396, 441

saddle point 426
sample 3, 6, 47, 52
   cause/effect 84
   incomplete 54
   labeled 97, 108
   likelihood 52, 68
   modeling 45
   specification 52
sampling mechanism 10, 47, 52, 67, 170, 275, 298
   Bernoulli 11
   Binomial 66
   double exponential 70
   dynamic 278
   Gaussian 80, 95, 228
   hazard function 92
   Laplace 86
   misclassification error 111
   negative exponential 11
   properties 66
   random function 86
   uniform 11
   universal 10, 52
scalability 332
search problem 285, 319
search space 285, 332
selection 286
   roulette 290
   tournament 290
self 326
self-organization 319
sensitivity 204, 309, 328, 385, 393, 404, 420
separability 234, 254, 304, 388

semantic   234
structural   234
shadowed set   139, 178, 226
   core   139, 178
   fuzzy set, transformation from   140, 226
   shadow   140
   three-valued logic   140
   uncertainty   140
similarity   199, 459
slack variable   260, 415, 422
sliding trick   428
social computation   275, 277, 278, 318, 325, 329, 334, 335, 343
solution
   approximation degree of   284
   feasible   284
   optimal   284, 419, 423, 425, 427, 429, 442, 443
sparseness   234, 248, 252, 256, 262, 267, 394
split clusters   215, 225, 239
state vector   279, 294
statistic   10, 52, 291
   extreme   53, 54
   independent   67, 94
   local sufficient   67
   monotone   67
   observable   67
   partition   68
   pivotal quantity   81
   pledge points   95, 108
   relevant   13
   sample mean   53, 54, 228
   separable   94
   sufficient   68, 69, 84, 298
      Gaussian   80, 228
      joint   87, 94
      negative exponential   69
      uniform   69
   well behaving   67
      Bernoulli   68
      double exponential   70
      negative exponential   68
      power law   68
      uniform   68
   well definite   67
statistical block   7, 61, 284
   equivalent   7

stimulus   314
stochastic matrix   279
stochastic minimization   266
stochastic process   278
   trajectory   279
stochastic vector   279
stopping rule   219, 220, 290, 332, 400, 438
structural optimization   333
structure   59, 61, 298, 344, 363, 389, 390, 434, 435, 437, 438, 446
support vector   109, 259, 260, 414, 428
   clustering   259
support vector machine   108, 310, 411, 412
   ϵ-tube   422
   C-SVM, see support vector machine, soft-margin classifier
   one-against-all   414, 415
   regression   422, 426, 428
   relevance-based   419, 420, 428
   single-class   259
   sliding trick   418, 426, 427, 430
   soft-margin classifier   110, 260, 415, 416, 419
   soft-margin coefficient   260
survival probability   325, 329
swarm intelligence   318
   ant colony optimization   319, 321
      daemon action   320
      pheromone   319
      preferential marking   319
   particle swarm optimization   322, 324
      acceleration   323
      inertia factor   323
      particle   322, 325
      position   323
      velocity   322
symmetry property   238, 437
synapse   see connection

target identification   325
test of hypothesis   119

Index   495

test set   295, 301, 363, 420
theorem
   Central Limit   237
   Glivenco-Cantelli   54
   perceptron   309
   probability integral transromation   53
   total probability   15
thought cycle   328
threshold   294, 295
   dynamic   328
tradeoff   *see* balance
training
   cycle   300
   error   260
   set   195, 295, 298, 301, 363, 420
transform
   Fourier   265
   linear
      orthogonal   235, 248
      orthonormal   236
   monotone   356
   nonlinear   256, 258, 416

wavelet   266
transition matrix   279, 297
transitivity   165, 437
triggering functionality   328
TSP   319
Turing machine   334
twisting argument   75, 301

uncertainty   113, 117
universe of discourse   16

validation set   295
variance   49, 235
Venn diagram   22
virtual distance   430
virtual shift   *see* support vector machine, sliding trick
volume transmission   329
Voronoi
   cell   198
   tessellation   197, 200

whitening   239, 419, 467
   local   251